Incorporating Cultures' Role in the Food and Agricultural Sciences

ELSEVIER
science & technology books

Companion Web Site:

https://www.elsevier.com/books-and-journals/book-companion/9780128039557

Incorporating Cultures' Role in the Food and Agricultural Sciences
Florence V. Dunkel

Available Resources:

- About the Author
- About this Book
- For Professors and Administrators Section
- For Policy Makers Section
- For Students Section
- Dancing Across the Gap: Journey of Discovery Film
- The Experiences of Linwood Tall Bull

ELSEVIER

ACADEMIC
PRESS

Incorporating Cultures' Role in the Food and Agricultural Sciences

Florence V. Dunkel

Department of Plant Sciences and Plant Pathology,
Montana State University, Bozeman, MT, United States

Academic Press is an imprint of Elsevier
125 London Wall, London EC2Y 5AS, United Kingdom
525 B Street, Suite 1800, San Diego, CA 92101-4495, United States
50 Hampshire Street, 5th Floor, Cambridge, MA 02139, United States
The Boulevard, Langford Lane, Kidlington, Oxford OX5 1GB, United Kingdom

Library of Congress Cataloging-in-Publication Data
A catalog record for this book is available from the Library of Congress

British Library Cataloguing-in-Publication Data
A catalogue record for this book is available from the British Library

ISBN: 978-0-12-803955-7

For Information on all Academic Press publications
visit our website at https://www.elsevier.com/books-and-journals

Working together
to grow libraries in
developing countries

www.elsevier.com • www.bookaid.org

Publisher: Andre Gerhard Wolff
Acquisition Editor: Nancy Maragioglio
Editorial Project Manager: Billie Jean Fernandez
Production Project Manager: Anusha Sambamoorthy
Cover Designer: Mark Rogers

Typeset by MPS Limited, Chennai, India

Dedication

Most authors and editors can remember where they were and who or what prompted them to write a specific book.

For me with this book, the moment was in 2011 in the office of P. Gregory Smith, US Department of Agriculture National Institute of Food and Agriculture (NIFA) Program Director for the Higher Education Challenge Grant Program. The gray carpet in his office, the sun coming in from the Potomac River side of the Waterfront Center, all seemed to be nudging me to the commitment to write a book. Greg heard my urgency in wanting to widely disseminate (shout loudly) the message of the key role culture plays in food choices, ultimately in food security and human health. He warned me, "A book is too slow, Florence."

I am dedicating this book to P. Gregory Smith, an inspiration and open ear for many years as the four competitive USDA NIFA grants provided the backdrop for these radical ideas to emerge. The fast way? Storytelling. Greg encouraged me always to listen to the ways of Indigenous people in Mali, in Montana, and my own people. Tell it as a story…the culture of food and agriculture. Just knowing Greg Smith sat way over there in the east in that gray carpeted office in Washington, DC, understood and cared about this urgent message, I will always be grateful.

Contents

LIST OF FIGURES xv

LIST OF TABLES xix

LIST OF CONTRIBUTORS xxi

PREFACE xxiii
Waded Cruzado

**LISTEN! A FOREWORD TO *RECOGNIZING CULTURE
IN FOOD AND AGRICULTURE*** xxvii
Hiram Larew

ACKNOWLEDGMENTS xxxi

Part I   **Fundamentals of the Culture and
            Agriculture Relationship** 1

CHAPTER 1 The Quiet Revolution: Where Did You Come From? 3
           Florence V. Dunkel

           Definition of Culture 8
               What Do We Mean by Indigenous? 10
               Ethno-Relativity 13
               Chapter-by-Chapter Summary 17
           References 20
           Further Reading 21

CHAPTER 2 Failures 23
           Florence V. Dunkel

           Case Study 1. The Chicken Award 25
           Case Study 2. The International War on Locusts and
           Grasshoppers 29

Case Study 3. Fry Bread 34
Case Study 4. The Use of River Systems in the
Western United States to Produce Food for the Nation.................35
Concluding Reflections on Failures 36
References 37

CHAPTER 3   Decolonization and the Holistic Process 41
*Clifford Montagne, Florence V. Dunkel, Greta Robison,
Ada Giusti, and Badamgarav Dovchin*

Introductions 43
Connection to Land Is a Prelude to Appreciating Wholeness.........45
Introduction to the Holistic Process.................................46
Decolonizing Methodologies: First Step Toward the Holistic
Process..49
    Identifying Communication Languages.....................51
    Establishing Ownership of Original Data or Products...............54
    Summary Reflections on Decolonization 56
    The Build Up of Colonization 56
    A Paradigm Adjustment for Reductionist (Western) Science 57
    Traditional Ecological Knowledge Becomes Visible During
    Decolonization...58
The Holistic Process..59
The New Paradigm for Teaching in the Food and Agricultural
Sciences..62
    Using the Holistic Process...62
    The Expansive Collaborative Model for Community
    Engagement and Service-Learning...............................63
Conclusion..66
References 68
Further Reading..70

CHAPTER 4   Immersion 73
*Ada Giusti, Florence V. Dunkel, Greta Robison,
Jason Baldes, Meredith Tallbull, El Houssine Bartali,
and Rachel Anderson*

Definition of Immersion..74
Case Study 1. An MSU French Professor Collaborates With
an Entomology Professor on a Malian Food and Health Project.......75
    Introduction 75
    My Naïve State 76
    Growing Awareness...76
    Applying Cultural Knowledge 77

Case Study 2. Arranging Successful Immersions at an
Institution Level 80
   Introduction 80
   Annual Undergraduate Immersions......................................81
   Evaluating Immersion Results..83
Case Study 3. Native Foods and Food Deserts 85
   Why Is the Mini-Immersion an Essential Part of
   Education?.. 89
   Mini-Immersion Process..90
   Native Plants and Native Foods on the Northern
   Cheyenne Reservation 92
   "Let's Pick Berries" Project 95
   Summary of Case Study 3..98
Are There Negative Consequences From Not Including
Immersion in General Education?.. 99
Concluding Thoughts on Immersion 100
References 102
Further Reading.. 105

Part II   Listening In and Between Communities ... 107

CHAPTER 5   Listening With Subsistence Farmers in Mali 109
*Florence V. Dunkel, Ibrahima Traore, Hawa Coulibaly,
and Keriba Coulibaly*

Introduction to Sanambele, Mali, and the Main Communicators 110
Reconciling the Role of Health in the Village's Food and
Agriculture System 115
Food Security, Stunting, Amino Acid, and Micronutrient
Deficiencies: What is the Culture-Smart Agriculture Answer
in Villages Relying on Grain-Based Diets? 118
The Gourd Story.. 124
Concluding Reflections... 127
References 128

CHAPTER 6   Listening With Native Americans 131
*Florence V. Dunkel, Linwood Tallbull, Richard Littlebear,
Tracie Small, Kurrie Small, Jason Baldes, and
Meredith Tallbull*

Phase 1: Awakening 132
Phase 2: Indigenous Teaching and Learning 136
   Beginning a Relationship With the Apsaalooke 141
   "Let's Pick Berries" Project 143

Phase 3: Linked Courses, Shared Curricula, and Classrooms 146
Closing Reflection.. 155
References 156
Further Reading... 157

CHAPTER 7   Listening Within a Bioregion... 159
Clifford Montagne, Badamgarav Dovchin,
and Florence V. Dunkel

Phase 1: Formation and Evolution of BioRegions.......................... 161
    Initial Visits to Mongolia.. 161
    Learning Through Crisis 164
Phase 2: BioRegions Program Matures to Include Annual
Work Visits and Fund Raising... 168
    Education.. 168
    Environment 169
    Health 171
    Boiling Duration and Water Quality for Tea 172
    The Arts and Traditional Knowledge............................ 172
    Festival of the Darhad Blue Valley 173
    Whole Community and Business.................................... 174
Phase 3: Deepened Listening Exchanges....................................... 176
    Reflections... 176
    Mongolian and Native American Students Working Together.... 177
    Including Health in Food and Agriculture 179
    The Yellowstone Connection.. 181
Outsiders and the Importance of Listening.................................... 183
Conclusion.. 184
References 184
Further Reading... 184

CHAPTER 8   Listening Over Power Lines: Students and Policy
Leaders 187
Hiram Larew, Florence V. Dunkel, Walter Woolbaugh,
and Clifford Montagne

Introduction 187
Case Study 1: Bringing USDA Into the university Classroom:
Two-Way Learning Through Mutual Listening 188
    Why Bother?... 188
    Getting Started 189
    Planning and Holding Classes....................................... 190
    Impacts 193

Case Study 2: Doing Intercontinental/Intercultural
Science With Middle School Students...194
    Finding Partners 195
    Science Trunks 196
    The Research Area 198
    The Research Process 199
    The Research Design and Data Collection 200
    Data Analysis and Conclusions...201
    Information Dissemination and the Global Research
    Symposium...203
    How to Accomplish This Activity in Your Area...........................204
    Tips and Suggestions 205
    Reflections..205
References 206
Further Reading..207
Power Line Resources 207

CHAPTER 9   Listening With Students 209
Greta Robison, Jason Baldes, and
Florence V. Dunkel

Case Study #1: Learn Communication
Patterns 211
Case Study #2: Recognize Unique Backgrounds
and Personal, Community Missions...214
Case Study #3: Connect Reality With Action 218
Case Study #4: Recognize Wealth of Insights Foreign
Students Bring to Classrooms and Policy-Making
Organizations 222
What Students Want 233
    Create Personal Connections 233
    Close the Gap Between Teaching Methodologies
    and Real-Time, Complex Issues..234
    Provide Cross-Cultural Immersion 236
    Encourage Passion 237
    Response From Florence Dunkel...238
Conclusion..240
References 240
Further Reading..241
Summary Illustrations..241

Part III   **Bridging the Gap Between Food and Agricultural Sciences and the Humanities and Social Sciences 247**

CHAPTER 10   Two Cultures and a Second Look: Humanities and Food Sciences 249
Florence V. Dunkel, Khanjan Mehta, Alison Harmon, and Ada Giusti

Case Study 1: Food, Storage, Marketing Research and Development of the National Food Quality Laboratory of Rwanda Combined Humanities With the Land Grant Mission 253
Case Study 2: Sustainable Foods and Bioenergy Systems: MSU 257
    The Design Team 258
    Curriculum and Learning Outcomes 260
    Lessons Learned in the First 5 Years..262
    Student Response, Graduates, and the Future 262
    How can a University Encourage Integration of the Humanities and the Food and Agricultural Sciences? 263
Case Study 3: Graphic Arts, the Humanities, and Entomology 264
Case Study 4: Humanitarian Engineering and Social Entrepreneurship Program: The PSU 269
    Introduction 269
    The Journey to Affordable Greenhouses 271
Closing Reflections 273
References 275

CHAPTER 11   Couples Counseling: Native Science and Western Science 277
Florence V. Dunkel, Jason Baldes, Clifford Montagne, and Audrey Maretzki

Case Study #1: A Change of Thought About Nutrition Education: Kenyan Women's Indigenous Nutrition Knowledge 284
Case Study #2: The Entomologist's Contribution to Early Childhood Nutrition 287
Case Study #3: Recognizing Indigenous Knowledge in the Academy 288
Strategies to Support Academics Interested in Indigenous Knowledge 290
Conclusion 293
References 293
Further Reading 295

CHAPTER 12 Putting It Together: The Way Forward 297
Florence V. Dunkel and Hiram Larew

Introduction 299
The Chapul Story 302
Other Young People Tell Their Stories 303
The Quiet Revolution 305
View From 30,000 Miles Above Academia 308
References 308

INDEX 309

The books' companion web site cab be found in the link: https://www.elsevier.com/books-and-journals/book-companion/9780128039557.

List of Figures

Figure 1.1   *Native Americans at Dartmouth*. A student at   12
Dartmouth College celebrates Indigenous People's
Day. As the United States celebrates Columbus Day,
many locations will be celebrating a parallel but very
different day: Indigenous People's Day—Lily
Rothman, Time.com, October 10, 2016.

Figure 2.1   Diagram of the chasm that exists people producing   24
primarily Western culture technical information and
Indigenous knowledge and ways to bridge the gap.

Figure 3.1   Participatory diagraming using Western culture science   53
and decolonizing methodologies. The chart with
numbers underlies the diagram below with
no numbers. The containers are bottle gourds grown
in the village and used for measuring. High lysine and
tryptophane foods are placed on the diagram from
left to right in descending content of these amino acids.
To avert stunting, a child must each day have 1 to 3
gourd-fuls of that plant as indicated by the diagram.

Figure 3.2   The Holistic Management Model (HMM) drawn from   61
Indian (Asian) tradition representing the cosmos
(https://en.wikipedia.org/wiki/Mandala) drawn by
C. Montagne, R. Kroos, L. Soderquist, B. Dovchin.

Figure 3.3   Basic components of learning communities in the   64
Expansive Collaboration Model using the holistic
process (NGO, nongovernment organization).

Figure 3.4   Simplified diagram used to begin a holistic process.   66
This is the detail used to obtain information for the
center concentric circle used in the diagram in
Figure 3.2. Once these details are determined by
the community, then the actions, testing questions,

tools, and whole to be managed can be described. The feedback loop operates on the entire system, on all parts, and is best illustrated as a third dimension.

Figure 5.1   Sanambele villagers reviewing draft of Carly Grimm's participatory diagram of high lysing high tryptophan foods grown/wild collected by villagers November 2014.   115

Figure 5.2   Hawa Coulibaly and Florence Dunkel in Hawa's cattle pasture, sanambele mali, 2009 by her fields.   117

Figure 5.3   Gathered in Dr. Dunkel's office at MSU, recent MSU Sustainable Foods graduate Carly Grimm (right) returns from her home in Wisconsin to meet with new Health Sciences graduate student, Jordan Richards (left) to explain how the revised lysine/ tryptophan chart for village mothers works so that Jordan can carry on the participatory diagramming process with the women to make the connection between the malaria season, the seasonable availability of these crops, and the importance of these nutrients to strengthen the immune system to build the body's defenses against malaria.   122

Figure 5.4   A large bottle gourd (callabasse): (A) in the edible phase, useful for cricket food; (B) dried callabasse after harvest cut in half to form a sphere similar to the callabasse used by Hawa Coulibaly in Sanambele to capture house crickets (used inverted, opening on ground) for feeding to her chickens and thereby significantly improving the growth and development of her free-ranging chickens; (C) homegrown kitchen utensils made from callabasse.   126

Figure 9.1   AGSC 465R students gather at the Dunkel-Diggs farm to discuss the concepts of Ayittey (2005) and practice participatory diagramming in the field and with farmer interviews.   242

Figure 9.2   Northern Cheyenne History course students at Chief Dull Knife College (CDKC), Lame Deer, MT meet with students in their linked course at Montana State University (MSU), AGSC 465R, who they challenged during the semester to learn how to make dry meat   242 and use it in a stew and to learn why this was important in Northern Cheyenne history. CDKC President Richard Littlebear joins with Elders, Kathy Beartusk, Mina Seminole, and George Nightwalker to test the dry meat stew created by the MSU students and their instructor, Dunkel.

Figure 9.3 Students in the linked course (Rachel Anderson, left; Gwen Talawyma, middle) and their Northern Cheyenne teaching assistant (Sierra Alexander, right) learn the dry meat cutting from Gwen, a traditional dry meat processor.   243

Figure 9.4 AGSC 465R students meet with Apsaalooke Kurrie Small, students' site mentor, a rancher, and former AGSC 465R student to listen to her debrief on the "Let's Pick Berries" project and to update her on their research at MSU (from left Tanner Mcavoy, Taylor Anderson, Danielle Bragette, Kyle Lavender, Kurrie Small, Florence Dunkel, Isaac Petersen).   243

Figure 9.5 Learning participatory diagramming developed their predecessors for the Sanambele village women to avert stunting in their children (Dunkel with two AGSC 465R students).   244

Figure 9.6 Village team, Karim Coulibaly (left) and Bourama Coulibaly (next), gather with AGSC 465R student Wendy Nickish (now a dentist in rural Montana) (far right), and her site mentor, Keriba Coulibaly (IER, USAID) (next left) to monitor local dry bed river in March 2009 for anopheline larvae whose adult stage can carry malaria.   244

Figure 9.7 From left, Keriba Coulibaly, Wendy Nickisch, and Bourama Coulibaly use equipment available in village to sample dry bed river pool for dry season populations of mosquitoes.   245

Figure 9.8 Dr. Florence Dunkel and Dr. Clifford Montagne engage with a transdisciplinary team of faculty (representing engineering, biology, and communications) on the University of St. Thomas campus, St. Paul, MT, November 2002 to learn to link the Holistic Management Model with service-learning   245

	methods and apply it in their classes at their respective universities and other US universities as they work together with villagers in Mali.	
Figure 9.9	Prairie turnips, a sustainable, wild-collected traditional food, introduced to AGSC 465R students by Linwood Tall Bull, elder of the Northern Cheyenne Nation, Busby, MT.	246
Figure 10.1	Graphic comparison of science and humanities.	252
Figure 10.2	Diagram illustrating the five goals of the World-Readiness Standards for Learning Languages from foreign language learning.	259
Figure 10.3	Diane Ullman (left) and Donna Billick (right), cofounders of the Art/Science Fusion Model for providing a visual linkage of the arts, humanities, and agricultural sciences.	264

List of Tables

Table 3.1   How Community Needs Are Addressed Using the   55
               HM Model in Generic and Summary Terms With
               Respect to the Village of Sanambele, Mali

Table 5.1   Growth-Limiting Essential Amino Acids Missing in   119
               Sanambelean Young Child's Diet

Table 5.2   Essential Amino Acid Needs for Children   123
               Ages 1−8 Years Old and Percentage of These
               Essential Amino Acids for One to Two Chicken Eggs

Table 10.1 Creative Works Associated With Teaching Entomology   265
               001 and Freshman Seminars at University of
               California Davis

List of Contributors

Rachel Anderson
The Madisonian, Ennis, MT, United States

Jason Baldes
Wind River Native Advocacy Center, Washakie, WY, United States

El Houssine Bartali
Department of Agricultural Engineering, Institut Agronomique et Vétérinaire Hassan II, Rabat, Morocco

Hawa Coulibaly
The Village of Sanambele, Bamako, Mali

Keriba Coulibaly
USAID, Sikasso, Mali

Badamgarav Dovchin
Renchinlumbe, Darhad Valley, Mongolia

Florence V. Dunkel
Department of Plant Sciences and Plant Pathology, Montana State University, Bozeman, MT, United States

Ada Giusti
Department of Modern Languages and Literatures, Montana State University, Bozeman, MT, United States

Alison Harmon
College of Education, Health and Human Development, Montana State University, Bozeman, MT, United States

Hiram Larew
U.S. Department of Agriculture (retired), National Institute of Food and Agriculture, Center for International Programs, Washington, DC, United States

Richard Littlebear
Chief Dull Knife College, The Northern Cheyenne Nation, Lame Deer, MT, United States

Audrey Maretzki
Department of Food Science (retired), Interinstitutional Center for Indigenous Knowledge, The Pennsylvania State University, State College, PA, United States

Khanjan Mehta
The Pennsylvania State University, State College, PA, United States; (currently: Lehigh University, Bethleham, PA, United States)

Clifford Montagne
Department of Land Resources and Environmental Sciences (retired), Montana State University, Bozeman, MT, United States

Greta Robison
Department of Earth Sciences, Montana State University, Bozeman, MT, United States

Kurrie Small
The Apsaalooke Nation, Baaxuuwaashe, MT, United States

Tracie Small
The Apsaalooke Nation, Baaxuuwaashe, MT, United States

Linwood Tallbull
The Northern Cheyenne Nation, Busby, MT, United States

Meredith Tallbull
The Northern Cheyenne Nation, Lame Deer, MT, United States

Ibrahima Traore
Lycee El Hadj Karim Traore, Bamako, Mali

Walter Woolbaugh
Manhattan Middle School, Montana Pubic Schools, Manhattan, MT, United States

Preface

Sustainable development is the pathway to the future we want for all. It offers a framework to generate economic growth, achieve social justice, exercise environmental stewardship and strengthen governance.

Ban Ki-moon, United Nations Secretary General

In 1974 the World Food Conference in Rome approved an ambitious declaration: that within a decade, the scourges of malnutrition and food insecurity would be removed from the face of the earth. The next year, the US Congress affirmed its commitment to that goal by passing the Title XII amendment to the US Foreign Assistance Act of 1961, establishing the Board for International Food and Agricultural Development, BIFAD.

While the lofty goals of the Rome Conference were not achieved within a decade, progress was made, but much work remains to be done. The commitment to freeing the world of hunger and malnutrition still weighs on our nation's conscience. In fact, with the recent passage of the US Food Security Act in 2016, the vision for a world free from hunger, malnutrition, and poverty endures.

BIFAD advises the US Agency for International Development (USAID) on issues as diverse as economic growth and trade; agriculture and food insecurity; democracy, human rights, and governance, among other things. BIFAD and the 1975 Title XII amendment of the US Foreign Assistance Act also recognize the critical role of land-grant universities in advancing science and technology that will assist in identifying and implementing projects to prevent famine around the globe. Those of us who have been blessed with the chance to work at land-grant universities understand the importance of converting bold ideas into new conditions that can transform our world.

The history and the mission of land-grant institutions is, at its root, a classic model for development of human capital. Thanks to the visionary work of

Justin Smith Morrill in the mid-19th century Congress approved an idea that was tremendously ahead of its time: the notion that, by establishing one public university in each state and territory of the Union, the sons and daughters of the working families of America would have access to higher education.

The passage of the Morrill Act in 1862 did not end Justin Morrill's work. His funding bills were killed time and time again until 1890, when the Second Land-Grant Act was passed, giving a boost of funding to established land-grant universities and offering unprecedented access to higher education to African-American students. The Morrill act even received new life over a hundred years later with the passage of the legislation that gave inclusion to a group of Native American tribal colleges in more than 12 states.

This last event is particularly pertinent to those of us in Montana. Thanks to its passage in 1994—and to a committee that was chaired by the late Michael Malone, then president of Montana State University—the Third Land-Grant Act now benefits tribal colleges all over the nation. Montana has eight land-grant campuses, the largest number of any state.

It is evident, therefore, that support for big ideas that can advance humankind and impel us toward solution-based changes is part of our DNA at land-grant and public institutions.

There are several important links between the history of this type of institution and the inspiration we need to tackle the urgent challenges faced by our most vulnerable populations around the world: the land-grant university was an innovative idea that challenged the social norms of the time, when higher education was largely restricted to the elite, particularly men. It potently carried the truth that we can change our most abject circumstances by working together, even (or perhaps especially) in times of great duress. And, finally, that we can promote positive social change by focusing some of our best efforts in advancing common men and women when we elevate their lot in life by enabling them to provide for themselves and their families, whether in America or in any corner of the globe.

A modern adaptation of these lessons is what I found when I arrived at Montana State University 7 years ago and had an opportunity to learn about the work of Prof. Florence Vaccarello Dunkel. She has dedicated her entire life not only to the study of agriculture and culture, but to that particular interstice between agriculture and human culture. Through patient observation and her love for interdisciplinary, Dunkel has actively taught generations of students how to communicate agricultural concepts to people of different ages and socioeconomic backgrounds in the Northern Cheyenne and

Apsaalooke' Indian nations in rural Montana and in communities in Africa, such as Sanambele, Mali.

Her work aims to be as disruptive as the history of public universities, which gave educational access that had been limited to the few up until that moment. Dunkel encourages us to move away from the traditional, hierarchical models that still undergird the dissemination of some scholarly work, no matter how well intentioned they might be. Her emphasis is on a holistic approach that asks us to listen rather than talk, to share knowledge back and forth between the academy and the communities they seek to help rather than simply delivering information to communities in need and expecting them to take ownership of something they had no role in creating. Her work is solid and strong because it is so difficult to do.

And yet, it is urgent work.

With the wisdom of decades of intercultural experience and passion for her subject, Dr. Dunkel does exactly that in *"Incorporating Culture's Role in the Food and Agricultural Sciences."* In this text, Dr. Dunkel weaves numerous voices from multiple cultures in a book that is as readable as it is vital in untangling the cultural obstacles to solving hunger around the globe.

A world free from hunger and malnutrition is possible, and it awaits us. Working together, we can and must end extreme poverty. We must remain committed to building communities in which inclusive and sustainable prosperity is a certainty for all. The publication of this exciting book is a positive step toward making that a reality.

Waded Cruzado
President, Montana State University,
Bozeman, Montana, United States

Listen! A Foreword to *Recognizing Culture in Food and Agriculture*

One word summarizes this book's powerful message: Listen! Regardless of who you are, where you live, your stage in life, what you believe in or seek, or what you know, the ability to truly hear and respect what others are saying is critical to your effectiveness in today's world. Using ears more, mouths less, is a talent that requires honing. It is a fundamental skill for the intercultural work that increasingly undergirds health, agriculture, and food security efforts in all countries around the world. The pedagogical "listening" case studies described herein are clearly focused on empowering students. Keep in mind, however, that the ability to listen can only be conveyed by those instructors who are able to practice what they preach; the book's message and methods are primarily relevant to teachers at the front of classrooms. And, as highlighted in Chapter 8, Listening Over Power Lines: Students and Policy Leaders, linking classrooms to policy makers in government, private sector, and other key stakeholders is a useful way to include real-time, headline, real-world insights into student discussions. Such links also provide nonacademic decision makers with insights about listening—a skill they need as well.

Listening implies an approach that focuses on building respectful partnerships, not hierarchical relationships, on working alongside each other rather than telling others what to do, and on being humble enough to realize that you have as much to learn and gain from others as you have to offer them. It centers on the principle that no one has all the answers, and no one is superior. You will learn more about such underlying principles in this book.

"Listen" is easy to say, much harder to do. But, if allowed to occur, it will be the key—the silver key—to success in linking sustainable health, agriculture, and food security policies to practice.

This book is carefully timed. Its rationale is pinpoint-focused on impending global food needs. Clearly, conventional approaches and their innovative application will continue to be critical in meeting the expected surge in want as the year 2050 approaches. But the magnitude of the challenge requires that all tools in the toolbox be considered. Wholly new paths and break-throughs that tap into largely untapped approaches will be needed. Paradoxically, many of those new approaches will be based upon traditional knowledge and insights gained over centuries, even millennia. By tuning in to all such quarters and to all experienced practitioners, we have the opportunity to (re)discover not only what agriculture means to others, but how we may move toward the goal of sustainable food security at home and abroad. In the coming years, our ability to innovate will increasingly synch with this willingness to respect the usefulness of many approaches.

There's no better place to explore new takes on agriculture than in the classroom. As this book points out, using class time to study the culture in agriculture as a way forward makes timely sense. And, the book points to mutually beneficial insights by both students and decision makers that emerge when they are given a chance to interact in the classroom. The search continues for food security responses that make a difference and that make sense; the teaching practices described in this book have uncovered helpful responses.

A culture-focused approach. Exploring the "culture" in agriculture is a powerful adjustment to the new teaching and learning approaches that are needed. The human dimension of food security—do behaviors contribute to food insecurity, and if so, how and why?—is often poorly understood or not fully considered, and students typically leave classes with a spotty or inadequate appreciation of how people affect and are affected by agriculture. Well-intentioned interventions are often undercut or rendered unstainable when a less-than-holistic approach is used. Said slightly differently, being deaf to the importance of culture puts development efforts at risk.

Students deserve to understand the power of listening as a must-hone skill for effectively working in agriculture. Food security policy makers also need to be given the chance to tune in—and chime in—as new teaching models that acknowledge the power of culture are brought online. This book highlights the role that culture plays in food security and other domains, and provides examples of how an appreciation for the role of culture can be incorporated into classroom discussions.

The intended audience. A wide range of stakeholders should find this book useful. First and foremost, the authors have assumed instructors as the primary stakeholders. The examples provided, and the insights and lessons learned that are described are geared toward enabling classroom leaders to more effectively facilitate discussions about food security. Students, too, should see themselves as central to the book's intentions; the book's goal is to inform and prepare students for their onward professional interests. Clearly, not all students who benefit from this book's coverage will choose to work in food security, but the skills that are promoted across the book's chapters will be professionally useful regardless of a student's eventual path. Decision makers, community members, policy shapers, donors, think tank staff members—generally anyone or any group in or beyond the United States who has a stake in agriculture, in teaching it and in practicing it—can usefully listen to the book's message. While these nonacademic stakeholders are less concerned with classroom enrichment, they all share an interest in effective programming—and in classrooms in which tools for effective programming are taught. Ultimately, of course, a wide array of public stakeholders—producers/farmers, processors, transporters, marketers, and consumers—all have a stake in ensuring that agriculture worldwide includes, rather than excludes, cultural dimensions.

A roadmap. This book isn't intended as gospel. All of the authors realize that the approaches they espouse will evolve and change. Instead, the reader should view the book as a roadmap of new terrain, and the work it encourages the reader to undertake as a journey. The book encourages a scaled-up, wider adoption of the described methods as well as an openness to revisions that will surely accompany experience. Policy makers, too, who influence or have a stake in educational models should consider the roadmap when strategies, plans, pedagogy, or other frames for learning or implementing are developed. And, early adopters should challenge themselves to develop and refine tools for assessing impacts as students are offered a deeper cultural awareness of food security: How does such an approach affect student learning, competencies, engagement, longer term employability, well-being, and productivity in the world of work? How does such teaching impact food security?

Learning from listening. All said, the roadmap described herein is ready for prime time and further testing. As the students who helped to develop it have done, those who try the new approach should maintain a reflective log of their experience so that learning from listening can be continuously folded into the model. And, lessons learned by the community of practitioners should be captured and shared.

But to circle back… As you take up this book, keep that first-among-words in mind: *Listen!* And enjoy.

Hiram Larew
US Department of Agriculture (retired),
National Institute of Food and Agriculture,
Director of the Center for International Programs,
Washington, DC, United States

Acknowledgments

There are scores of people to whom I owe a great measure of gratitude, beginning with the US Department of Agriculture (USDA) National Institute of Food and Agriculture (NIFA) program officers, particularly P. Gregory Smith and Hiram Larew, who over a decade ago began to listen and come to understand there was something about culture's role that needed to be accounted for in the agricultural sciences in order to make progress in solving complex issues of food and health.

Since 2012, when my colleague of many decades in entomology, natural product toxicology, and fellow activist, Sonny Ramaswamy, became Director of USDA NIFA, the book's message became urgent. Sonny completely understood the ideas I was trying to convey. We had discussed them for years before his appointment by President Obama. I thank Sonny for his steady encouragement, his advice to focus on stories, and his directive to let the book speak clearly and concisely, but quickly publish.

Thanks are due to my colleagues in the Department of Entomology (dispersed in 2004) and my new colleagues in the Department of Plant Sciences and Plant Pathology who consistently remind me that I am odd to be considering the cultural aspect of our science. I appreciate their gentle and their not so gentle, but steady reminders that cultural aspects of entomology and plant sciences are borderline tolerable, but not a substitute for "pure science." John Sherwood, my department head during more than a decade of book preparation and the writing phase, has come to understand the relation of "pure" plant science to food security and health disparities, particularly the cultural part of it.

I am incredibly grateful to May Berenbaum, department head of Entomology at University of Illinois, fellow of the American Association for the Advancement of Science, member of the National Academy of Sciences, and past president of the Entomological Society of America. Little does she know that she has often brightened my days with her wry humor published in the *American Entomologist*, in her book, *Bugs in the System*, and in replete form in her book, *Buzzwords*. Humor is the best prescription for academic knots. Entomology remains a male dominated discipline at the higher levels. The ability to laugh at one's oddness, just being female for starters, is essential for working in entomology, especially, I found, when, in 1988, I became the first-female-in-50-years to head nationally a Department or group of Entomologists. To be able to share one's own cultural differences in humorous way is a recipe for diffusing conflict and confusion. Preparing for the unexpected, approaching it with humor, is May's great lesson. I used it when after completing the keynote address for 900, mostly male engineers in Seoul, South Korea. A typical Korean dinner and entertainment was planned to honor me but when the planners saw that Dr. Dunkel was a woman suddenly new plans on the spur of the moment had to be made. My first meal of sushi was delightful but there was really no entertainment except a group of engineers, and rural development specialists watching a Westerner learn to eat sushi for the first time.

Editors at Elsevier, on the other hand, had no problem understanding cultures' roles in science—be it food or agricultural sciences or agricultural engineering. To have this group of cheerleaders, who immediately understood what I was trying to convey about food and culture, was a weekly boost for me, even if I was apologizing about missing yet another deadline. Billie Jean Fernandez deserves a huge shout out as does Nancy Maragioglio. Billie Jean kept me organized, confident, and enthusiastic about the book, and Nancy was the best for reigning me in or asking me to explain the backstories. My wish is that she is my editor forever and that the team remain to organize me for the rest of my life.

To all the Montana State University (MSU) students who first listened to these stories and then contributed to them I am forever grateful—students in both my University Core courses AGSC 465R Health, Poverty, Agriculture: Concepts and Action Research and in BIOO 162 the Issues of Insects and Human Societies. My colleagues and site mentors on several continents and from several Native American Nations, Dr. Richard Littlebear, Meredith Tall Bull, Linwood Tall Bull, Jason Baldes, Tracie Small, Kurrie Small, Alma Hogan Snell, have made the process of saying what needed to be said a joyful, caring process. For all their guiding and patience while I made cultural errors, I am thankful. The two MSU faculty for whom I owe a great debt of gratitude are Dr. Ada Giusti, an MSU professor of French language and literature who helped me see the people side of the world—who guided me out of the strictly science-oriented approach to people, and Dr. Clifford Montagne, a soil scientist who enlightened me with the holistic process.

And now, I gratefully acknowledge my MSU colleagues who are coming to understand the message of the book and are turning this understanding into action in our new 2017 USDA NIFA grant creating an internship program for Tribal College students focused on health, food sovereignty, agriculture, and traditional wealth: principle investigator, Dr. Holly Hunts, consumer economist; Dr. David Sands, microbiologist; and Dr. Edward Dratz, nutritional biochemist.

My parents, Mildred Nageli Behr and Vincenzo Vaccarello, who created a home environment filled with playing of music, doing art, learning about other cultures, welcoming people of other cultures into our home and family, and in sharing their own cultures were a foundation that led directly to this book. The long dinner conversations, as a teenager, usually included both arguments about the chemistry, biochemistry, physiology, of the human body, and the day's new observations of the local flora and fauna on our farm, as well as stories from several cultures represented at the table that, as a young child, included my grandparents. A special thanks go to my tolerant Sicilian family who thought I was the most buried-in-books person they had ever known, and who not only did not tell jokes but did not understand the jokes her cousins told.

Special thanks go to Robert Diggs, my husband and first level reviewer who works at any hour and on copy not fit for any other eyes to see and is brutally honest, but at the same time a strong supporter. The book could not have happened had it not been for his unwavering support for the late, but delicious suppers I served and that he had, at times, to step in to create, and for the weekends I spent buried in my computer.

And I thank my eight children (Anne-Marie Pfaff, Alec Dunkel, and Lynn Dayan) and grandchildren (Roman and Anna Pfaff, Annika and Aidan Dunkel, and Aria Dayan) who sacrificed time without Mom/Grandma and who have taught me to think of things other than my intense cultural missions and who draw me quickly into their world. It is these three children, now all parents, who as teenagers, I would, in addition to packing them up and going to see their Grandmas, I would fly them to my workplace in Rwanda and when they arrived, I would take a break and we would go to the rainforest to visit a family of mountain gorillas that Dian Fossey had befriended. Or I would ask them to come meet me in the People's Republic of China to learn to know my Chinese colleagues there. I remember giving them several yuan and suggesting they practice their Chinese in the market around the corner and buy something. My son and two daughters still take joy immersing their own children in other cultures, visiting my son-in-law's Swabish German family farm, or traveling to another continent to live for several weeks as a family, or, as my younger daughter did, chose to buy their home in a school system offering Chinese and then homeschool her own daughter in Chinese and guide her into Chinese immersion for her first year in public school while she, the mom, continued learning to speak Chinese herself. And special thanks to my daughter, Anne-Marie, who traveled with me to my father's village to continue my search for who I am and where I came from—a culture that I was protected from knowing much about, but, I am learning, is a lovely culture with their own agricultural system and food unique to their village, and to their festivals.

I am grateful to all who contributed to this book in writing in helping me tease out this important message.

Thank you all from the bottom of my heart.

Fundamentals of the Culture and Agriculture Relationship

CHAPTER 1

The Quiet Revolution: Where Did You Come From?

Florence V. Dunkel

Department of Plant Sciences and Plant Pathology, Montana State University, Bozeman, MT, United States

CONTENTS

Definition of Culture 8
What Do We Mean by Indigenous? 10
Ethno-Relativity 13
Chapter-by-Chapter Summary 17

References 20

Further Reading .. 21

Into my classrooms and interactions with students, I bring a strong dose of African sense. Thirty-three years I have struggled as a person of Western culture origins to understand sub-Saharan Africa while building long-term relationships with Africans and helping other outsiders understand the wealth of Africa (Dunkel et al., 2013; Dunkel and Peterson, 2002; Dunkel, 2004; Dunkel et al., 1986, 1987; Lamb and Dunkel, 1987). A quarter century ago, I began a similar learning odyssey with the Native American Plains people of the Northern Rockies (Chaikin et al., 2010; Weaver et al., 1995). Some of these students took my courses and some were awarded assistantships to work in my research laboratory at Montana State University (MSU). Some of these Native Americans traveled with me and other faculty to Africa. Seven of the Africans moved to Bozeman, Montana, for 2 years. During those years, from both African and Native American colleagues, students, and other friends from these cultures, I learned a similar lesson: know where you have come from and listen to hear where your students, colleagues, administrators, and policy makers you interact with have come from.

The sub-Saharan Africans enjoy telling me the story of the African stork that reappears many times in literature and art in Africa. The basic image is an imposing bird with a huge wingspread. I have often imagined it was like the wings of pair of California condors I found in 1967 as I hiked in a canyon near my uncle and aunt's cabin in the Sierra Madre Mountains. The condors were surprised and showed me their gigantic wings, 7 m wingspread, as they fled into the sky. The African image and story comes, I think, from a smaller bird, probably a Marabou stork. These Marabou storks frequented a place where we often stayed overnight while visiting the Akagera National Park during my work in Kibungo Prefecture in eastern Rwanda. It was not so much the wings of this African stork that impressed me, or its very large size (about the size of a goat). It was its huge eyes and enormous beak that caught my attention.

3

Incorporating Cultures' Role in the Food and Agricultural Sciences. DOI: http://dx.doi.org/10.1016/B978-0-12-803955-7.00001-8

A symbol that is often depicted in African art is of this majestic bird in flight going toward some future place with its egg (of course very large) in its beak. As it did this, carrying the fragile egg ever so gently, the stork is depicted looking backward! From the Africans' use of this symbol we learn that the stork carries in that one egg all its future generations, but is depicted with its head looking backward. The symbol reminds us of the importance of being aware of from where one has come. In the Euro-American culture we still see a stork symbol used to signal caring for future generations or to bring the new generation somewhere, but the concept of simultaneously looking backward, of respecting Elders' wisdom, is lost.

Listening for the wisdom of one's Elders, for the stories of one's culture, and knowing the cultural practices of your own people is important. Knowing the traditional wisdom and the traditional ecological knowledge (TEK) of one's people and carrying it forward into the future with new challenges is a sound way to prepare to survive new challenges (Norberg-Hodge, 1991).

Each meeting of a Native American Elder for the first time begins, it seems, with the question: Where are you from? Sometimes this lesson is learned in an embarrassing process. I tell myself and others this story to remind myself of this important lesson and share it with others so that they might learn without the embarrassment.

In February 2013, Kurrie Small and Tracie Small, my former and, at that time, current Apsaalooke students in AGSC 465R Health, Poverty, Agriculture: Concepts and Action Research invited me to attend one of the monthly meetings of Apsaalooke Elders in which these Apsaalooke sisters were using the skills I had taught them. They were learning from their community what its members considered their key to a happy, fulfilled life, in other words, their desired quality of life. I suddenly found myself in an immersion experience surrounded by 36 people, all Apsaalooke, all speaking in Apsaalooke. I was 3 hours from my own office on the MSU campus, but I was able to discern what was taking place only by carefully watching body language and voice intonations, not by the words being spoken. In the middle of the morning, Tracie suddenly motioned to me to come up to the front of the room. "I want to introduce you," she said. But, I said, "I can't speak Apsaalooke." I silently vowed to myself to do better with this. I had been working together with my Apsaalooke students on the reservation since 2009 and rued the fact that I had not yet at least learned how to introduce myself in Apsaalooke. "Just tell the Elders where you are from," she said. I thought for a moment. What does Tracie mean? All the Extension presentations I had given flashed before my mind and I quickly recovered to do my usual introduction, "I am an Associate Professor of Entomology at MSU in Bozeman." I looked at Tracie and her sister Kurrie for verification and perhaps to translate what I said. They were both negatively shaking their heads and smiling, "No, no, they want to know *where* you are from—your family." Now I understood. They wanted to know my tribe, my people, my land, what and where. It was then I told them about my father's people and our village in Sicily and my mother's people from Switzerland and Germany and the land they immigrated to in Minnesota to farm. Certainly, with one exception, never in a university Extension service presentation had I ever revealed this information about myself. The one exception was when I returned to my own family farm community in central Minnesota. Then it seemed relevant that I was of the Behr-Krueger family and thus part of the German immigrant community called Salem in Zion Township in Stearns County, Minnesota.

When working in Africa, either Morocco, Niger, Mali, or Rwanda, villagers would usually ask me where I was from. *Only when I was in a "safe" location, i.e., where being an African, or, as in my case, an "almost African" was a plus* would I describe the island in the middle of the Mediterranean where my father's family had lived for hundreds of years with no record of living elsewhere. Thinking back, those were times my colleagues or the family I was visiting would express a signal of relief, "Well," they would say, "You are almost African. No wonder you like to work here in Africa." The Euro-American culture when I was growing up in Kenosha, Wisconsin, a working class, 50% Sicilian/Calabrian community, considered Sicilians "almost African." It was not a compliment then and so I intrinsically knew that it was best not to remind people from where I came. It was only in the safety of an African village, with no "white person" for miles around that I first felt comfortable admitting that, "Yes, I am Sicilian" and feeling deeply very proud of it. I am not sure of the exact time and place this occurred. It happened subtly, in one after another of many immersion experiences. Over decades, I often found myself surrounded by sub-Saharan Africans, first the Rwandans and then the Malians. I came to admire their views of life and respect the depth of their traditional knowledge and their art. When they would pose the inevitable first question, where are you from, I soon learned that the most interesting and positive response was to share my Sicilian origins. To be remotely related to these people I admired seemed to give me a safety zone to proudly say "what I was." In earlier decades of my life, I would answer this question apologetically or just say I was "Italian" which is what my dad said "never to say, because I wasn't Italian." In fact, Italians were a rival country and in 1860, just before my grandfather was born, the Italians fought the Sicilians and village-by-village took over our country, Sicily. So strong were the anti-Italian feelings that my grandfather, years later, forbade my aunt from marrying my uncle when it was discovered he had some Italian lineage.

Not until 1989 did I visit the village of my father for the first time. Most of the family who had immigrated to the United States or who were "first generation" like me thought it not important, some thought it even dangerous. So my cousin (a full Sicilian from this village area) and I (a half-breed) took the plunge, the first of our family to return to the village. We traveled with our husbands for protection. When I worked in Morocco, I felt even closer to "my people." Everyone seemed to look just like my cousins, my father, and grandparents. Father would often kiddingly say to me as a child that when he was my age and travel in the wagon to the Mediterranean Sea coast near Agrigento with his dad that if it was a clear day, he could see North Africa. As I worked in North Africa in the Moroccan field project with Matmora (underground) grain storage (Bartali et al., 1990) and shared meals on the farm in their homes, I imagined this was like my grandmother's childhood farm home. In the small towns, like Settat, Morocco, I imagined this is like the village where my dad grew up, similar wagons, similar square mudbrick houses. In April 1989 when I actually visited Aragona for the first time, it was a surprise to see the "out-of-context" shiny Mercedes trucks for collecting garbage navigating the narrow treeless streets beside the mudbrick adobe buildings that my dad so often described where he played and went to school. It looked like Morocco on the outside. Father's house had the street name painted on the brown mud corner of the square row houses. Just like in Morocco the street name was faded and flaked off, almost too worn to be read. Still in Aragona there were no restaurants or hotels, but the little shops had beautiful jewelry and glassware from Sicily and the European mainland.

These are my people and the land to which I belong to. The village records of our family go back hundreds of years. But thousands of years ago, this area of the world had a thriving port, Agrigento, just a morning's journey by wagon south of our village. There is no story of our family living elsewhere. How long did we occupy this village? As I wandered through the streets where father played, running with his friends, and walking out to his grandparents' farm with his sister and brothers, it seemed like every shop had a sign with our family names on it. I was home. Finally, I could tell people when they asked me where are you from, I could truthfully and specifically answer.

One of the important life lessons I learned from these greetings with the Apsaalooke and in African countries is that the Elders care little about where I worked, what my company or university name was, what my title or discipline was, or even what town I was presently living in. The Elders were asking me who were my parents and grandparents? Who were my people? What was my ancestral land? To what ancient people of the world do I belong? Once these fundamental questions are answered, the Elders usually remind me, "How can you know where you are going if you don't know from where you came?"

In learning where you yourself are from sometimes brings up multiple origins as it did in my case, along with pain and tears. Honesty in explaining where you are from to others and especially to yourself is critically important. In fact, it is essential to recognizing culture and incorporating cultures' role in one's classroom, student activities, administrative, corporate and staff responsibilities.

In 2009 when we were in the throes of creating and filming *Dancing Across the Gap* (Chaikin et al., 2010), Michele Curlee at Chief Dull Knife College (CDKC), Lame Deer, Montana, shared an insight that instantly helped me understand myself, my students, and others. Michele was the dean for Academic Affairs at CDKC. She is of French origin but part of a Northern Cheyenne family by marriage. To help her counsel students at this tribal college, Michele formulated this story using what is currently understood scientifically about global movement of tectonic plates. In a scientific venue, tectonic plate movement helps us understand earthquakes, volcanoes, hot springs, and other evidences of unstable land formations. This concept is also helpful to understand mixed cultural origins. Michele's students often have part Cheyenne, part Mexican origins. Some have black and/or white origins, along with their Native American heritage. "Pay attention," Michele counsels students, "to the origins within you." She continues, "Just like the tectonic plates in the earth, if you let one of those cultures dominate or suppress the other, expect a volcano or geyser eruption when you least expect it to happen." In Cheyenne culture, this could express as an extended "crazy period." It could be a period of substance abuse, people abuse, depression, or even high productivity. "Let no culture within you subsume the other. Give space and time to each of the cultures within you, to avoid the earthquakes and volcanoes of life," Michele explained. Since Michele's instructive story, a new understanding of tectonic plates has been revealed by Western science. From subsurface, oceanic research, scientists have recently found that tectonic plates moving over one another create geothermal energy. There may be a positive part of Michele's story yet to be told, about the cultural energy or resilience one can harvest from being a half-breed or other combination of cultures.

Sometimes we are too tied to the present, the things of Western culture success, the newest technologies, and we forget from where we have come. We forget to ask our Elders to tell us the stories of our own people. Like me, we forget until it is too late and all the storytellers have lost their voices and we must grovel in the remnants of centuries for the tiniest thread of the story of where we came from, and if lucky, discover a small kernel of wisdom from our people to share with new generations.

Tool #1. Know and understand from where you have come, your own culture(s).

So the first tool, to put in one's tool kit for incorporating culture into the food and agricultural sciences is to recognize culture, your own culture, your own people: to know from where you came. What was the culture we the professor, instructor, scientist, administrator, or policy maker came from and how can we show that we are proud of this heritage, whatever it may be. How can we recognize and celebrate from what cultures each of our students, patients, employees, or clients come? What are the opportunities and the barriers to learning and teaching that each of these cultures create?

DEFINITION OF CULTURE

What do we mean by culture? Culture has an ancestor in the Latin cultura, meaning "cultivation." Both land (agricultura) and people can be cultivated (cultura animi). Marcus Tullius Cicero, a Roman philosopher, politician, lawyer, political theorist, consul, and constitutionalist, known as Rome's greatest orator, has been credited with suggesting that "cultivation of the soul" can be "improved by labor, care, or study" just as cultivation of the land can be improved (https://www.vocabulary.com/dictionary/culture). From its origins, therefore, the term culture was closely related both to the land for growing food and to people and their innermost being.

The first known use of the English word culture was in the late 15th century as a noun *cul·ture \ kəl-chər*. In Middle English, the meaning was cultivated land, cultivation, from Anglo-French (Oxford Dictionary, 2016). *From this cropping systems term, later arose these three broader definitions for the noun:*

1. The beliefs, customs, arts, etc., of a particular society, group, place, or time
2. A particular society that has its own beliefs, ways of life, art, etc.
3. A way of thinking, behaving, or working that exists in a place or organization (such as a business).

But this is just the beginning of the diversity of this word. "Culture" just as a noun is a confusing word. The Merriam-Webster offers six definitions for it (including the biological one, as in "bacterial culture"). The problem is that "culture" is more than the sum of its definitions. If anything, its value as a word depends on the tension between them (Rothman, 2014).

In 2014 the noun *culture* became the "Word of the Year," the most frequently looked up word in the Merriam-Webster Dictionary (http://www.merriam-webster.com/words-at-play/2014-word-of-the-year). This reflects the interest of the Millennial Generation and the classroom syllabi that they were given as well as to the conversation at large, appearing in headlines and analyses across a wide swath of topics. Culture conveys a kind of academic attention to systematic behavior and allows one to identify and isolate an idea, issue, or group (Rothman, 2014). One can speak of a "culture of transparency" or "consumer culture." Culture can be either very broad (as in "celebrity culture" or "winning culture") or very specific (as in "test-prep culture" or "marching band culture").

Culture can be used both as a noun as above and as a verb, cul · ture \ˈkəl-chər\ *meaning to grow under controlled environment conditions.* Synonyms include crop, cultivate, grow, dress, promote, raise, rear, tend. Related words include breed, produce, propagate; plant, sow; gather, glean, harvest, reap; germinate, quicken, ripen, root, sprout. In this book, we return to the concept of human cultures' once close association with the term *agricultura* as it originated over 2000 years ago with the Romans and to the concept of culturing the land and associated water to produce food, a concept that evolved thousands of years earlier.

We will explore culture as the integrated pattern of human behavior that includes thought, speech, action, and artifacts, and depends upon the human capacity for learning and transmitting knowledge to succeeding generations particularly as it is intertwined with the culture of, or production of, food.

We began this chapter by suggesting that it is important to keep one's eye on the wisdom of the past while one creates a new educational system for future generations. We have also suggested that subsuming one of one's own cultures will result in volcanic, earthquake type explosions at some later point. Are these new ideas? No. When my family arrived on the boat from Sicily in 1910, they settled first in a Sicilian enclave on what is now Cabrini Street in Chicago near Hull House. One of the first things the older boys of the family said to their mom, Rosaria, was, "We are in America now, don't ask us to wear the garlic clove around our neck to ward off colds and other illnesses." This was the only story of the rapid assimilation of my family that survived intact. We all noticed, though, that our parents learned English and our grandparents never spoke English. Along with this rapid assimilation by the younger generations came a shedding of the "old ways," the Sicilian culture. The dominant culture at that time in Chicago was determined by the Northern Euro-American settlers who had preceded the Sicilians. Sicilian culture had a negative connotation even within our family. It was subtle, not verbalized, just acted upon.

Jane Addams (1912) and her colleagues who started Hull House and managed it when my family lived in the neighborhood saw this process happening and took action. For instance, once a week she invited the Sicilian mothers in the neighborhood to Hull House to demonstrate their household arts. It is likely that my grandmother Rosaria, daughter of a sheepherder, demonstrated carding and spinning wool during those events. Rosaria was also well known for her traditional herbal knowledge and was often called upon in the neighborhood to help with respiratory infections. Maybe she also shared these skills at the Hull House gatherings. Other Sicilian women demonstrated needlework and culinary arts at the Hull House Sicilian Day events. These were celebrations. The Sicilian homecrafts were celebrated. Daughters of these women were invited to the social event to learn from the elder women of the neighborhood. On these days at Hull House, Jane Addams was living her deeply felt philosophy of celebrating the rich culture of the Elders, the mothers of the neighborhood, in this case, those who came from Indigenous villages in Sicily with their families. Years later when these boys and their one surviving sister married and had children, the Elders continued to speak Sicilian (a strong dialect of Italian, not recognizable as Italian). The first generation children, my cousins, and I, though, were not allowed by our parents to learn our Indigenous language. Unfortunately, the fundamental part of our culture among us immigrants, our Sicilian language with regional dialects, is now lost. Jane Addams has some key ideas clearly shared in her book over a century ago, ideas that we might use now in higher education to recognize culture's role and comprehensively and inclusively internationalize.

What Do We Mean by Indigenous?

We have just introduced the term Indigenous. Since this concept is woven through most chapters of this book, we pause a moment now to explore its complexity and form a working definition of the concept and its importance to the food and agricultural sciences.

Indigenous may mean either "produced, growing, living, or occurring naturally in a particular region or environment," or "*innate, inborn.*" It comes to English from the Late Latin *indigenus*, which is from a Latin noun for "native" (*indigena*). The word has been in use since the early 17th century; our earliest evidence comes from Michael Stanhope's 1632 work, *Cures Without Care* (Merriam-Webster, 2016). In botany and other organismal sciences, we use the term "center of origin," meaning the location where this species evolved. How can this be applied to humans when there is no species distinction? Do we need a new term? Do we want to refer to the people who have lived together for some arbitrary period of time, such as 500, or 400, or 300 years in one place on a specific piece of land? "Indigeosity" is about a special and deep relationship with the land. It is a spiritual relationship, because it is the land that defines an Indigenous people. In fact, that particular land that defines the indigenousness of a specific group of people is actually responsible for their survival. The land, water, and air related to that location is used for a long period of time (how long?) as the sole supplier of the Indigenous people's food, shelter, and clothing.

We are all Indigenous to the planet Earth. This book puts a strong emphasis on learning food ways and agricultural production by people who are indigenous to much smaller portions of the land and water of the Earth. In this book, we reflect on the issues of Indigenous people, and, indeed, this book consists of contributions from Indigenous people in Europe, Africa, the United States, and Asia, all of whom are involved in the food and agricultural sciences. Why have we devoted so much space in this book on food and agriculture to Indigenous people? This book is for those who work with the land and water and the food that the land and water produces. By the time the reader comes to Chapter 12, Putting it Together: The Way Forward, it should be perfectly clear why we have emphasized understanding Indigenous approaches to food and agriculture.

Many of this book's key concepts are highlighted with stories from Indigenous people in Mali, Mongolia, and Montana. The relationship to the land and its environs is easy to define, but now we must decide on a time period. The Bambara farmers and trades people called Sanambeleans whom we will meet in Chapter 5, Listening With Subsistence Farmers in Mali, are indigenous to the spot they now occupy between Dialakoroba and Bougoula, Mali. Three hundred years ago, about 1710, the Sanambeleans (defined in a patriarchal sense, so just the men and their immediate families) emigrated from Bougoula (3 km NW) and established a new village. They called it Sanambele. Meanwhile, about 1710 the Northern Cheyenne were living in long houses in a sedentary lifestyle, living near where the border of North and South Dakota meet. To use 300 years as the benchmark for indigenous, the Cheyenne are newcomers to the High Plains of the United States. They are immigrants from the Dakotas. The Cheyenne are indigenous to Minnesota River Valley 500 years ago, 400 years ago to the Dakotas, 300 years ago to the Dakotas also. Three hundred years ago the Sanambeleans lived in Bougoula. Five hundred years ago, the Vaccarellos likely lived in Aragona, Sicily. Four hundred years ago, the Vaccarellos were definitely living there in the same village where they live today. The village of Aragona was founded 200 years before Sanambele.

Considering the human species, *Homo sapiens* is 200,000 years old and its immediate predecessors are millions of years old, the concept of attaching a time period to the definition of Indigenous does not work well. Defining Indigenous as the First People, first humans to occupy a location is also somewhat difficult. It is actually the concept of indigenousness that is so specifically important. Indigenous people have an intimate, spiritual relationship with the land. Their land is the source of their food, shelter, wealth, health, and livelihood. It is this definition that we will focus on in this book.

FIGURE 1.1
Native Americans at Dartmouth. A student at Dartmouth College celebrates Indigenous People's Day. As the United States celebrates Columbus Day, many locations will be celebrating a parallel but very different day: Indigenous People's Day—Lily Rothman, Time.com, October 10, 2016.

> The interaction between indigenous and non-indigenous societies throughout history has been complex, ranging from outright conflict and subjugation to some degree of mutual benefit and cultural transfer. A particular aspect of anthropological study involves investigation into the ramifications of what is termed first contact, the study of what occurs when two cultures first encounter one another. The situation can be further confused when there is a complicated or contested history of migration and population of a given region, which can give rise to disputes about primacy and ownership of the land and resources (https://en.wikipedia.org/wiki/ Indigenous_peoples). Given the extensive and complicated history of human migration within Africa, being the "first peoples in a land" is not a necessary precondition for acceptance as an indigenous people. Rather, indigenous identity relates more to a set of characteristics and practices than priority of arrival. For example, several populations of nomadic peoples such as the Tuareg of the Sahara and Sahel regions now inhabit areas where they arrived comparatively recently; their claim to indigenous status (endorsed by the African Commission on Human and Peoples' Rights) is based on their marginalization as nomadic peoples in states and territories dominated by sedentary agricultural peoples (https://en.wikipedia.org/wiki/ Indigenous_peoples).

In this book, we will be calling upon a number of Indigenous people from various cultures in Asia, Africa, and the United States to compare and contrast with my own observations, just one generation from an Indigenous community in southern Europe (Figure 1.1). These people from Mali (Bambara), Mongolia, the Eastern Shoshone, the Apsaalooke, and the

Northern Cheyenne will be sharing with us suggestions for additional tools to help us recognize and incorporate culture's role in the food and agricultural sciences. In most cases, we ourselves have been in conversation with each other for many years, with some of the voices shared in this book, for over a decade. The knowing and understanding of from where you have come, your own culture(s) is(are) captured in the following excerpt from a university senior's reflective log during the first week of class as he recognized the need to come to terms with from where he had come.

> Week 1. AGSC 465R Health, Poverty, Agriculture: Concepts and Action Research. Undergrad, Honors College student's reflective log. *"Entering these readings* (Bennett, 2004, Savory and Butterfield, 1999, *Helena Norberg-Hodge, 1991*), *it becomes apparent that the development process in communities must start with a demonstration and reflection of self. Approaching intercultural situations as cultural also demands attention to detail and a real evaluation of one's motivation for being in a community. Grounded in the holistic process and intercultural competence (a la Bennett), the importance of this self-reflection and an understanding of the context become paramount. Further, I think language is incredibly important even within dialects of the same lexicon. Last semester in Dr. C.'s sociolinguistics course we talked about the idea of social capital + language and how it is related to a habitus or a collective of our…"—C.J. Carter Honors College student majoring in Film and Photography at MSU.*

Ethno-Relativity

In 2007 I learned a great lesson about creating a classroom which incorporates cultures' role from my teaching assistant. In January, I invited Keriba Coulibaly to be the teaching assistant of my course BIOO 162CS The Issues of Insects and Human Societies. Keriba is one of the Malians, an agronomist, who came to live in Bozeman, Montana, and take courses at MSU for 2 years. He has been a major translator and interpreter of culture's role in food and agriculture and will reappear several times in this book. The course that Keriba assisted in was a University Core course in Contemporary Issues in Science that had evolved from a University Core course in Multicultural/Global Dimensions. The course dealt in large part with insects for food and feed (Dunkel, 1996; Defoliart et al., 2009) as well as sustainable protection of food sources from insect competition. The course retained a strong flavor of intercultural understanding in the concepts and content shared. Keriba was my graduate student and had been trained in English, the holistic process, and laboratory bioassays using insects and traditionally important natural products. Keriba was of the Bambara ethnic group. He grew up in a village in western Mali. Bambara was his mother tongue. So the first day of class with a group of 40 students seemingly eager to hear each of us rather diverse instructors greet them. I turned to Keriba to begin the greeting. Immediately his presence filled the room. I will always remember this moment. I had taught this course already for 18 years, had received a couple of teaching awards, but did not, I feel, ever come close to the presence that Keriba seemed to create in a moment in that classroom. He began the daily greeting process in Bambara. "Ini tilin (Good Afternoon) Au ka kene? (How are all of you?)" Then he continued with the specific caring phrases for youth "Ma ka kene? (How is your mother?)" "Fa ka kene? (How is your father?)" And the specific caring phrases for our beyond traditional age students "Somogou ka kene? Iche ka kene? (How is your wife/husband?)" "Demisinu ka kene? (How are your children?)" Just in those few moments of greeting, Bambara style, students knew this was to be a class in which each of them as an individual mattered and as a group we would care about each other. Keriba returned to Mali September 2007 as scheduled, but he changed the way I teach forever. It was amazing to my students and to me as well that such caring statements could be part of sidewalk greetings or of the beginning of class for the day.

Inclusivity begins with recognizing where you are coming from and where your colleagues and students and administrators are from. If one does not recognize at all (or subconsciously denies) that there are other cultures, one will not "see" other cultures. This "ignoring" of another's culture results in pulling others into one's own culture. This is one of the positives about immersion in another culture. In immersion, you are expected to respond the way someone in that culture might respond, not your own. It may be opposite from the way you learned, but one quickly learns what is considered norm for the culture in which you are immersed.

The opening of one's awareness to the importance of cultural differences, deeper than physical traits, language, cultural icons, or religious backgrounds, is an essential first step toward inclusivity. This is the beginning. This is the awakening process, but still in a world view of ethnocentricity. Two of the forms of ethnocentricity that emerge often are: (1) "I see that there are differences, but my own culture is still the best"; and (2) "there are differences between my culture and others, but basically we are all quite similar."

Milton Bennett (2004) has been a leader in both describing and measuring this process of cultural awakening as a continuum. Ethno-relativity is the next step in intercultural competency following recognizing other cultures that Bennett describes. In this broad category of response, one recognizes subtle, subsurface differences between cultures and accepts these profound differences as ok, just different. In advanced levels of ethno-relativity, one adapts to eventually move seamlessly between different culture systems. Using categories or "boxes" to describe a behavior pattern such as response to other cultures has shortcomings. Although Bennett describes the process in a linear and stepwise path, such a categorical view of intercultural development does not seem to fully describe the process of becoming ethno-relative in actual practice.

Tool #2. Strive for an ethno-relative approach.

In practice, we have noticed that development of intercultural competency may be a nonlinear process. There may be a tendency in most situations to be adaptive, but certain experiences or combination of circumstances may stimulate a person to choose another form of intercultural response. It is possible that the unconscious-directed responses operate at a different level of intercultural development than do one's conscious-directed responses in an intercultural situation. From Paul Rozin, the world authority on the "disgust factor," food and agricultural scientists have learned that disgust becomes imbedded in our psyche at an early stage and it is almost a psychological contamination factor that prevents otherwise logical, well-educated persons from considering certain normal-to-billions food (Rozin and Fallon, 1987; Fallon and Rozin, 1983), such as insects, as nutritious, delicious, and sustainable ingredients (van Huis et al., 2013).

Why is a discussion of cultural origins and responses to cultures different from one's own relevant here? Why is it especially relevant in incorporating cultures' role in the food and agricultural sciences? The easiest way to answer this question is with a story.

> Amy Wright, a Tlinget from Hoonah, Alaska, and a 2010 graduate of MSU's College of Liberal Studies, was clearly interested in sustainable food production among her people, particularly in relation to berry picking, salmon berries, and other berries of the area when she arrived at MSU in Bozeman. Amy was also very clear about why she did not choose to major in Horticultural Science after being a student at MSU for a year. She confidently said, "I really didn't see that the College of Agriculture was going to give me the things I needed for life. It seemed clear that Horticultural Science was about Plants and Science and NOT about people. I was worried that the Horticulture Program would put me in a greenhouse for the rest of my life."

Amy was the first student who explained to me the "border crossings" that she must make every time she comes to class. It was not until she was a junior that she discovered a class without border crossings and a place that connected people, science, and the opportunity to explore an issue that was important to her own Tlingit people, but it was too late then to change majors. Amy found a safe place to explore/research a topic that was important to her own self and her people. In the process, she taught me how to articulate the border crossing concept to other professors. Because her research paper then became a permanent document in the resources of my course for students that followed, Amy also opened the door for other Native Americans to reflect on the importance of their own food systems. It is possible that her paper influenced the Small sisters (Kurrie and Tracie) as they together opened holistic process conversations with their people (Apsaalooke) to learn about their desired quality of life (Small et al., 2012, 2014). The details of this project are found in Chapter 6, Listening With Native Americans. Now, 4 years later, Amy has returned to Hoonah, Alaska, with more knowledge about managing Salmon berries and the Small sisters have guided a *Let's Pick Berries Project* that has emerged from within their community to address issues of Youth/Elder connections, nutrition and exercise, and revitalizing their culture (Small et al., 2012, 2014).

Why are Native American students not choosing to major in the food and agricultural sciences? What did we learn from Amy and our other Native American students about the teaching techniques that caused her to feel more at home? Some examples are recognition in the classroom that traditional ways have value; serious appreciation of "other-than-scientific ways of knowing"; and simply the arrangement of a classroom where the professor is part of the circle of learners, rather someone in the front of the room dispensing knowledge.

Another example of knowing where you are from and from where one's students, colleagues, customers are from is simple, but difficult for Euro-American cultures to implement. It is about voice patterns and politeness in conversation. Sierra Alexander is a Northern Cheyenne and at the beginning of the semester she was very quiet in classes. When I asked Sierra to explain to non-Native students in our class preparing to visit her reservation the oral tradition and speech-hesitation followed by thoughtful-speaking of the Northern Cheyenne, she looked puzzled. It was not until that moment that Sierra realized her quietness in classes was a normal Northern Cheyenne response. It was the professors who were causing her quietness by not providing the required hesitation time for her culture. So much rich communication can be lost when one does not know from where one comes. A non-Native classmate of Sierra's summed it up in her reflective log, "*...so articulate, ...poetic and inspiring dialogue!...the long tradition of oral storytelling creates a skill of hesitation-yet-thoughtfulness in...speech.*"

For many years it has been common for textbooks (Kittler and Sucher, 2008) to fail to include insects when describing food from the world's various cultures. Even when textbooks or other documents recognize the use of insects in other than Euro-American cultures, the statement is often made that in the United States edible insects are not used. Research in just a dozen years of the *Food Insects Newsletter* (DeFoliart et al., 2009) revealed 65 species of insects recorded being used as food by just the 565 federally recognized Native American nations occupying the United States. Two billion people of the world use insects deliberately as the normal part of their diet. To have omitted this protein source for one-third of the people of the world sends an important message about the ethnocentricity of the editors and authors. The Food and Agricultural Organization (FAO) of the United Nations has publicly recognized insects as one of the world's sustainable protein sources. This recognition has been summarized in the document, *Edible insects: Future prospects for food and feed security*. Within the first 24 hours of release of this document (van Huis et al., 2013), the most comprehensive treatment of this food source, there were 1 million downloads of the report. Thus far there have been 6 million downloads worldwide (van Huis and Dunkel, 2016). In 2016 two textbooks were published in the United States that followed on this seminal report of FAO: one recognized edible insects as one of the worldwide sustainable protein sources for people (Naduthur et al., 2016); and the other entire textbook was devoted to the industrial process of using insects as food ingredients (Dossey et al., 2016). Even with this clear recognition by Western cultures of the acceptance of other cultures' food sources, skeptics remain in other organizations with a global food focus. That is the urgency of this current book: to attempt to decolonize the language of the food and agricultural sciences by providing examples of the importance of raising awareness and the tools to decolonize language and learning structures.

Honestly knowing the nature of one's own cultural filter and how one consciously and unconsciously responds in an intercultural situation will create a broader, wider field in which to learn and to teach. Being open to and excited about the rich resources that a student of a minority culture brings to the classroom, the playing field, the dorm, or any venue on campus enriches any institution of higher education. Being open to a similar cultural richness in a food and consumer science company boardroom or at a government policy meeting or a funding decision-making table can also unleash creativity. It deepens the learning experience and the development of innovative foods to feed the world if the freedom to express these cultural resources is provided.

Chapter-by-Chapter Summary

In this book, we detail the tools used in the birth of a number of programs that evolved in the food and agricultural sector of higher education, programs that incorporated cultures' role in their programs. Most of our examples are drawn from two programs which are examples of the Quiet Revolution now underway in food production and consumption and its related teaching and learning processes (Martin, 2014; Nabham, 2014; Roos et al., 2010; van Huis et al., 2013). We will illustrate how two small-scale, discovery-based, mentored research programs linked undergraduates in Montana with Malian scientists and subsistence farmers in West Africa (www.montana.edu/mali) and Mongolian seminomadic communities (BioRegions International, 2014). Each program grew into an *interdisciplinary, multiinstitution, international service-learning, action research teaching and learning* virtual center, and a BioRegions program. Both programs began by recognizing and incorporating cultures' role in the issues of food production, food consumption, nutritional education, and related health issues. The food and agricultural sciences continue to be the heart of these programs. Those looking in from the outside may say this is a "soft science program, simply sociology, cultural geography, and anthropology." These are examples of programs that are "making a difference in their communities" because the "hard sciences" are *not* isolated from the "soft sciences" (NCR, 2009). Both programs used similar tools which they adapted from the same source, Savory and Butterfield (1999) and their approach to range management called Holistic Thought and Management. In our research programs, we call our modifications the Holistic Process. But first, we must remember to ask the *most* basic question: From where have you come?

These fundamental parts of knowledge, "hard science," "soft science," and the humanities, are essential for acquiring new knowledge. They form a filter through which the individual acquires new information and need to be presented as an integrated body of knowledge with these several parts not in competition with each other, but respectfully complementing each other. Traditional wisdom is also an important reservoir of knowledge that often does not appear in textbooks in the hard or natural sciences, the soft or social sciences, and the humanities. This book delves deeper into TEK and examines the importance of recombining these separate products of Western culture's scientific processes and scholarly activity with the knowledge base, TEK, that is created by the process of Native Science (Cajete, 1999).

Although this is an interesting intellectual challenge, we pose the hypothesis that this recombination of Western Culture Science and Native Science is essential for our food system, our environmental health, our personal health, and for our very existence as human beings on Earth.

In Part I of the book, we present our view of the Fundamentals of the Culture and Agriculture Relationship. Chapter 2, Failures, will remind us of failures, how to recognize them and to move beyond failure. Chapter 3, Decolonization and the Holistic Process, will provide an in-depth explanation of the holistic process and guide the reader to recognize colonizing methodologies and how to decolonize those approaches in many venues that include the classroom, mentoring sessions, even in room arrangements, and interview formats—basic concepts to avoid failure. Chapter 4, Immersion, will explore examples of several types of immersion and present a strong case for the importance of immersion.

The heart of the book is found in Part II: Listening In and Between Communities. This part of the book is replete with examples of how the holistic process and decolonizing methodologies can take place in practice. Chapter 5, Listening With Subsistence Farmers in Mali, brings the reader to Mali, mainly to a subsistence farming village in sub-Saharan Africa. Chapter 6, Listening With Native Americans, brings the reader to two Plains Native American reservations in Montana, the Apsaalooke, and the Northern Cheyenne. Chapter 7, Listening Within a Bioregion, is set in Mongolia among the seminomadic herders of the Darhad Valley. In this chapter Clifford Montagne shares the stories of how the holistic process guided the development of various projects within the Darhad Valley program, part of BioRegions International. In Chapter 8, Listening Over Power Lines: Students and Policy Leaders, we walk with the reader through the process of how the university classroom as well as middle school classrooms on three continents can be connected with each other *and* with the USDA National Institute of Food and Agriculture (NIFA) food and agricultural policy makers in Washington, DC. Chapter 9, Listening With Students, is replete with voices of Millennial Generation students from many nations and representing many aspects of the food and agricultural sciences.

This book will conclude with a call to action, Part III: Bridging the Gap Between Food and Agricultural Sciences and the Humanities and Social Sciences. In Chapter 10, Two Cultures and a Second Look: Humanities and Food Science, we specifically examine the gap between the food and agricultural sciences and the humanities. Chapter 11, Couples Counseling: Native Science and Western Science, recognized another gap which is overlain on the issues discussed in Chapter 10, Two Cultures and a Second Look: Humanities and Food Science. If Native Science and Western Science were people in a marriage relationship, we would recommend marriage counseling. Native Science and Western Science when intertwined complement each other well, each contributing essential understandings to the whole of place-based, culture-based knowledge. Chapter 11, Couples Counseling: Native Science and Western Science, aims at replacing the skepticism and fear of those practicing each process with the rewards from achieving respect and understanding between the two parts of this couple. Chapter 12, Putting it Together: The Way Forward, provides pathways for putting these ideas together and creating a new way forward, a way that incorporates cultures' role in the food and agricultural sciences.

Incorporating cultures' role in the food and agricultural sciences is at the intersection of teaching, scholarship, and engagement. In large part the nudge and push to get this book written has come from the Millennial Generation who were quite clear that this is what they are about "connecting the dots between our technical education and real people in the real world." This connection process so many of the millennials speak of is not about a time and place *after* graduation but for action taken, active learning, *during* their degree. This book and the pedagogy that led up to it is meant to address the millennial's and the new generations' requests about changing the teaching process, a Quiet Revolution.

References

Addams, J., 1912. Twenty Years at Hull-House With *Autobiographical Notes*. The MacMillan Company, New York, NY. (c. 1910)

Bartali, H., Dunkel, F.V., Said, A., Sterling, R.L., 1990. Performance of plastic lining for storage of barley in traditional underground structures (Matmora). J. Agric. Eng. Res. 47, 297–314.

Bennett, M., 2004. Becoming interculturally competent. In: Wuzel, J. (Ed.), Toward Multiculturalism: A Reader in Multicultural Education, second ed. Intercultural Resource Corp, Newton, MA, pp. 62–77.

BioRegions International. 2014. <http://bioregions.org/mission/>.

Cajete, G., 1999. Native Science: Natural Laws of Interdependence, Natural Laws of Interdependence. Clear Light Publishers, Santa Fe, NM, 339 pp.

Chaikin, E., Dunkel, F., Littlebear, R., 2010. Dancing Across the Gap: A Journey of Discovery. Montana State University. 56 minute documentary aired Montana PBS. Northern Cheyenne TV.

Defoliart, G., Dunkel, F.V., Gracer, D., 2009. The Food Insects Newsletter: Chronicle of a Changing Culture. Aardvark Global Publishing Co, Salt Lake City, UT, 414 pp.

Dossey, A.T., Morales-Ramos, J.A., Rojas, M.G. (Eds.), 2016. Insects as Sustainable Food Ingredients: Production, Processing and Food Applications. Academic Press, San Diego, CA.

Dunkel, F., Wittenberger, T., Read, N., Munyarushoka, E., 1986. National Storage Survey of Beans and Sorghum in Rwanda. Miscellaneous Publication. Minnesota Experiment Station, University of Minnesota, St. Paul, MN, 205 pp.

Dunkel, F.V., 1996. Incorporating food insects into undergraduate entomology courses. Food Insects Newslett. 9, 1–4.

Dunkel, F.V. 2004. Discovery-Based Learning for Undergrads: Facilitating Farmer-to-Farmer Teaching and Learning. 2004–2007. USDA CSREES Higher Education Challenge grant.

Dunkel, F.V., Peterson, N. 2002. Discovery-Based Learning for Undergrads: Collaborative Research Support Programs. 2002–2005. USDA CSREES Higher Education Challenge grant.

Dunkel, F.V., Coulibaly, K., Montagne, C., Luong, K.P., Giusti, A., Coulibaly, H., et al., 2013. Sustainable integrated malaria management by villagers in collaboration with a transformed classroom using the holistic process: Sanambele, Mali and Montana State University, USA. Am. Entomol. 59, 15–24.

Fallon, A.E., Rozin, P., 1983. The psychological bases of food rejections by humans. Ecol. Food Nutr. 13 (1), 15–26, <http://dx.doi.org/10.1080/03670244.1983.9990728>.

Kittler, P.G., Sucher, K.P., 2008. Food and Culture, fifth ed. Thomson Higher Education, Belmont, CA.

Lamb, E.M., Dunkel, F., 1987. Studies on the Genetic Resistance of Local Bean Varieties to Storage Insects in Rwanda. Miscellaneous Publication. Minnesota Experiment Station, University of Minnesota, St. Paul, MN, 122 pp.

Martin, D., 2014. Edible: An Adventure Into the World of Eating Insects and the Last Great Hope to Save the Planet. New Harvest, Houghton Mifflin Harcourt, New York, NY, 250 pp.

Merriam-Webster, 2016. Americans Celebrated Indigenous Peoples' Day. <https://www.merriam-webster.com/news-trend-watch/americans-celebrated-indigenous-peoples-day-20161011> (accessed 16.12.16.).

Nabhan, G.P. (Ed.), 2014. Savoring and Saving the Continent's Most Endangered Foods. Chelsea Green Publishing, White River Junction, VT.

Nadathur, S., Wanasundara, J., Scanlin, L. (Eds.), 2016. Sustainable Protein Sources. Elsevier Publ. Co. Academic Press, Boston, MA.

NCR, 2009. Transforming Agricultural Education for a Changing World. National Research Council of the US National Academies. The National Academies Press, Washington, DC, 194 pp.

Norberg-Hodge, H., 1991. Ancient futures: Learning from Ladakh. Sierra Club Books, San Francisco, CA, 204 pp. (pp. 1−71 and 101−192).

Oxford Dictionary, 2016. English Oxford Living Dictionary. <https://en.oxforddictionaries.com/definition/culture> (accessed 07.11.16).

Roos, N., Nurhasan, M., Thank, B., Skau, J., Wieringa, F., Khov, K., et al., 2010. WinFood Cambodia: Improving Child Nutrition Through Improved Utilization of Local Food. Poster for the WinFood Project, Denmark, Department of Human Nutrition, University of Copenhagen.

Rothmam, J., 2014. Cultural Comment: The Meaning of "Culture." The New Yorker. <http://www.newyorker.com/books/joshua-rothman/meaning-culture> (accessed 07.11.16.).

Rozin, P., Fallon, A.E., 1987. A perspective on disgust. Psychol. Rev. 94 (1), 23−41, <http://dx.doi.org/10.1037/0033-295X.94.1.23>.

Savory, A., Butterfield, R., 1999. Holistic Management: A New Framework for Decision Making. Island Press, Washington, DC. (Chapters 9 and 10)

Small, T., Berg, A., Dauw, C., Dunkel, F., 2012. Revitalizing berry picking in the Apsaaloooke community: preserving traditions and improving community health <http://www.montana.edu/mali/pptsaspdfs/BergAndrewCrowTeamPoster.pdf>.

Small, T, Dunkel, F., Robison, G., Nyman, T., Setzer, D., Killian, C., et al., 2014. Let's Pick Berries: Addressing Apsaalooke needs for nutrition, plant propagation, community engagement, and youth-elder connections through the holistic process in a community-based, service-learning course. Poster presentation at the national meetings of the North American College Teachers of Agriculture. Bozeman, MT. <http://www.montana.edu/mali/pptsaspdfs/Ap%20-%20Robison%20Greta%20Poster%20-%20Cultural%20Nutritional%20Significance%20Traditional%20Berries%20Apsaalooke%20People.pdf>.

van Huis, A, Van Itterbeeck, J., Klunder, H., Mertens, E., Halloran, A., Muir, G., et al., 2013. Edible insects: future prospects for food and feed security. FAO Forestry Paper 171. Food and Agriculture Organization of the United Nations, Rome, Italy, 187 pp.

van Huis, A., Dunkel,, F., 2016. Edible insects, a neglected and promising food source. In: Nadathur, S., Wanasundara, J., Scanlin, L. (Eds.), Sustainable Protein Sources. Elsevier Publ. Co, Boston, MA, pp. 341−355. (Chapter 21)

Weaver, D., Phillips, T., Dunkel, F., Nance, E., Grubb, R., 1995. Dried leaves from Rocky Mountain plants decrease infestation by stored product beetles. J. Chem. Ecol. 21, 127−142.

Further Reading

Dunkel, F., Clarke, S., Kayinamura, P., 1988. Storage of Beans and Sorghum in Rwanda: Synthesis of Research, Recommendations and Prospects for the Future. Miscellaneous Publication. Minnesota Experiment Station, University of Minnesota, St. Paul, MN, 74 pp.

CHAPTER 2

Failures

Florence V. Dunkel

Department of Plant Sciences and Plant Pathology, Montana State University,
Bozeman, MT, United States

This chapter is about the Gap, a gap that faces us all—one in our understanding that is as deep as your misassumptions and my biases, one that is as wide as our cultural backgrounds and tendencies, one that is as pervasive and pernicious as everyone's good, but poorly informed intentions. This chapter is about the millions of lives, billions of dollars, and entire careers of scientists that fall in the Gap (Fig. 2.1). Western culture science has the technical knowledge to put poverty in museums, to eliminate cerebral malaria, and other global diseases vectored by insects, and to eliminate stunting that affects 40% of the children in parts of the world, and yet, we have not accomplished this. Why not? Because as scientists most of us have not seen the gap, and so we do not manage our research to address this gap. Some scientists see the gap, but do not have the skills to bridge the gap. WHY?

What is this Gap? The gap is a hiatus, a chasm that exists between two groups who wish to communicate with each other. Often neither group envisions or engages in bilateral communication, where both parties are enabled or willing to truly hear what the other group is conveying.

This situation is as relevant to food scientists as to those working with diet-related disease, environmental conflict, and nearly every imaginable scenario where two parties need to effectively communicate. Along with this gap come high levels of stress and other forms of mental and emotional unrest in the world and barriers related to world peace. This chapter is about recognizing the gap in effective communication and the rest of the book is about building skills to bridge the gap. I want to share with you my passion for overcoming that communication gap—a vision to make use of my own and others' (sometimes embarrassing) failures and to begin the process of improvement, of incorporating cultures' role in the food and agricultural sciences.

CONTENT

Case Study 1. The Chicken Award 25

Case Study 2. The International War on Locusts and Grasshoppers 29

Case Study 3. Fry Bread 34

Case Study 4. The Use of River Systems in the Western United States to Produce Food for the Nation 35

Concluding Reflections on Failures 36

References 37

23

Incorporating Cultures' Role in the Food and Agricultural Sciences. DOI: http://dx.doi.org/10.1016/B978-0-12-803955-7.00002-X

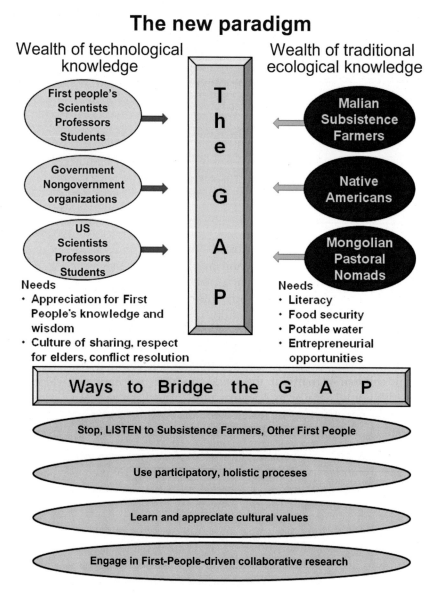

The new paradigm

Wealth of technological knowledge

First people's Scientists Professors Students

Government Nongovernment organizations

US Scientists Professors Students

Needs
• Appreciation for First People's knowledge and wisdom
• Culture of sharing, respect for elders, conflict resolution

The GAP

Wealth of traditional ecological knowledge

Malian Subsistence Farmers

Native Americans

Mongolian Pastoral Nomads

Needs
• Literacy
• Food security
• Potable water
• Entrepreneurial opportunities

Ways to Bridge the G A P

Stop, LISTEN to Subsistence Farmers, Other First People

Use participatory, holistic proceses

Learn and appreclate cultural values

Engage in First-People-driven collaborative research

FIGURE 2.1

Diagram of the chasm that exists people producing primarily Western culture technical information and Indigenous knowledge and ways to bridge the gap.

Each of the four case studies in this chapter focuses on facilitating communication at a different organizational level. The first case study is about how I learned an important life lesson from a woman who has never had any formal education in a school. The second case study takes place at an international level, between nations. The third case study is from the Plains Native Americans in Montana. In the final case study we will explore food production and consumption patterns among the Native Americans and the Euro-Americans related to water in the western United States. The final case study is from the Plains Native Americans in Montana.

CASE STUDY 1. THE CHICKEN AWARD

The Global, Participatory Integrated Pest Management (IPM) Collaborative Research Support Program (CRSP) began in 1993. My role was to focus on insects destructive to staple crops postharvest in Mali, specifically at the smallholder subsistence farmer level. To facilitate communication with these villages, we had taken on a Bambara tutor and most of us acquired a 0+ level of competency in this primarily oral language. To make up for our lack of relevant language skills, the project was based on participatory diagramming. To be exposed to the best practices in participatory process at the beginning of the project, I went to London to visit the International Institute for Environment and Development. There I learned about Chambers and his group who focused on empowering the smallholder farmers globally (Chambers et al., 1989). I left the Institute with a half meter high stack of reports filled with illustrations of participatory diagrams hand-drawn by groups of farmers on several continents—maps, Venn diagrams, matrixes and pyramids of constraints, transects, system diagrams, and timelines. Gueye (1993) was the editor of one of those dozen books I carried to Mali. In these reports was the secret to appreciative listening, listening to learn. But we were about to reach beyond just learning. We were about to undertake participatory *research*. In the United States and Mali, with my US and Malian colleagues, all agricultural scientists, we tried out the participatory process in weeks of practice sessions. We were, I thought, well prepared for communicating with this new-to-us group of smallholder, subsistence farmers in Mali.

It was the sixth year of our collaborative research support in a set of small Bambara farming villages each with a population of about 500–1000 located in south central Mali. I had been working in other Bambara villages in the north central part of Mali in prior years. We had been using participatory diagramming, but the topic of the discussion and the diagramming that followed was always agricultural, crop related. One of the best examples of this narrowed focus is the "cowpea story." Cowpeas had become my focus because neither of these farmers' main grains, millet, nor sorghum had major postharvest insect problems due in large part to their kernel size (millet) or their antibiosis compounds (sorghum). One of the best hedges against the "hunger season" (period when these grains run out just before the new harvest), it seemed, was cowpeas. Cowpeas, *Vigna unguiculata*, are high in one of the essential amino acids, lysine, missing in corn, their third main grain. Villagers, however, told us that it was impossible to store cowpeas more than two months after harvest when they will have been entirely consumed by the larva of a bruchid beetle, the cowpea weevil, *Callosobruchus maculatus*.

My colleagues and I at University of Minnesota had just finished a project on storing the dry edible bean, *Phaseolus vulgaris*, in Rwanda. We published results of many sensory evaluations (Edmister et al., 1986) and bioassays with the bruchid beetle, *Acanthoscelides obtectus*, that attacks *P. vulgaris* after harvest including some results using neem kernel extract as a growth regulator for the beetle (Dunkel et al., 1990, 1995; Sriharan et al., 1991). The sensory evaluations included preparing and cooking the beans exactly the way the rural women did it: rinsing in water, discarding the rinseate, and then boiling for an hour or more. It was a logical decision, we thought, to look to neem kernel extract as a way to solve this long-term storage issue with another species of bruchid beetle on a related legume, cowpeas in Mali. Neem trees, *Azadirachta indica*, seemed to grow everywhere we worked in Mali, in the villages and in the capital city, Bamako. Making the extract from the seeds did require a cold manual press, but the press that worked well was produced in Mali and sold for about 200 USD. The neem tree had green leaves year-round, but produced the seeds to make the kernel extract once, sometimes twice a year in this area. Based on this information, we launched an in-depth research program in the laboratory at Montana State University (MSU) and in the laboratory at l'Institut d'Economie Rurale (IER) Sotoba (Bamako) (Gamby and Dunkel, 1997; Jenkins et al., 2003). At the time this seemed a logical solution to the cowpea issue, but it wasn't.

The root of our failure was that we were oblivious to the diverse practices of cowpea storage in Malian homes. We did not listen for reference to or ask about specific local cultural practices of storing cowpeas and preparing cowpeas for meals. Villagers shared with us stories of poor storability of cowpeas and how prized they are for special meals, but, looking back, there was a hesitancy to share any "better ideas" with us, relatives of former colonizers (Europeans). Looking back very carefully, it is not that we were not listening, it was that the Malians were not saying and we did not at all detect their hesitancy. Not hearing any better suggestions to our proposed plan, we developed a set of drawings to convey how the cowpea weevil uses the cowpea for a safe, protected incubator filled with food to nourish its young. The series of drawings then explained how to triage cowpeas after harvest, make a water solution of the neem kernel (the nut inside the seed) made from local tree seeds, and treat the best quality cowpeas for long-term storage. We then instructed the villagers to shake the cowpeas in the solution and let them dry. We knew there were 135 or more insect repellent, oviposition deterrent, and growth regulating compounds in the neem kernel extract (Biswas et al., 2002) and we had tested it in the laboratory (Dunkel and Gamby, 1997). It worked well in preventing the development of cowpea weevils, but we neglected to cook the cowpeas Malian style and taste them.

For several years, we provided each Peace Corps trainee in Mali with a packet of these laminated cards. The drawings had been prepared in a participatory way, but not with the villagers: that could have been the missing link, but we did not get feedback directly from the Peace Corps Volunteers after their training was complete, only from their trainers. We received no negative feedback from the two Malian trainers, and no stories of adoption of the long-term cowpea storage process being used. Here again, the politeness to former colonizers, Europeans, may have influenced the exactness of the response. We had neglected to just listen to the villagers after establishing a culturally safe environment where trust and respect thrived. By 2006 middle school students at the International School of Bamako had undertaken publishable scale experiments with neem kernels they collected from just outside their classroom. These city students collaborated with rural students in a town called Kati, just north of Bamako. Those rural students discussed their classroom research with their parents and grandparents and learned that they used ashes and onion skins mixed with the cowpeas during long-term storage to preserve this food source almost until the next cowpea harvest. Malian parents and grandparents were not hesitant to share the truth about neem with their children. Their children did not represent former colonizers (Europeans). The key piece of information missing was that these village women did not rinse their cowpeas before cooking them, and after our laboratory experiments, we did not cook the cowpeas in a Malian way and taste them. The key piece missing was that these Rwandan women did rinse their beans before cooking (Dunkel et al., 1995), and the Malian women did not. Hence, the bitter notes of the neem kernel extract were offensive to the Malians and so protection with neem kernel extract was not adopted. For the Rwandans, it didn't matter, their traditional practices took away the coating of neem along with the bitter flavor.

In 2005 MSU students in the Mali Extern Program made a simple suggestion, "Let's just try the holistic process," they proposed. I was skeptical, but open to a new approach since what we had tried with typical Western science methods did not address the cowpea problem in practice, locally. The first step in the holistic process is to listen. Since these were only Bambara speakers with whom we were listening, we had to assemble a team of scientists who were native Bambara speakers to listen with us. Our listening team was a group of social scientists and plant protection specialists all from l'Institut d'Economie Rurale (IER). We returned to the farming villages where we had been collaborating for six years. We sat down with villagers in focus groups: women scientists and women MSU students with village women farmers; and male scientists and male MSU students with men farmers in the village.

We learned many new things, for instance, from whom village women liked to receive new information; and that wells and reliable access to water was a constraint to the good life; (Kante et al., 2009). From one village, Sanambele, we learned that the two main constraints to achieving the "good life" was first, their children dying from malaria and secondly their children being "hungry." We visited several villages each day. The day we visited Sanambele, we received a message while at our next destination that it was very important on our return to pass by the village of Sanambele the same day. We did, hot and tired, and as we drove by the edge of the town on the main dirt road between villages, in the distance we could see a woman smiling broadly and walking quickly toward us, almost running, with a live chicken held high in each hand.

It was the chicken award. We had bridged the communication gap. To sit down and listen, to listen as we did for hours while the women discussed the village concerns among themselves, and the men separately and then together and then to listen intently to their decision about the main village concerns was of high value, worthy of the awarding of two live chickens.

Years later, the level of trust and respect between the Malian villagers, our students, and me, had risen considerably. We had just completed a 4-year grant from United States Agency for International Development (USAID) for a country-based model of graduate training. Keriba Coulibaly was one of the mid-career scientists chosen to receive this training managed by us at MSU. Coulibaly is a common Malian name and Keriba has no known relationship to Hawa Coulibaly. He had listened carefully to our many years of frustration trying to "solve" the cowpea storage issue. One day, he shared with me that his mother and grandmother actually used shea butter to coat the cowpeas for long-term storage. They made the shea butter from a water extraction of the crushed nut of the shea tree and used the butter for cooking. It worked well for Keriba's family. Keriba and I tried the method in a sophisticated bioassay analysis in my laboratory at MSU, evaluating more than a thousand eggs of the cowpea weevil. It was clear, the coating of shea butter stopped embryogenesis in a dose-dependent response. Later that year, Hawa Coulibaly, the woman in Sanambele who presented me with the chicken award in 2005, showed with us how she had been using a local mint, *Hyptis spicigera*, in her cowpeas and was able to store them safe from the cowpea weevils for a year.

Clearly, long-term cowpea storage was not a mystery for many Malian villagers—the mystery was malaria, and to a lesser degree, "hunger," or, in nutritional science terms, malnutrition. We had failed to understand what the villagers we collaborated with really wanted to improve in their lives. We changed our focus to malaria. We listened and nudged the villagers to take action to eliminate the mosquitoes, once they understood the intertwined life cycles of the protozoan that causes malaria and the anopheline mosquito that transmits it. Now, with two exceptions in 9 years, children no longer die from cerebral malaria in Sanambele. Now we work together exchanging ideas, focused on how to keep malaria locally eliminated and to avert malnutrition with just the local, village-grown foods. We continually use the holistic process.

The life lesson I learned from Hawa was ask basic questions to learn about the core values of people, and also to listen for minute cultural differences. I, a professor with several formal degrees and many books and publications to my credit, had learned these listening skills from a woman who had no formal schooling. Hawa had politely waited until I showed deep listening behavior, then she immediately, publically rewarded this listening behavior. Specifically, we set in motion a communication process that led to local elimination of deaths, mainly in children, from cerebral malaria. The approach that we need to move beyond the practice of superficial listening can be learned from Indigenous people.

CASE STUDY 2. THE INTERNATIONAL WAR ON LOCUSTS AND GRASSHOPPERS

The widespread existence of stunting (kwashiorkor) in West Africa is a travesty. The importance of the first 1000 days, and defining of perfect weaning foods, has been a serious focus in the United Nations (UN) Food and Agricultural Organization (FAO) since at least 1983. An even greater tragedy is the fact that all the time of field and laboratory scientists taken to develop a diet, as well as that of policy makers developing strategies, to solve this issue of stunting, has not in 35 years led to a solution for stunting.

I pose my hypothesis to you that this travesty of the continued existence of stunting in West Africa is tied directly to the "War on Locusts and Grasshoppers[1]" launched by Western culture scientists to take out a prime food source, a traditional snack food—edible insects. This case study is tied to the inability of nutrition scientists to teach nutritional biochemistry to

---

[1]Locust is a term used for insects in the family Acrididae that can have multiple generations per year, in contrast to grasshoppers that are in the same family, but are univoltine or have only one generation per year. Some grasshoppers can become locusts such as *Oedaleus senegalensis* depending on environmental conditions. Locusts go through phase changes, i.e., they change color, shape of the pronotum, and change in behavior, from solitary to gregarious. In the gregarious phase locusts form bands when they are immature, before their flight wings mature and when they are adult form swarms. When swarms of locusts, a million or more individuals, catch the prevailing winds, they can travel hundreds of miles and darken the sky as they travel. The arrival of these swarms is often referred to as a plague.

those without a formal education. This case study is also related to malaria because stunting and micronutrient deficiencies weaken the adaptive and innate immune systems, respectively. Suppression of the cultural aspects of food is not only in large part responsible for malnutrition in Africa, but, in part, for the millions of deaths of young children from malaria as well. These mistakes are similar to my failures in understanding of adoption of long-term protection of cowpeas with neem kernel extract as described in Case Study 1 in this chapter, but on an international scale.

War, terror, plague, destruction are not the usual terms one associates with food, but they have been common terms used to describe the outbreaks of locusts in Africa in European and American news media. In contrast, my own introduction to locusts in Africa was the exclamations of joy of young Rwandan children with their plastic bags collecting swarming red locusts, *Nomadacris septimfaciata*, under the streetlights in Kigali. I purchased a package for the equivalent of 1 USD and brought them to the hotel chef to prepare from me that evening. It was my first taste of freshly sautéed-in-butter locusts. I thought I was eating soft-shelled crab. The year Warren Rush (USAID) shared this copy of this Unclassified Department of State Incoming Telegram, dated July, 1986, with *The Food Insects Newsletter*, I perfectly understood (DeFoliart et al., 2009, p. 21).

> Subject: Locust and grasshopper treat—Burundi….Criquet migrateur (migratory locusts) began entering Burundi in May and June from Tanzania on the east. They are now to be found in Ruyigi, Kankuzo, Muyinga, Ngozi, and Kayanza Provincees. No alarming damage has been reported to date. The only defensive measures reportedly taken to date by local residents in affected areas has been to gather them up and eat them.

Rush also provided *The Newsletter* with an article from a Cape Town, South Africa, newspaper (Yeld, 1986) reminding us that the next outbreak of locusts in South Africa might be a good source of protein for food or animal feed. South Africa spent 2.8 million USD on pesticides to fight the locust war in 1986. Funds could have been spent, Director of the Endangered Wildlife Trust and research entomologist, Dr. John Ledger, proposed paying the unemployed to collect the locusts making it safer on the environment and provide needed nutrition.

By 1986 there was a well-developed infrastructure among international aid agencies such as USAID/AELGA (United States), GTZ (Germany), PRIFAS (France), to manage the locust and grasshopper populations in Africa. As an entomologist working on the ground in 1987 in Niger with the USAID locust/grasshopper management program (Radcliffe et al., 1990), it seemed to me like these national aid organizations were vying for recognition for the best destruction of locusts and grasshoppers. The 1987–89 locust "plague" cost the international community 600 million USD and took 5 years to bring under control (DeVreyer et al., 2012). At first the "war on locusts and grass-hoppers" was fought with synthetic, commercial pesticides, organochlorines, such as dieldrin and aldrin, and organophosphates, such as malathion. These were all neurotoxins for insects. Because insects and humans share the same chemicals used for neurotransmitters, such as acetylcholine, these neurotoxic insecticides were also toxic to humans. The dose makes the poison. When an airplane carrying an entire cargo of malathion crashed over Senegal (West Africa), contaminating the soil and likely harming the humans, especially children, the weapons in this insect war turned to biocontrol agents and bio-rational (mainly plant-based) approaches (Showler, 1995), international aid organizations outlawed the neurotoxin approach and turned to biocontrol. The intent, however, remained the same: kill the grasshoppers and locusts and dispose of them. More bio-rational approaches emerged after the pesticide-carrying plane crashed, but these approaches did not address the fundamental problem of food nutrient dense food availability in West Africa: stunting of humans. To avert cognitive and physical stunting, nutrient dense foods need to be supplied to young children, i.e., sufficient quantities of each of the 10 essential amino acids for children and all of the required micronutrients on a daily basis. In this region, grasshoppers and locusts were the nutrient dense, traditional food, not the grain and leaves the insects were eating.

Locusts have cycled in high and low populations and have created swarms that catch prevailing winds across Africa for millennia. Africans know how to use this food source and how to preserve it for long-term storage.[2] International interventions can lead to failures in food security. We need to ask ourselves and each other in our daily work of research, teaching, and liv-ing: have I explored all the interconnections of my daily work, have I asked the cultural questions? In 1987 when I arrived in Niger to help with the grasshopper and locust problems in the cereal cropping systems, I already knew the unmistakable delicious taste of locusts, but my team and I were not aware of the seriousness of the stunting issue in West Africa and its con-nection with the high density of essential nutrients in grasshoppers and locusts. We had come to Niger to transfer traditional knowledge about

---

[2]Story told by unknown visitor in Algeria during the 1891 invasion of the desert locust, *Schistocerca gregaria*. "The natives are well disposed to carry out orders for the destruction of the locusts, since they use them for food. Around Tougourt every tent and house has prepared its store of locusts, on the average about 200 kg to each tent. Sixty camel loads (9000 kg) are the quantities of locusts accumulated daily in the Ksours of the Oued-Souf. They are a valuable resource for the poor population. To preserve them, they are first cooked in saltwater, then dried in the sun. The natives collect and prepare such considerable stocks that apart from their own needs, they have some for trading on the markets of Tougourt, Temacin, etc. I have in my hands now two boxes of freshly prepared locusts and I convinced myself that they are quite an acceptable food. The taste of shrimps is very pronounced; with time they lose their quality (DeFoliart et al., 2009, p. 392)."

the repellent properties of the seeds of a subcontinent Indian tree, neem, *Azadirachta indica*, that grew abundantly in Niger. Applying it to sorghum and millet seedlings to prevent damage with a safe-for-humans locally grown tree that did not kill the grasshoppers or render them inedible seemed like a sound research choice (Radcliffe et al., 1990).

Not until I lived and worked in villages in Mali did I put this story of the grasshoppers and locust and the international pest management effort together with an interesting statistic from sub-Saharan Africa that had not changed in about 40 years, the percent of children in West Africa that were stunted, both cognitively and physically. In rural West Africa, this metric has hovered around 40% without much change. The diet for rural smallholder farmers is grain-based. When we did the calculations from tables in the literature and what we observed children eating in the households where we stayed over a period of 10 years, it was no surprise that stunting was so prevalent (Stein et al., 2012). From the time of weaning, diets for children are based on millet, sorghum, and corn without a daily supply of foods rich in two of the essential amino acids, tryptophan and lysine. We also observed that the sources of these essential amino acids in the village were chicken, fish, sheep, goats, cattle, cashews, milk, peanuts, eggs, and grasshoppers and locusts. After years of conversations with the villagers and sharing meals with them, we came to understand the reasons for each of these sources of nutrients not being provided to the young children. Reasons ranged from: it's a man's cash crop, to eggs are a "foreign" food, to those foods are reserved for feast days and weddings and funerals. The only daily complete source of tryptophan and lysine for the young children was grasshoppers and locusts. And, not surprisingly, those insects were considered by the Malians, "kids food." After beginning work in Mali (1994) I asked most Malians I encountered what were their views of this "traditional kid's snack." Those who grew up in rural locations all said "sure, they are good, we all ate them as kids."

In Sanambele, Mali, the village where I spent most of my time while in Mali, the young boys had even developed "hunting bags" made from the polyester woven bags used to deliver salt to the village. Once home from the hunt, the boys one by one impaled the hoppers on a stick and roasted them over their dad's charcoal burner used for the tea-making process. This grasshopper roasting must have been an important activity for the dad to lend the charcoal burner to his children. Making tea is a man's thing and to be the tea maker in the family is an important responsibility. Women don't use this, only the big fires for cooking in big pots.

Toddlers and girls would join the boys around the burner and the hoppers would be shared.

Locust populations are cyclical, partly due it is thought to cyclical, multiyear weather patterns. The "war" reemerged in 2004. The UN FAO "appealed to the international community for 100 million USD to help contain this locust invasion, the 'worst' which West Africa has seen for 15 years" (UN Integrated Regional Information Network, 2004). Still the value judgment from a Western culture point of view that locusts were not food prevailed.

This focus on killing the locusts over successive years was a serious mistake. To remove a valued, traditional food source for young children in a country where the very nutrients that were missing from their diets were available in the insects that were being removed is a tragedy. Removal of this food source for young children contributes to structural, muscle, and bone developmental delays as well as poor brain development, greater risk for malaria and other infectious diseases, and in some cases, death itself.

Why did this happen? Was it European and Euro-Americans' odd ideas of what is food and what is not? Did no one mention to the entomologists who organized the "War on Locusts and Grasshoppers" that this was an important and traditional food source? While the UN FAO was tirelessly trying to develop weaning foods for children to avert stunting, the international aid organizations of France, Germany, England, and the United States were systematically removing the best weaning food for children in this part of the world. This was a grand international failure to recognize culture and to incorporate it into local, national, and international food and agricultural programs.

The tragedy, however, continues. In 2009 data obtained by PLAN Mali and the health center in Sanambele revealed that with young children 0—36 months, only 23% of their very young children had or were at risk for kwashiorkor (stunting) in contrast to the national level of 40%. Until this year, there was a positive trend. In 2009, though, mothers and grandmothers announced to me that they had now forbade their children from harvesting grasshoppers and locusts in the fields. Cotton was now being grown in those fields around the village exactly where the children hunted hoppers. Cotton was grown in Mali organically, at least experimentally, but it was the organophosphate pesticides of parathion and other pesticides that were being used in the fields around Sanambele, according to pest management advisors. The mothers and grandmothers had given good advice to the children. But now, the children were at an even greater risk for stunting.

By 2012 the "war" seemed to be over.[3] The UN FAO called to Rome, the world's experts in edible insects for a consultancy. We came and shared our knowledge. I specifically shared this story of Case Study 2. In 2013 report

---

[3]Although Niger and Mali were put on alert for desert locust invasion in 2012 coming from Libya, the FAO Commission for Controlling the Desert Locust in the Western Region (CLCPRO) provided only $300,000 to tackle infestations in Libya. FAO-Rome added $400,000 more to address the problem (FAO, 2012).

#171 from this conference was published (van Huis et al., 2013). There was a global reaction. Within 24 hours there were 1 million downloads of the document. The first international conference on edible insects, "Insects to Feed the World" convened the next year in Ede, the Netherlands. By 2016, this initiative of FAO, part of the Forestry Department, was canceled. The primary donor country, the Netherlands, had done their part. It was time for another country to contribute to this scholarly effort related to food and agricultural sciences. No country stepped forward to continue the effort begun by FAO with funding from the Government of the Netherlands to end the "war on locusts and grasshoppers" while simultaneously, as an unintended benefit to significantly reduce malnutrition and malaria in Africa. The war on locusts and grasshoppers continues.

Being aware of cultures' role in the food and agricultural systems in which one works and advises others is the key to avoiding failures such as the one revealed here. This is a story that college students should know. This happened and is still happening because food and agricultural scientists forgot and continue forget to ask the cultural question or do their homework or are unable to resolve a deep-seated "food disgust factor" (Looy et al., 2013).

CASE STUDY 3. FRY BREAD

The removal of traditional foods from a culture intimately connected with the land and a specific landscape is a serious action. We saw this in Case Study 2, undertaken with the best of intentions, that of saving the grain for feeding the local population. In this case study we explore the history of fry bread, what the failure was, and how this failure contributed to diet-related ill health.

After the removal of the Native Americans from the Great Plains in the United States using a genocide process by the Euro-American settlers was nearly complete, reservations were created. Hunting and gathering in the typical ways for the Native Americans was no longer possible. In a weak humanitarian action, the US government organized emergency supplies until the reservation inhabitants could be taught to raise crops in the way that the Euro-Americans did. Surplus food supplies were provided to reservation families. These were called "commodities" and included flour, oil, salt, baking soda, and sugar. With the familiar foods unavailable, the bison, wild berries, prairie turnips for example, mothers were hard-pressed to create interesting and attractive food to replace the traditional "comfort foods" that were nutrient rich. Over time, mothers developed a combination of the commodities that when cooked in the same metal pots used in the past to boil bison and vegetables in water could be boiled in oil to make to make a palatable food.

When sprinkled with sugar, it served as a way to stave off the hunger from the missing nutrients. Thus, fry bread originated, replacing bison and other natural, local, traditional foods such as wild berries with a gluten-based, high omega-6, low omega-3 food. Salt and sugar were the attractive parts of the fry bread. As a new, favored dietary component, fry bread contributed to diabetes, celiac disease, and other forms of diet-related ill-health. In addition, reservation life was measurably more sedentary than the nomadic life, further exacerbating the likelihood of these diet-related conditions such as diabetes, celiac disease, and heart disease when fry bread became a typical part of the diet.

Soon, fry bread became the "new" traditional food. Indian tacos which are fry bread with a toping of salsa, meat, and sour cream are thought by many Euro-Americans unaware of this story to be a Native American traditional food. Indeed, fry bread and Indian tacos continue to be a mainstay food at celebrations for the public, such as powwows, and at gatherings within the Native American communities in the western United States. This is another example among numerous cases involving food scientists, plant scientists, animal scientists, and policy makers making well-meaning resource decisions without full appreciation of the cultural background.

CASE STUDY 4. THE USE OF RIVER SYSTEMS IN THE WESTERN UNITED STATES TO PRODUCE FOOD FOR THE NATION

As opportunities for "new" land to raise crops in the eastern United States diminished, the Great Plains were considered by Euro-Americans an ideal option to provide more locations to raise row crops and European cattle. The push was for food and land to create it. The new settlers had no long-term, multigenerational, revered, and traditional relationship with the land, the soil specifically, and the water resources. Water and soil were merely a way to create a commodity that could be turned into cash.

Raising alfalfa using irrigation in a low rainfall, arid environment promised rapid results for addressing the need to feed both the growing human population and European cattle requirements. Other crops and food production systems, such as sugar beets and dairy cows requiring silage from crops such as corn, were all new cash crops or food items that required supplemental water in an arid ecosystem such as the Great Plains. Settlers chose to initiate farms and ranches near rivers in the Great Plains in the late 1800s and early 1900s. Soon their collective economy was dependent on the use of these natural waterways (Carr and Hawes-Davis, 2000).

As a result of this introduction of water-based, industrial agriculture, there are now some river segments, such as the Wind River in Wyoming (Big Horn River in Montana), that have not been able to maintain an in-stream flow, or sufficient water flow to support native fish populations that live within the natural waterways (Flanagan and Laituri, 2004; Miller et al., 2015). There has been a standoff of Euro-American farmers who have invested their lives in developing the land and water diversion processes to produce lucrative crops, and the Native Americans, Eastern Shoshone, and Northern Arapaho, who revere the water as sacred and want to protect the ecosystem as a wild fishery and wilderness area. Resource managers for decades have been questioning whether Euro-American resource management adequately constitutes an effective model for managing resources sustainably (Ruppert, 1996).

Thanks to rising respect among resource managers for the traditional ecological knowledge of Indigenous people in the Great Plains, younger generations of Euro-Americans are realizing that ignoring or **under**utilizing of crops grown successfully there on the same land for centuries was a mistake. These same resource managers can be encouraged to identify and understand cultural values of water as an important first step in collaborative resource management. If these resource managers are able to communicate this understanding to farmers who lack the Indigenous perspective, conflict and lengthy resolution in court can be prevented. The remedy is emerging, but it is important to know and remember the mistakes that have been made lest they be repeated.

CONCLUDING REFLECTIONS ON FAILURES

In this chapter we have tried to convey passion for improvement, not anger. We wanted to explain our vision to learn from these failures and to strengthen the process of improvement by incorporating cultures' role in the food and agricultural sciences. Probably the most important failure that many of us share is simply the continuing thought of "fixing something." Poverty, particularly as observed in material resource poor countries like Haiti, Mali, and other Sahelian countries, and both rural and urban India, pulls at our heartstrings. For some, particularly the nongovernment organizations, it has been a continuing business growing in diversity and reach. Yet, as eloquently demonstrated to us in the film *Poverty, Inc.* (Miller et al., 2015) and the book, *Africa Unchained* (Ayittey, 2005), we need to let Haiti recover from the earthquake on her own terms and Africa develop on her own.

Our job is to listen, but to listen not just in the ordinary places. As was shared by then President Obama in his farewell speech in Chicago, Illinois, "For too many of us it's become safer to retreat into our own bubble, whether in our neighborhoods, or on college campuses, or places of worship, or especially our social media feeds, surrounded by people that look like us and share the same political outlook and never challenge our assumptions. In the rise of naked partisanship and increasing economic and regional stratification, the splintering of our media into a channel for every taste, all this makes this great sorting seem natural, even inevitable" (Obama, 2017). We can begin to counter the acceptability of this great sorting by looking for information in not the usual places and listening deeply to not the usual voices.

Listening, deep listening, contemplative listening is a solution that will help remediate failures we have identified in this chapter and prevent many future failures. It is a positive sign that calls to engage in contemplative listening are coming now from a wide variety of scholars (Cavanaugh, 2017) as well as a former US President. True contemplative listening is hard work. It requires that we separate personal needs and interests from those being expressed by the speaker. In Chapter 6, Listening With Native Americans, we present the importance of the "pregnant silence," or the wait period before responding that we learn from Native Americans and other Indigenous people. Incorporation of this skill into our institutions of higher education is essential for societies to address the challenges we now face. It is this form of training that must permeate educational arenas. We must reach beyond inquiry, analysis, and critical and creative thinking to understand the value of significant amounts of silence.

Chapter 3, Decolonization and the Holistic Process, will provide readers with the fundamental process of decolonizing one's words and actions and to use the holistic process to insure that we are in tune with the desired outcomes and qualities of life of our communities of focus. We will begin to build these skills, first with decolonization, unchaining, and mindfulness which is the prelude to using the holistic process.

References

Ayittey, G.B.N., 2005. Africa Unchained: The Blueprint for Africa's Future. Palgrave, Macmillan, NY, 483 pp.

Biswas, K., Chattopadhyay, I., Banerjee, R., Bandyopadhay, U., 2002. Biological activities and medicinal properties of neem (*Azadirachta indica*). Curr. Sci. 82 (11), 1336–1345.

Cavanaugh, J.C., 2017. Point of view: You talkin' to me? Chronicle of Higher Education January 27, A48.

Carr, D.G., Hawes-Davis, D., 2000. Wind River. High Plains Films. 35 minutes.

Chambers, R., Pacey, A., Thrupp, L.A., 1989. Farmer First: Farmer Innovation and Agricultural Research. Intermediate Technology Publications. London, UK/Short Run Press, Exeter, Great Britain, 218 pp.

DeFoliart, G., Dunkel, F., Gracer, D. (Eds.), 2009. Chronicle of a Changing Culture: The Food Insects Newsletter. Aardvark Global Publishing, Salt Lake City, UT, 414 pp.

DeVreyer, P., Guilbert, N., Mesple-Somps, S., 2012. The 1987–89 locust plague in Mali. Paris School of Economics. G-MonD Working paper number 25. <http://www.parisschoolofeco nomics.eu/IMG/pdf/WP25-GmonD-locust-plague-mali-june2017.pdf> (accessed 28.02.17.).

Dunkel, F., Gamby, K., 1997. Postharvest monitoring and pest management. IPM CRSP. Fourth Annual Report 1996–1997. Virginia Tech Management Entity. 282–289.

Dunkel, F., Sriharan, S., Nizeyimana, E., Serugendo, A., 1990. Evaluation of neem kernel extract (Margosan-O) against major stored insect pests of beans and sorghum in Rwanda. Proceedings of the 5th International Conference on Stored Product Protection. Bordeaux, France, vol. I, pp. 527–536.

Dunkel, F., Serugendo, A., Breene, W., Sriharan, S., 1995. Influence of insecticidal plant materials used during storage on sensory attributes and instrumental hardness of dry edible beans (Phaseolus vulgaris L.). Plant Foods Hum. Nutr. 48, 1–16.

Edmister, J.A., W.M. Breene, Z. Vickers, A. Serugendo. 1988. Changes in the Cookability and Sensory Preferences of Rwandan Beans During Storage. University of Minnesota, USAID, Government of Rwanda: Minnesota. Agr. Exp. Sta. Misc. Publ. 47-1986. 390 pp.

FAO, 2012. Niger, Mali on alert to desert locust risk. Food and Agricultural Organization, Rome. <http://www.fao.org/news/store/en/item/146885/icode/> (accessed 19.02.17).

Flanagan, C., Laituri, M., 2004. Local cultural knowledge and water resource management: the Wind River Indian Reservation. Environ. Manage. 33, 262–270.

Gamby, K., Dunkel, F., 1997. Variation de la composition biologique et de l'efficacité de l'huile de grains de neem (NKE) en fonction du temps et de la température de conservation. Proceedings of the Joint Congress of the Entomological Society of Southern Africa (11th congress) and African Association of Insect Scientists (12th congress), Stellbosch, pp. 124–125.

Gueye, M.B. 1993. Rapport de l'atelier regional de formation de formateurs sur la method accel-eree de recherché participative (MARP), Dakar, Senegal. Program des Zones Arides, International Institute for Environment and Development, London, UK.

Jenkins, D., Dunkel, F., Gamby, K., 2003. Storage temperature of neem kernel extract: differential effects on oviposition deterrency and larval toxicity of Callosobruchus maculatus (F.) (Coleoptera: Bruchidae). J. Environ. Entomol. 32, 1283–1289.

Kante, A., Dunkel, F., Williams, A., Magro, S., Sissako, H., Camara, A., Kieta, M., 2009. Communicating agricultural and health-related information in low literacy communities: a case study of villagers served by the Bougoula commune in Mali. Proceedings of the Annual Meetings of the Association for International Agricultural and Extension Education, San Juan, Puerto Rico, 24 May 2016. Used as a case study by USAID Feed the Future. MEAS Case Study Series on Human Resource Development in Agricultural Extension, Case Study #11, January 2013.

Looy, H., Dunkel, F., Wood, J., 2013. How then shall we eat? Insect-eating attitudes and sustain-able foodways. Agric. Hum. Values 31 (1), 131–141.

Miller, M.M., Scionka, S., Mauren, M.R. K., Fitzgerald, J.F., 2015. Poverty, Inc.: Fighting poverty is big business. But who profits the most? 91 minutes. <http://www.povertyinc.org> (accessed 19.02.17.).

Obama, B., 2017. Farewell speech. Chicago, Illinois. <http://www.nytimes.com/2017/01/10/us/politics/Obama-farwell-address-speech.html?r = o> (accessed 22.02.17).

Radcliffe, E., Dunkel, F., Strozk, P., Adam, S., 1990. Antifeedant effect of neem, Azadirachta indica A. Juss., kernel extracts on Kraussaria angulifera (Krauss) (Orthoptera: Acrididae), a Sahelian grasshopper. Tropical Agric. 68, 95–101.

Ruppert, D., 1996. Intellectual property rights and environmental planning. Landscape Urban Plan. 36, 117−123.

Showler, A.T., 1995. Locust (Orthoptera: Acrididae) outbreak in Africa and Asia 1992−1994: an overview. Am. Entom 41, 179−185.

Sriharan, S., Dunkel, F., Serugendo, A., 1991. Efficacy of neem kernel extract (Margosan-O) for management of stored product insects in Rwanda and evaluation of sensory properties for humans. In: Ahlem, S. (Ed.), Proc. Int. Neem Conf. East West Center, Univer. Hawaii, Honolulu, HI, 30 May 1991.

Stein, C., Dunkel, F., Kone, Y., Coulibaly, K., Jaronski, S., 2012. Potential approach to regulate and monitor moisture for *Brachytrupes membranaceus eggs*. < http://www.montana.edu/mali/pptsaspdfs/SteinrCarissaCricketsinSanambelePoster.pdf > .

UN Integrated Regional Information Network, 2004. West Africa: Locust Invasion. AfricaFocus Bulletin. < http://www.africafocus.org/docs04/loc0409.php > .

van Huis, A., Van Itterbeeck, J., Klunder, H., Mertens, E., Halloran, A., Muir, G., et al., Edible Insects: Future Prospects for Food and Feed Security. Food and Agricultural Organization of the United Nations, Rome, Italy, FAO Forestry Paper #171, 187 pp.

Yeld, J., 1986. Eat locusts, don't poison them. Argus July 7, 1986, Cape Town, South Africa.

CHAPTER 3

Decolonization and the Holistic Process

Clifford Montagne[1], Florence V. Dunkel[2], Greta Robison[3], Ada Giusti[4], and Badamgarav Dovchin[5]

[1]Department of Land Resources and Environmental Sciences (retired), Montana State University, Bozeman, MT, United States, [2]Department of Plant Sciences and Plant Pathology, Montana State University, Bozeman, MT, United States, [3]Department of Earth Sciences, Montana State University, Bozeman, MT, United States, [4]Department of Modern Languages and Literatures, Montana State University, Bozeman, MT, United States, [5]Renchinlumbe, Darhad Valley, Mongolia

CONTENTS

Introductions 43

Connection to Land Is a Prelude to Appreciating Wholeness 45

Introduction to the Holistic Process .. 46

Decolonizing Methodologies: First Step Toward the Holistic Process 49
Identifying Communication Languages 51
Establishing Ownership of Original Data or Products 54
Summary Reflections on Decolonization 56
The Build Up of Colonization 56
A Paradigm Adjustment for Reductionist (Western) Science 57
Traditional Ecological Knowledge Becomes Visible During Decolonization 58

The Holistic Process 59

We have now seen the gap and explored its impact globally and in specific locations in Chapter 2, Failures. As agricultural scientists and policy makers, and as citizens of the Earth, we have responsibility for the 2 billion people in food insecure households throughout the world. This includes 40% of children in West Africa who suffer from physical and cognitive stunting because they are subsisting on a grain-based diet without their traditional, nutritionally dense snacks—grasshoppers and locusts. In East Africa, with an attempt to alleviate hunger, crop scientists pushed farmers of Kenya to switch from traditional grains to corn-based diets, deficient in two essential amino acids, lysine and tryptophan (Dunkel et al., 2016). The result today is malnourishment across sub-Saharan Africa because we have nudged these peoples to abandon traditional sources of essential amino acids and micronutrients. Because we, food and agricultural scientists, are responsible for this tragedy, we also have some responsibility for the 450,000 children who will likely die this year from cerebral malaria due in part to their compromised innate and adaptive immune systems caused by the inadequate content of these amino acid and other micronutrient deficiencies in these children's diets. There is a good news. We are now recognizing how including cultures' role in our learning process as scientists, policy makers, and citizens, we can bridge the gap and contribute to solving these issues.

Food insecurity and insidious diseases like malaria can be solved. We have the technology, but without the skills to bridge the cultural gap, conventional technology will not transfer. In this chapter, we present the most difficult of

Incorporating Cultures' Role in the Food and Agricultural Sciences. DOI: http://dx.doi.org/10.1016/B978-0-12-803955-7.00003-1

The New Paradigm
for Teaching in the
Food and
Agricultural
Sciences............... 62
*Using the Holistic
Process* 62
*The Expansive
Collaborative Model for
Community Engagement
and Service-Learning* . 63

Conclusion 66

References 68

Further Reading .. 70

the cultural technologies: decolonizing methodologies and the holistic process. Just as good pipetting techniques are necessary to conduct polymerase chain reaction (PCR) and microarray studies, these cultural technologies require specific skills: careful, active listening; patience; and development of an ethno-relative worldview.

The most basic of these skills is recognizing that there are other cultures. Then, when one recognizes the diverse value unique to each culture, it is possible to develop an ethno-relative worldview, a decolonized worldview. A simple example of this developmental process works like this:

- One can *deny* that there is any such thing as other cultures. With food this may be expressed as "Insects? I do not eat insects. No one eats insects. They are not food."
- One can *defend* one's culture as the only approach to a topic. With food this may be expressed as "Insects? Oh sure. I have heard that there are people in the world who eat insects, but these are inferior, aboriginal cultures. We don't eat insects and that is the way in most all the world. Insects are disgusting."
- One can *minimize* cultural differences. With food this may be expressed as "Insects? There are a few people all over the world who eat insects. If they had better food opportunities, I am sure they would not choose to eat insects."

These responses are examples of expressions of ethno-centric world views (Bennett, 2004). There are varying degrees expressed here of recognizing one's own culture, beginning with denial that there even are other cultures. The following are examples of ethno-relative world views:

- One can *accept* cultural differences, seeing one's own culture as one of many. With food this may be expressed as "Yes, my culture, Euro-American culture, has not had a strong history of eating insects. I know that other cultures in the world have. In fact, it is interesting that all the other main US cultures, African American, Hispanic, Asian American, and Native American all have sub-cultures within them that enjoy insect cuisine."
- One can *adapt* to cultural differences in a nonjudgmental and authentic way. With food this may be expressed as, "You know, I have never tasted insects before, but I am looking forward to tasting how you prepare them at home in Guangzhou. Hmm-m-m. These ants have more of a lemon flavor than I expected. They might be quite tasty on a pork stirfry."
- One can be so *integrated* into a variety of cultures that it is hard to recognize which culture one belongs to. With food this may be expressed as, "Are you having escamole (ant eggs) with dinner this evening. I am excited to taste them. Will you be preparing them as a main course or an hors d'ourves?"

We have presented Bennett's (2004) categories as one way of explaining intercultural competency. This process of development begins with recognizing that there are people with world views other than one's own. This was a simple set of examples from one category of one of the most important part of the paleo-diets that helped us humans develop large brains and literally become human. Had we, as food scientists and entomologists, been more interculturally competent in the 1970s, 1980s, and 1990s, we would likely have not declared war on grasshoppers and locusts, one of the most complete proteins and micronutrient food sources traditionally used by children in West Africa. The human neurotoxins used to kill off this food source were expensive and not without danger. This killing of a normal-for-those-cultures food source led to considerable time spent monitoring what could have otherwise been spent harvesting, drying, and distributing the "meat."

This is just one example of how the global agricultural and food production system has been colonized, with subsequent expansion of the gap between European/Euro-American cultures and Indigenous cultures. Colonization was directed by human decision making. To reduce these gaps, human decision making and the resulting actions must become open and transparent between these Western and Indigenous cultures. Decision-makers must operate within a climate of trust and understanding rather than power and domination when interacting with other cultures. To begin the process, people must engage on a personal level. The wealth of personal experience and skills can be tapped by first engaging with people as people where listening with respect, recognizing one's own and others cultures is a fundamental prerequisite.

INTRODUCTIONS

We have all come from somewhere and have a personal history. It is important to begin to exchange that personal history information whether in the extended family, a classroom, a research team meeting, a board meeting, or in a community. It is the first step toward trust and open, honest engagement. To share our culture, we must recognize that we have a culture and it is one of many cultures, all of which are equally relevant. Communication failures often occur when an important question is not asked: From where have you come? At times, the answer to this question brings out colonization experiences, and reveals failure to self-determine, or self-actualize. Knowing how to use the holistic process and decolonizing methodologies is a fundamental way to avert failure in culture and agriculture relationships. Beyond avoiding failure, there are wealth of ideas, plant materials, sustainable protein sources, nutraceuticals, and storage and processing ideas that can save human lives and fresh water, conserve land, and reduce production of greenhouse gasses. Indigenous peoples of the Earth have been quietly collecting this wealth of knowledge and storing it in their oral, story-based archives.

To access respect, and protect this wealth, several important concepts must be learned and subsequent actions taken. In this chapter, we explore the interconnections of decolonization, traditional ecological knowledge (TEK), reductionist (Western) science, and the holistic process. Decolonization is essential, because it allows for an ethno-relative worldview and an ability to see, hear, and appreciate TEK. Holistic management (HM)—most often referred to as "the holistic process"—is equally important. Although its application can appear relaxed and informal, it does not mean it is less credible or effective.

In this chapter, Dr. Clifford Montagne provides insights into his own life pathway, one that eventually led to our collaboration and development of a new teaching process that creates an intellectual environment that allows learners and guides to incorporate cultures' role in the food and agricultural sciences and the related sciences as well. Before we introduce this learning environment, Cliff provides a bit of his own personal history.

How does one choose a pathway through life? For me, Clifford Montagne, I have always felt fortunate to grow up with my particular set of parents. My mother came from the plains and mountains of Wyoming, where she grew up riding horses and skiing mountain trails. My father fled the confines of the East Coast to become a geology professor at Montana State University. Along the way both parents got to know the Murie family, wildlife biologist Olaus and his wife Mardie and Olaus' brother Adolph and wife Weezie. These two Norwegian-origin biologists worked throughout the American West and Alaska to document wildlife and their interrelationships with the landscape. I remember camping with my family at the Murie ranch in Jackson Hole when my father was doing field geology for the US Geological Survey. The Muries went on to play key roles in the founding of the Wilderness Society and garnering key national legislation to recognize the value of wild places. My father, and then my wife Joan, both went on to both become Presidents of the Montana Wilderness Association. While growing up I spent nearly all summers in our family mobile field camp, traveling around Wyoming and the Greater Yellowstone Ecosystem while my father did his geology mapping. All this while I was reading cowboy and Indian lore books and trying hard to grasp the feel of the 'Old West' where in my mind, Indians had lived with tremendous skill sets, legends and adventures close to the land; and where the cowboys trained horses and punched cows across the open range.

Living in this idyllic style, I also remember wondering about issues on a global scale, like the Cold War nuclear threat, the migration of refugees out of Hungary, the poverty of post-World War II Europe, and that great mass of people in the more densely populated parts of Asia. These years were also the time when science was being promoted in the US as a panacea, with medical science breakthroughs bringing longer lives and more comfort, and also the feeling of a crusade in which we could all participate and perhaps even win, and with the ever present arms and space races. In high school, I realized my cowboy dream with a full-time ranch job in the summer and some weekends during the school year. This is where I became acquainted with basic skill sets of animal handling and hard physical work. This experience gave me great confidence in knowing that I could support myself with my hands and that I was more like a member of the physical-labor class than the son of a university professor. In college, geology became my major because I believed in the power of science but wanted to be out of doors as much as possible. A student field trip to Guatemala and El Salvador brought me face to face with cultures rich in tradition and poor in material wealth, and for the first time I saw people living in the street, horses and mules being worked nearly to death, and a landscape of erosion and devastation. We encountered rural people living a scramble of seemingly endless work, but with smiles of happiness on their faces.

CONNECTION TO LAND IS A PRELUDE TO APPRECIATING WHOLENESS

Allan Savory and Jody Butterfield (1999) coined the term "Holistic Management" and described it as a process to begin to appreciate one's wholeness as part of one's connection with the land or water from which one's food is produced. In their description of the process, it is first necessary to see the wholeness of the physical landscape with which one is most associated. Second, it is essential to understand one's own innermost motivations, and interconnectedness with that landscape. Recognizing the interconnectedness of the landscape and its role in creating one's desired quality of life is the beginning of appreciating wholeness. Watching for changes in one's motivations, changes in the landscape, and recognizing the factors that initiate these changes are equally important. Cliff continues his story. . . .

From 5th grade I grew up next door to the Charles C. Bradley family in Bozeman, Montana. Charles was also a geologist. Dorothy went on to be elected as the youngest member of the Montana State Legislature and participate in landmark environmental policy legislation. Charles became my graduate advisor. Dorothy continues to guide and cajole me to this day with encouragement for the pathway my wife Joan and I are following with dedication to the love of and appreciation for landscape and the people inhabiting landscapes. The landscape appreciation comes from parents interested in geology and wild places, the influences of several prolific and well-published ecologists, naturalists, and geologists.

The first key to using the holistic process is to understand your own connection to the landscape and to be able to see that landscape as a source of food and shelter whether it is a natural landscape or a cityscape. The second key to using the holistic process is to see yourself and other humans as part of the landscape and essential to each other. Both Cliff and I came to understand this fundamental urge to put the landscape together with "our tribe," Cliff as a soil scientist and me as a protozoologist/entomologist, fascinated by the "little things that run our world." Cliff walks us through his own awakening.

> Throughout this entire growing-up period in the 1950s and 1960s, I was feeling the lack of being part of a tribe. I envied people living in a closely connected village, something I had no experience with. I was part of the science and technology-based 'modern' generation where kids were expected to leave home at 18 and spread around the country, where the rise to job success often meant successive moves and change of neighborhood. I felt a need but could not express it.
>
> By the mid-1980's, I had become a soil science professor heavily engaged in teaching while facing the challenges of developing a credible research program. I had the technical knowledge and therefore power to prescribe landscape management practices which would conserve environmental quality but might limit people's livelihoods. I felt the need to connect my natural science knowledge more directly with the human condition. Growing up closely associated with nature, I also, felt the power of the great landscapes of the West where so-called natural processes seemed more powerful than human influences. This gave me a confidence that nature, and her processes, would hold the keys to human survival and well-being. Hence, I was well set up for learning from nature and applying such lessons to human situations.

INTRODUCTION TO THE HOLISTIC PROCESS

It was the synergy of a prepared mind, such as Cliff's, superimposed in real time on just the right combination of guides and companions that brought the holistic process into the food and agricultural sciences at Montana State University. A critical mass had accumulated. It was the start of the process to rediscover the interconnectedness of natural systems, particularly related to the use of rangeland and water resources to produce food.

> In the late 1980's, Allan Savory was touring around the West presenting workshops and talks about his 'Holistic Resource Management' approach. Those working in range science appeared to be the most interested in his work. In spring 1988, I attended a lecture Savory gave for the College of Business at Montana State University (MSU), sponsored by a rancher

Warren 'Busk' Jones of the Twodot Land and Livestock Co near Harlowton, MT. Jones, trained in economics at Dartmouth College, was serving as the first full time visiting professor in the David B. Orser Distinguished Visiting Professorship, an endowed chair in the College of Business at Montana State University-Bozeman.

I sat in my seat and trembled with excitement as Savory explained how decision makers could build trust and mutual understanding through sharing, learn from natural processes, make simple inventories of their resource base, vision better futures, and decide on the most effective tools. Savory's approach required some paradigm changes, and that is where I first heard about the concept. I expressed my excitement to him and he invited me to attend an introductory course in Albuquerque.

The occurrence of the Yellowstone fires, through which I had driven to attend the course, was a useful example which Allan Savory cited as he taught the Holistic Management class in Albuquerque. The fires were so intense for multiple reasons. One was the happenstance of a very dry and warm summer, leaving Yellowstone's forests vulnerable to fire. Another, now widely recognized factor, was that ever since a series of catastrophic forest fires early in the 20'th Century, the prevailing paradigm in the United States was to prevent forest fires. This led to huge efforts and expenditures to guard against forest fires by increasing public awareness and preparedness, increasing surveillance, and increasing and using human, chemical, and mechanical means to extinguish wildfires as quickly as possible. A result was encroachment of trees into grasslands as well as buildups of forest fire fuels in existing forest stands. In Yellowstone, the lodgepole pine forests were especially thick and there were more trees invading grasslands. Without fire suppression, the forests and invaded grasslands would likely have had much less fuel ready to burn, and the fires would likely have been less intense and less widespread. Allan Savory would say that in the Yellowstone case, human mismanagement had led to a higher than usual buildup of fuel. A more holistic view might have recognized forest fire as a natural process that plays important ecological roles. Fire can be used as a positive tool in management of forests and grasslands once people decide on the kind of ecological condition they would prefer. This realization would have had to first been made at the policy level to make a difference. And as the case with many situations, it only took a few key people to point out the fallacies of the 'prevent all forest fires' paradigm, for private, and then public, and finally agency practices to change. Today, across North America, wildfire remains a prominent issue, but fire is now widely recognized for the positive roles it can play.

Wildfires in Yellowstone National Park may seem to be a long way from food and agricultural production, but fires intimately affect the foundations of both.

Soil content and water availability, both of which are dependent in part on fire, are the fundamental building blocks for producing food. Without a holistic approach, those fundamentals of food production are often forgotten, just as the fundamental connection of food to culture is often ignored.

> The US Forest Service is a branch within the US Department of Agriculture charged with maintaining healthy and productive forests and grasslands within the National Forest System. These lands are key to providing food and fiber, as well as critical 'ecosystem services' like clean water for agricultural crop irrigation, domestic use, and hydroelectric power in the United States. A sister agency, the Bureau of Land Management within the Department of Interior, has a similar mission, but is more focused on rangelands than forest lands. The great watersheds of the US, tend to originate in higher elevation National Forest land, and then spread out into great valleys and plains which provide much of the American agriculture base.
>
> Not long after this introduction to holistic management, range scientist Roland Kroos moved to Montana. Roland had been a range scientist employed by the Soil Conservation Service (SCS), now Natural Resources Conservation Service (NRCS). In 1985 he resigned his government position to receive extensive training in holistic management from Allan Savory. Roland was certified as an educator at the non-profit organization started by Savory called Holistic Management International. In 1988 Roland moved to Montana where he started and continues to operate a ranch consulting service based on the holistic process.

A critical mass assembled in Montana, including Roland, Professor Brian Sindelar and some Bozeman area ranchers and graduate students who had begun working together to promote the practice of HM. In addition to the group in Bozeman, ranchers in central and eastern Montana gathered together.

> At about this time I drove to Post Falls ID to attend a workshop in Systems Thinking for Agriculture put on by Washington State University. The main instructor was Kathy Wilson, co-author of a newly published book on systems thinking applied to agriculture (Wilson, Morren 1990). Kathy's materials paralleled Savory's approach. With help from my colleagues mentioned above, I worked on a synthesis obtained by explaining the many formally named steps of Holistic Management within the so-called Soft Systems Circle of Kathy Wilson. Armed with these new approaches, I converted my Soil Conservation course to a new course in 1990: Holistic Thought and Management, which continues today. The course used holistic management as a template, so it was guided as much by the students as by the professor. It was highly participatory with individual and small group discussions and projects as major components of the course. Formal

lectures and testing received much less emphasis. Students conducted individual 'holistic management personal plans' on topics of personal relevance. Many students benefited from the opportunity to apply principles of holism to their personal and upcoming professional lives. With this approach, the course continues to be rewarding to teach.

By 1996, a perfect opportunity arose for Cliff to put into practice the holistic process in a community in tune with their own culture and their interconnectedness to the landscape. It was a subsistence herding community in Mongolia, but the landscape was similar to wild Montana.

After spending a year in Japan and returning multiple times, I visited Mongolia (1996) and became entranced with both the landscape and its people. The landscape seemed similar to the Western US which I loved, and the people were the first people I had encountered who seemed to be tuned into their own culture and landscape, but were having challenges emerging from the Socialist Period when Mongolia was colonized by communist Russia. Tim Swanson, Bozeman businessman, mayor of Bozeman and Board Chair of the Greater Yellowstone Coalition, and I were in Hokkaido, Japan visiting a landscape with similarities to wild Montana (believe it or not), when he suggested an organization to forge links and connections between similar yet differing 'bioregions'. The concept of a bioregion had just been presented by a guest speaker and book author at the Greater Yellowstone Coalition annual meeting. This led to creation of a loose affiliation of student and faculty at MSU and the conduct of annual 'work trips' to Mongolia's Darhad Valley. The underlying premise of BioRegions, both an informal association at MSU and the formal non-profit BioRegions International, speaks to the benefit of learning, stimulation and paradigm change through outside visitation. BioRegions became a vehicle to practice implementation of Holistic Management. The non-profit BioRegions International works parallel with the informal MSU BioRegions Program.

DECOLONIZING METHODOLOGIES: FIRST STEP TOWARD THE HOLISTIC PROCESS

Understanding one's interconnectedness with the landscape is basic to the holistic process. Also basic to the holistic process is to bring to consciousness one's own desired quality of life. If one is working with other than one's own community, it is essential to know the desired quality of life of the community, group, or person with which you are interacting. The fundamental basic step after this "knowing procedure" is to decolonize one's interactions with the community, that is, to refrain from imposing one's own values on the community.

To begin the difficult process of first identifying colonizing actions and verbiage and then undertaking the even more difficult process of *de*colonizing one's actions and verbiage, we present this example. During a cab ride in the United States, I noted that the cab driver was not a Euro-American and looked like he had emigrated from Africa. From where, I asked? Ethiopia. It was the weekend of Christmas and so we quickly moved beyond a weather discussion to comments on major religions, to ethnic-specific calendars, and how they work. I quickly discovered from the "taxi-professor" that his country has its own calendar, its own official national, but not European colonial, language in addition to its 83 different tribal languages, and a national tradition of helping each other. When I expressed amazement, the Ethiopian cab driver explained his country was one of two in Africa that had never been colonized. How did this happen? "Well," he said. "The king was strong and said to the people of Ethiopia that we must all help each other. We must care for each other." I asked how the king able to bring this about. Without hesitation, the Ethiopian pointed to his ear and said, "hearing." Communication! Thinking about the holistic process, I asked: "What is the one most important thing that makes Ethiopians happy?"

"Eating together. We even have a word for eating together. In my language, Amharic (Amharigna) one says, 'Enebla' which means 'Come. Let's eat together.' To eat alone is considered undesirable behavior. No matter if the person is a stranger or if there is only one person and it is a very young child nearby, it is still considered better to eat together than it is to eat alone." So we find in some cultures that the first step in understanding, in living peacefully, and in decision making is listening and that listening is best done when sharing a meal, no matter how simple. The close connection of hunger, food, the act of eating together, communication, and peace has been part of that Ethiopian culture for almost a millennium. Peace and food go hand-in-hand for the Ethiopians. And further, what is defined as food, as we learned in Chapter 2, Failures, is also defined in a cultural sense.

This simple vignette exemplifies decolonizing methodologies and illustrates some essential components of this approach. These include posing nonjudgmental questions, and responding with appreciation for the information shared and for specific cultural values. Such an approach can open doors for further discussion, and a deeper exploration of the topic if both parties were so inclined.

The next level of basic decolonizing methodologies that need to be addressed are:

- Identifying communication languages.

- Recognizing cooperation-based groups versus competition-based groups.
- Achieving participation of all, Indigenous and non-Indigenous, peoples in all research phases, including the planning, methods development, data collection, data analysis and interpretation, in writing publications, in authoring written and oral presentations, and in sharing the knowledge gained.
- Establishing ownership of original data or products.

Identifying Communication Languages

Of course one of the most important factors in decolonization methodologies is establishing the language to be used. Speaking or writing to an Indigenous colleague in the language of their colonizers could convey a strong message of continuing colonization. Similarly important is the use of singular versus plural pronouns in any common language. Indigenous peoples' societies are usually based on a cooperative sharing of resources versus a competitive sharing of resources. This was the double problem facing Carly Grimm, an MSU Honors College senior double majoring in Sustainable Foods and in French Language and Literature. At the beginning of the semester, she had to write a letter of introduction of herself to her site mentors in Mali. This group of mentors included the President of the Women's Association in a small farming village in Mali fluent only in Bambara and the principal of a high school in Bamako, the capital of Mali. The principal was fluent in French, and Bambara. We were at the course retreat for the weekend so there were 10 other students and instructors to help. Carly struggled for a while and then asked all of us for help. Because she was writing to a community that is based on cooperation, sharing of resources, use of the singular pronouns sounded colonial. All the rest of us were stumped, except for Aedine Ndi Peyou, a student from Cameroon. Aedine smiled. Yes, the letter sounded odd to her. Aedine spoke French, English, and some Ewando. In a short time, she and Carly reconstructed the sentences to avoid use of the words "I" and "my." This was an interesting exercise for all of us, college seniors, and professors, constructing a letter without beginning a sentence with the word "I," one of the first steps in decolonizing behavior.

Carly had a good start with her "decolonized" letter and her research flourished. Later in the semester, she was faced with sharing the nutritional calculations she had made with the Sanambele village women, which led to a further "decolonizing" methodology, using images or pictures to communicate where a shared spoken language is not available.

Since there was no common language and no way to transfer the spoken word to this community whose adult members did not read or write, Carly turned to participatory diagraming, a collaborative graphic representation created together in a group with each member having an equal role is what is added to or subtracted from the diagram (Chambers et al., 1989). Carly used images to answer the community's request to reduce hunger in the children. From living in the village with Bourama's family, we knew twice a day the typical meal was tou (a stiff, not liquid, grain base made from millet, sorghum, or corn) dipped handful by handful into a bowl of sauce and then eaten. Breakfast was a grain gruel, sometimes with dry milk and sugar. We were able to estimate how much tou and sauce, the toddler, newly weaned, usually ate. An earlier student had calculated the essential amino acids in these servings of the grain part of the meal (Turley, 2011). We knew that both lysine and tryptophan were lower than recommended by FAO/WHO standards. Carly then had to communicate with the village mothers to find out which foods they grew in the village and used to make sauce for the tou. The next calculation was simple, to look up which plants would make up the difference. Carly calculated from nutritional tables of FAO/WHO 1981 (Makkar and Becker, 1996). It was a big challenge, though, in a village with no graduated measuring cups to translate these numbers into a form these village women could use. Finally, we settled on a small cup-size callabasse, a gourd grown in the village to make home-grown food preparation utensils of various sizes. Carly calculated how many of the small callabasse cups would be needed of each sauce ingredient for each young child each day to meet the recommended tryptophan and lysine intake to avert stunting. Then she arranged drawings of these food plants and food insects in descending order of content of these two amino acids (Fig. 3.1). For contrast, Carly added the house cricket and the termite, both of which occur naturally in the village and are common food insects in other parts of the world. Voila! A quantitative food choice chart had been created without numbers or words (Fig. 3.1).

The next step was a critical review by the women farmers themselves. Carly's site mentor hand-carried the painting from the classroom at MSU to the village in Mali. There were a few requests by the village women to include additional plants used for sauce. The women reminded us that although Carly's painting included cashews, they were a man's crop and not available for the women to use to improve their children's diet. The unforgettable moment was the comments about the insects. Termites we learned were so small that they were too difficult to harvest for human food so they were fed to the chickens raised in the outside-the-village fields by the men. The women understood the high value of the house crickets for overcoming their young children's malnutrition, but grasshoppers were children's snacks not crickets. During the back and forth conversations that ensued over cell phones, Skype, and email with the help of our site mentor, the women learned that at MSU and in the United States, we do eat crickets incorporated into a variety of different foods. At MSU, we even have an annual luncheon with over 800

Participatory Diagramming to Eliminate
Malnutrition in Sanambele, Mali

Carly Grimm,[1] Florence Dunkel,[2] Hawa Coulibaly,[3] Yakouba Kone,[4] Ibrahima Traore,[5] Salifou Bengaly[6]

Food Source	Protein (g/100g)	Tryptophan (mg/100g)	Serving size (g) for full daily requirement - tryptophan	Lysine (mg/100g)	Serving size (g) for full daily requirement - lysine	Sources
Amaranth leaves	10.7	229.5	26.14	705.55	63.78	Ghaly & Alkoaik 2010 ; Rodríguez et al 2011
Bambara groundnuts	17.7	192	31.25	1141	39.44	FAO/WHO 1981
Cashew	17.4	378	15.87	942	47.77	FAO/WHO 1981
Chicken	20	205	29.27	1590	28.3	FAO/WHO 1981
Eggplant fruit	1.2	12	500	63	714.29	FAO/WHO 1981
Fish	18.8	211	28.4	1713	26.27	FAO/WHO 1981
Orthoptera (crickets and locusts)	67.6	695	8.63	5240	8.58	Rumpold & Schlüter
Mango	0.6	12	500	65	692.31	FAO/WHO 1981
Milk	3.5	48	125	268	167.91	FAO/WHO 1981
Moringa	25.1	176.7	33.96	453.8	99.16	Makkar & Becker 1996
Okra fruit	2.1	12	500	70	642.86	FAO/WHO 1981
Okra leaves	4.4	40	150	217	207.37	FAO/WHO 1981
Peanut	25.6	305	19.67	1036	43.44	FAO/WHO 1981
Isoptera (termites)	39.7	1430	4.2	5420	8.3	Rumpold & Schlüter 2013, Kinyuru et al 2013

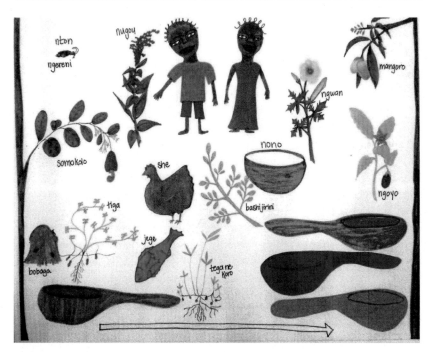

FIGURE 3.1

Participatory diagraming using Western culture science and decolonizing methodologies. The chart with numbers underlies the diagram below with no numbers. The containers are bottle gourds grown in the village and used for measuring. High lysine and tryptophane foods are placed on the diagram from left to right in descending content of these amino acids. To avert stunting, a child must each day have 1 to 3 gourd-fuls of that plant as indicated by the diagram. (Grimm et al., 2014).

people eating insects. After a few more semesters went by with the student successors of Carly, we saw the food choices of the women in the village shift relatively quickly to make use of higher quality tryptophan/lysine-containing foods already grown in the village and available at the start of the malaria/ hunger season and to introduce foods to provide vitamin B12 for their young children. Interestingly, the malaria season is the same as the wet season, so it corresponds with the planting growing season. With one cropping system per year for grain, this means the part right before the harvest is when food supplies might run out and incidence of malaria is the highest it is all year. If the details in this story seem excessive in comparison to the simplicity of the output, consider the changes that occurred in food choices mothers in that village made once they understood what they needed to know about the underlying nutritional biochemistry.

"Hunger," kwashiorkor, stunting can disappear if we take the time and energy to engage with the community-at-risk, use the holistic process (Table 3.1), and practice decolonizing methodologies. This story of place-based, culture-based nutritional science and crickets is continued in Chapter 5, Listening With Subsistence Farmers in Mali.

Establishing Ownership of Original Data or Products

When I try to explain decolonizing methodologies to anyone or when I prepare myself for a meeting on the Native American reservations or anywhere in Africa, I remind myself or my colleagues about the ownership of original data and products. I tell the following story. It was 2003. I had invited Clifford Montagne to join me in Mali because his approach to working with communities using the holistic process seemed to be that missing part of the puzzle I detected in doing field research in Mali with the small-holder farming villages and the national agricultural research organization (l'Institut d'Economie Rurale [IER]) there. I needed an appropriate, respectful way of introducing Cliff to this village, a way that followed the principles of the holistic process of Savory and Butterfield (1999). When we arrived in the village and entered the courtyard at the home of Bourama Coulibaly and his family, our host and usual meeting place, we were offered two of the only chairs in the household. After the formal introductions by one of our collaborators from IER, I mentioned that we might explain the whole farming system of Sanambele and their hunting/gathering locations and transportation opportunities to Cliff by drawing a map. I took out a 3 m × 3.3 m white paper and laid it on the ground in the middle of all of us and offered a pencil and black marker to Bourama. After much collaboration among the men

Table 3.1 How Community Needs Are Addressed Using the HM Model in Generic and Summary Terms With Respect to the Village of Sanambele, Mali

HM Model[a]	Community Need	How This Happens Generically	How This Happened in Sanambele to Reduce Stunting and Micronutrient Deficiency[b]
Step 1: Define the "Whole"	Trust and acceptance	Sharing culture in daily life activities, learn about "whole" situations[c]	Focus groups not aimed only at crop production and storage Homestays of USA/Malian students/faculty/scientists involved with village
Step 2: Identify tools, test for sustainability	Shared collaborative vision	Learning together about problem situations, finding win–win solutions[c]	Wordless, numberless but quantitative tables of nutrients available locally vs needed for normal growth/development
Step 3: Identify tools, test for sustainability[d]	Creative solutions	Brainstorming, testing for sustainability[d]	Handicraft Enterprise initiated and thrives on local creativity Nutritional diagrams with local crops developed along with Women's Association
Step 4: Plan actions, use[e]	Action and feedback loop	Action plans, agreeing on how and when to monitor and adjust[e]	Children's height, weight, age compared Acceptance of new, but local foods by mothers, children

[a]In practice, the holistic process does not always occur in a linear path and steps may occur simultaneously.
[b]For chronological history of how Sanambele addressed its holistic goal of stunting and micronutrient eradication in collaboration with transformed classrooms, see Chapter 5, Listening With Subsistence Farmers in Mali.
[c]Quality of life is discussed at this time.
[d]Present resource base and future resource base are described and discussed at this point.
[e]Ecosystem Processes and Human System Processes are identified and used in monitoring at this point.
Adapted from Dunkel et al., 2013. Sustainable integrated malaria management by villagers in collaboration with a transformed classroom using the holistic process: Sanambele, Mali, and Montana State University, USA. Am. Entomol. 59 (1), 45–55.

and women farmers, a map was created. The map was without words, of course, since all but one of the farmers assembled did not read or write.

We took a photo of the map, but left the original map with Bourama and he kept it safely for many years afterward. It seemed at first this was simply a symbolic act of friendship and respect, but, after many more years of working together, about 17 years, I have thought that Bourama deemed it his responsibility to keep the map safe. When we would return to the village, Bourama would retrieve the map and lay it out for us to verify that he had responsibly preserved it. This was, in itself, an awesome task. The map was made of paper and the only storage for it was the earth floor of Bourama's mud hut. The rain comes each year, usually every day from end of June until the beginning of October, sometimes in torrents. Paper not protected from this rain disintegrates. This story helps me remember the importance of always leaving original data with the village.

Summary Reflections on Decolonization

Throughout this chapter, we have shared specific examples of small group interactions with indigenous communities that illustrated decolonization methods. It is important to note that the term decolonization is also used on a much larger scale. For instance, *Decolonizing Methodologies: Research and Indigenous Peoples* (2012), authored by Linda Tuhiwai Smith, is replete with passion and formal examples of colonizing and decolonized methodologies in research. Smith writes not against research or sharing knowledge with Indigenous peoples but for new ways of knowing and discovering, and new ways to think about research with Indigenous peoples. Another large-scale example is George Ayittey, a macroeconomist and a Ghanaian, who wrote *Africa Unchained: Blueprint for Africa's Future.* This book about decolonizing sub-Saharan Africa presents the strength of the "Atinga," the subsistence farmers of the region, respecting the ability of the Atinga and the power of the informal sector to solve the current issues of this continent. It is the "Atinga" with whom Carly learned to communicate complex nutritional data all the while respecting social practices in raising food, wild collecting, and in serving food (gender- and age-related food dominance).

If this is decolonization, how did colonization begin?

The Build Up of Colonization

Summarizing the history of colonialism is a difficult task because each culture's story and interaction with colonialism is unique. The summary we provide later is meant to give general context to Western science and colonialism's relationship. Every culture has been impacted by colonialism differently and we encourage the reader to specifically explore colonialism's impact on every culture with which one works.

In Europe, Descartes described the human body as a mechanical system, paving the way for a view of nature as a mechanical system apart from humankind (Capra 1982). Sir Isaac Newton followed Descartes in developing the laws of motion, again using the mechanical paradigm supported with replicable linear mathematical relationships. These paradigms, in which humans manipulate nature to gain power and material gain, helped Europeans increase agricultural production and accumulate wealth, which supported further scientific exploration and technology development. With increased technological capabilities, European countries became powerful enough to colonize other lands. Colonization was accompanied by agricultural development, often focused on introducing domestic animals like cattle along with crop farming. Agricultural development, pushed by technological applications, tended to overlook the inherent knowledge and skills (TEK) of small farmers and those pursuing subsistence activities like hunting, gathering, and fishing. Colonial agriculture often involved commercial firms which tended to encourage large scale mechanized production to gather agricultural products for export back to Europe or other centers of wealth and power.

Colonialism, which was based on a top-down power structure, led to fragmentation of landscapes, ecosystems, and societal organizations. Complex human-ecological systems were simplified as local peoples were exploited and native species reduced or eliminated. Resources and power became more concentrated in the wealthy nation states, and more intensive natural resource and agricultural development occurred to support increasing global populations.

A Paradigm Adjustment for Reductionist (Western) Science

In the 19th and 20th Centuries formal nation-to-nation colonialism largely ceased, but the former colonies faced huge issues of leadership and governance, which often remained power-dominated with top-down decision-making. This encouraged continued cultural and environmental exploitation. Also during these centuries, in Western cultures, scientific discoveries became compartmentalized. Interconnections between the whole biosphere or even the ecosystem were lost. Western culture science earned the adjective reductionistic. As science further developed, thinkers became increasingly aware of the complexities of nature. These complexities came to be seen as webs or systems of interconnections (Capra 1996). The resulting synergistic processes produced results which could not be predicted by knowledge of just the component parts. This view, termed 'systems thinking,' allowed science to move towards learning how the complex systems of nature operate to achieve a steady state of production (a certain level of emerging properties) before additional perturbations push the system to readjust by going through a period of chaos to emerge at another level, with an adjusted set of emerging properties.

As an illustration, consider the story of the US Great Plains grasslands (Manning 1995). In the early colonial period, Europeans reduced the bison population and displaced the Native American populations with ranching and farming. Crop farming required bare soil, so when years of low precipitation and high wind came along, crops failed, the topsoil was eroded, and the system of supporting small scale crop farming collapsed. Replenishment of the soils through windbreaks and cover crop plantings revitalized the degraded landscapes and in some cases they have been repopulated with native species of plants and animals.

Aldo Leopold's Sand County Almanac (1949), brings the human dimension into ecological science. In his chapter The Land Ethic, he writes: 'In short, a land ethic changes the role of Homo sapiens from conqueror of the land-community to plain member and citizen of it. It implies respect for his fellow-members, and also respect for the community as such.' (p204). Dan Dagget (2005) reminds us that humans and landscapes have evolved together, and that human roles should fit within, rather than outside, of nature.

Traditional Ecological Knowledge Becomes Visible During Decolonization

TEK is the product of the process of doing Native Science. Native Science (Cajete, 1999) is a process of learning through direct experience. It requires a connection with heart and mind, interrelationships, and a participatory approach. Through ever-changing relationships and the ever-evolving techniques of Native Science, TEK is built and changed by this process over multiple generations. TEK accumulates within cultures living in a particular place as the culture adapts and adopts as it learns how to provide for living necessities, generation after generation. During colonization when one culture attempts to supplant another culture, TEK is suppressed or devalued. TEK is easier to see, and for some, is only possible to see, after one's approach is decolonized, that is when one's worldview is ethno-relative.

Yarrow, *Achillea millefolium*, is an example of a plant whose usefulness is described in TEK many places in the world. Colonization has resulted in the knowledge being devalued, and now largely forgotten, though useful and needed, particularly in Western culture medicine. Yarrow is among the most widespread and widely used medicinal plants in the world. It has been popular for millennia as a treatment for wounds and infectious diseases, as well as for many other conditions (Applequist et al., 2011). This knowledge was obtained from Native Science. Western Scientists are urgently searching for antibiotics that will aggressively stop the MERSA strain of *Staphylococcus aureus*, *Listeria* the causative microorganism in Legionaire's Disease, and antibiotic resistant strains of biofilm-forming *Listeria monocytogenes and Listeria innocua* obtained from food-processing environments. Antibiotic properties of yarrow have been known for many decades through Native Science (www.herbvideos.com), Only recently have Western culture scientists confirmed the bioactivity of yarrow extracts against *S. aureus* (Tajik et al., 2008) and *Listeria* (Jadhav et al., 2013). It is often Native Science that "finds" and Western Science that verifies. This was the case with the revealing of yarrow's usefulness.

In a way, Native Science and Western Science have one basic similarity, which is accumulation of data. Generally, the timeframe of data accumulation in Native Science is hundreds of years, sometimes millennia. The timeframe of data accumulation in Western Science may be 24 hours to 10 years.

Data testing, verification, and replication all occur in these respectively diverse timeframes as well. TEK and Native Science (Cajete, 1999) evolved within their cultural and environmental contexts. In contrast, Reductionist or Western science has capabilities to isolate variables and develop more mechanistic solutions. This enables people to exploit natural and human resources to produce concentrations of wealth, energy, and power, but often overlook or choose to ignore "external effects" on culture and environment. Both scientific processes are needed to solve complex issues such as food insecurity and to reduce the ecological footprint of food and agricultural production. We explore more of these two science processes in Chapter 11, Couples Counseling: Native Science and Western Science.

Both Native Science and Western Science are equally valued in the holistic process.

THE HOLISTIC PROCESS

The holistic process methods can be described as follows in a formal, linear logic (Dunkel et al., 2013). This is what underlies the informal, nonlinear process that occurs in actual practice. These are categories of action, not steps followed in a sequence. For the approach to work, there must be spontaneity:

- Describing the "whole to be managed" and creation of the holistic goal.
- Identifying tools and testing questions.
- Establishing the feedback loop.
- Creating focus groups.
- Convening stakeholder meetings.

In the later example, the first step was the focus group. Most of the other steps happened simultaneously and are still happening now 12 years later.

In 2005 MSU students, faculty, and National Agricultural Research Organization scientists in Mali were aware that they had experienced some failures in their research and outreach programs over the past 7 years. Postharvest techniques for storing cowpeas, which provide one of the missing essential amino acids in children's diets, were not being adopted. Regional access to the Internet for the world's cropping information and pest management ideas were used in unintended ways. For example, we found that this technology was mainly being used by local men farmers to connect with family living outside the village, or to pornographic sites.

We, the outsiders decided to start a holistic process by holding gender- and age-based *focus groups* in 11 village clusters where we had been collaborating, and where one of those Internet centers was located. Much to the surprise of the outsider stakeholders (IER, US Agency for International Development (AID), and MSU faculty), the main way the farmers preferred to receive information was from their village chief. In fact, the Internet center was lowest in priority as an information source (Kante et al., 2009, 2011). We also discovered that cerebral malaria and "hunger" were standing in the way of one village's desired quality of life (*Holistic Goal*). About 8% of their young children died each year from malaria and 23% had or were at risk for stunting due to malnutrition. Over the next several years in this village, we informally learned what *tools* villagers recognized to help achieve their holistic goal. We then offered diagrams so that they might understand the biocycle of the mosquitoes that vector malaria. We also provided diagrams of the biocycle of the microbes that cause malaria and who use the body of the mosquitoes as an incubator. *Testing questions* arose from this information transfer. All these years, *stakeholder meetings* were happening simultaneously as well. A *feedback loop* developed when villagers showed us where the mosquito larvae lived in the dry season and we showed villagers how to identify the anopheline larvae whose adult form transmits the microbe causing malaria. Now this monitoring loop operates informally in bimonthly or weekly interval, depending on the time of year. When villagers find the number of larvae in their sample reaches our agreed upon action level, they treat the pond with an insect growth regulator we taught villagers to make themselves from leaves of a local tree, *Azadirachta indica* (Loung et al., 2012). The success has been rewarding. As cerebral malaria and children's deaths from it were eliminated since 2008 in this village with only two deaths in the meantime and those were due in part to complications from kwashiorkor (stunting) (Dunkel et al., 2013).

The HM model encourages decision-makers and all other stakeholders to communicate openly with each other to define the "Whole to be Managed" and create a "Holistic Goal" to describe a very specific future condition for the resources within the whole. With a consensus-driven vision, decision-makers and stakeholders collaborate to identify "Tools" to apply, but first the tools are evaluated for sustainability through a series of "Testing Questions." Included in the testing questions are also system diagnostic queries encouraging participants to identify and focus on causes of issues rather than becoming constrained by focusing on effects. There is also an exercise to identify "Chains of Production" within the whole and to realize that *the weakest link must be strengthened first*. The process includes the vital "Feedback Loop" in which participants agree on specific ways to monitor the system and readjust when boundary conditions or other factors change, as they always do in complex systems.

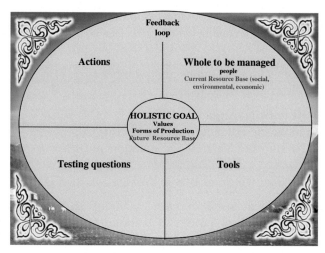

FIGURE 3.2

The Holistic Management Model (HMM) drawn from Indian (Asian) tradition representing the cosmos (https://en.wikipedia.org/wiki/Mandala) drawn by C. Montagne, R. Kroos, L. Soderquist, B. Dovchin.

Cliff, along with Animal and Range Sciences professor Brian Sindelar, consultant Roland Kroos, and graduate student Charles Orchard, developed a circular diagrammatic version of the HM model (Fig. 3.2) that incorporated it into a systems thinking context. The circular depiction gives learners a simple visual way to organize holistic thinking to include the whole, the goal, the tools, the testing questions, planning/actions, and the feedback loop. It is a reminder that the process can be entered at any point rather than requiring a linear approach, and it can be continually repeated as boundary conditions change and more knowledge is gained. The circle also suggests that this is a systematic process, which must also be systemic and originate from within the community.

Decisions and control and power come from within rather than being dictated from the outside.

A mandala is useful in visualizing this process. The mandala is a circle diagram from the Indian (Asian) tradition representing the cosmos. Clifford Montagne, Roland Kroos, Lora Soderquist, and Badamgarav Dovchin created the Holistic Management Mandala (HMM) to represent the main components of the HM model. It is a culturally significant visual presentation signifying that the holistic process is continual as conditions keep changing. Although the mandala is in the form of a circle, it is not a stationary circle simply whirling in place like a car wheel spinning in the mud, rather it is like a spiral moving and changing indefinitely as a dynamic ever-changing process.

The HMM circle has four quadrants and an inner circle (Fig. 3.2). The inner circle is the Holistic Goal. The quadrants are:

- The Whole to be managed.
- Tools to move the Whole toward the Holistic Goal.
- Testing Questions to diagnose the situation and select sustainable Tools.
- Plan and Action, which are best to do after considering the other parts.

An over-all component used in each quadrant is the Feedback Loop (Monitoring). In any holistic process, it is important to have a way to reexamine results and adjust components conditions change.

THE NEW PARADIGM FOR TEACHING IN THE FOOD AND AGRICULTURAL SCIENCES

Using the Holistic Process

The methods of HM and decolonization just presented formed the heart of four courses at MSU. Students engage with the topic through:

- Directed readings and content modules that are prepared and guided by the coinstructors.
- Reflective writing in which students bring course content into the context of their own set of values and concerns while applying knowledge from their personal and professional or disciplinary outlooks, these writings are often shared within the class.
- Small and whole group discussions in class or on-line (the in-class discussions often start with a group of two or larger, with groups then sharing with the whole class and thus eventually a "whole group" discussion). Who participates in each discussion group varies over the semester so participants have opportunity to interact with people with similar and different backgrounds, values, skills, and expertise.
- Whole class group case study based on a real community and location in which students "role-play" the various decision-makers and other stakeholders as they go through the steps of using the HM process for learning about key issues and making decisions.
- Individual Holistic Management Plan in which students select an issue of personal concern and use the HM process to make progress on solving the issue.

- Guests who share "real-world" examples in an interactive presentation and discussion format.
- Field trips to land-based locations where the instructors have long-term relationships built on respect and trust, such as subsistence farming villages in Mali (Chapter 5: Listening With Sugsistence Farmers in Mali), as well as Native American reservation communities (Chapter 6: Listening with Native Americans), and semi-nomadic herding communities in Mongolia (Chapter 7: Listening Within a Bioregion) to engage with decision-makers and other stakeholders and/or to practice HM techniques such as monitoring for various components of environmental quality, food insecurity, or disease management.

These student opportunities to bring natural science together with the human dimension and use this "trans-disciplinary approach" to engage with people in ways which respect and consider both environment and culture (psychology, value-sets, world views, traditions, economic necessities, political realities) allow them to learn the importance of having "definite outcomes" for stages of problem solving. They learn the importance of understanding "cause" versus focusing on "effect." They learn ways to diffuse contention through inclusion and participation and how this can help them as citizens within a participatory democracy.

Evaluation of student's ability to use the holistic process is accomplished through:

- Performance on several simple "content quizzes."
- Participation in the discussions, writing assignments, case study, and field trips.
- Quality of the Personal Holistic Management Plan.
- Ability, on the final exam, to use the HM process to understand and suggest solutions to a land- and culture-based issue.

The Expansive Collaborative Model for Community Engagement and Service-Learning

When I learned of Montagne's course in Holistic Thought and Management, this seemed like the answer to the failures we were experiencing. We worked together with faculty and students from across the United States to apply these concepts with collaborators from Mali, West Africa, and the Northern Cheyenne Nation (Chapter 5: Listening With Subsistence Farmers in Mali and Chapter 6: Listening With Native Americans). In 2005 a watershed moment occurred when two of Cliff's students who became Mali externs nudged me to use the holistic process. This immediately reversed the learning and research process (Kante et al., 2009, 2011; Dunkel et al., 2013). At the same time, the French Language and Literature students of Dr. Ada Giusti became Mali externs and brought to the Mali extern program a deeper appreciation of cultural nuances. Specific literature written by Malian authors was introduced by Ada into our predeparture training along with stricter language requirements (Dunkel and Giusti, 2012). Cross-fertilization of learning with the Northern Cheyenne students added a profound appreciation for the family and TEK (Chaikin et al., 2010). The holistic process helped disciplinary professionals and their students learn to listen with village citizens in Africa and discover simple solutions to food and health issues (Dunkel et al., 2013).

The Expansive Collaborative (EC) Model (Fig. 3.3) emerged from a combination of the holistic process and my work with the global, participatory Integrated Pest Management Collaborative Research Support Program (IPM CRSP) (Norton et al., 2005). The IPM CRSP began in 1993. By 2000, it was clear that the new Millennial Generation students were ready for experiences to address their concerned engagement about the Gap: failures of aid to material poor communities. Based in the same communities where the IPM CRSP was engaged, I initiated the Mali Extern program at Montana State University and it spread via USDA National Institute of Food and Agriculture (NIFA) funding in collaborations other universities in California, Virginia, and Minnesota, and a tribal college (Dunkel et al., 2011). Mali externs, students who engaged with communities in Mali through a service-learning process, had their needs met for involvement at a grass roots level. Students actually became engaged with the community. Community leaders became site mentors for the students and coinstructors in the course. Still, the

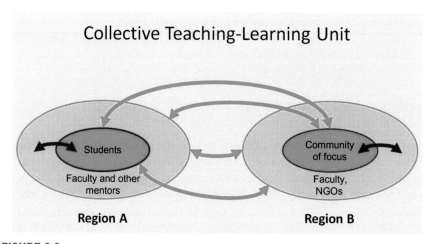

FIGURE 3.3

Basic components of learning communities in the Expansive Collaboration Model using the holistic process (NGO, nongovernment organization) (Dunkel et al., 2011).

participatory IPM process we used led to some failures as described in Chapter 2, Failures.

This EC model requires two main groupings of higher education institutions (in Region A and Region B) linked in what should be entered into as a long-term relationship (Fig. 3.3). Two or more culturally distinct regions are needed in the model to set the stage for cognitive dissonance followed by perspective transformation. We found this was an ideal environment for students trained in the food and agricultural sciences (as well as in health/cell biology/neuroscience and engineering) to understand social consequences of technological change.

By 2007, the Mali Extern program was both addressing needs of the Millennial Generation students for meaningful engagement and needs expressed by subsistence farming communities. Nudged again, this time by an MSU campus leader preparing for her career in industrial engineering and economics, we developed a course to explore new concepts being articulated by economists working in material resource poor countries (Ayittey, 2005; Easterly, 2006; Calderisi, 2006). Combining the holistic process, the Mali Extern Program, and the challenge from Nobel Prize recipient, Mohammed Yunus that poverty could be put in museums (Yunus, 2003, 2007), AGSC 465R Health, Poverty, Agriculture: Concepts and Action Research was created. The course was built on the development of the EC Model (Dunkel et al., 2011) and has been tested by 220 undergraduate and graduate students. Nationally, 131 students participated in the Mali Extern program, until Islamic militants became active in Mali and US student travel to Mali was prohibited by US universities. Thanks to increasing sophistication of communication electronics, students in AGSC 465R maintain weekly communication with Malian communities through this course.

The EC Model was initially designed by Dunkel in 2002 and revised in 2004 and 2007 (Dunkel and Gamby, 2007; Dunkel and Montagne, 2006, 2009; Dunkel et al., 2007) (Fig. 3.3). This Model provides a methodology to learn how to decolonize one's words and actions, to practice the holistic process. In so doing, the EC Model promotes education for sustainable development. Suggestions of Easterly aligned with the holistic process encouraging outsiders not to come with a plan, but to let the community develop their own plan in their own timeframe (2006). Because the EC Model has members of the community-of-focus acting as on-site mentors, there is opportunity for guided interpretation of cultural dissonance. Usually, this interpretation leads to a transformative effect in both faculty and students (Kiely, 2005; Mezirow, 2000).

Evaluation of student's ability to use the holistic process is accomplished in four ways. First the students create their own holistic goal statement

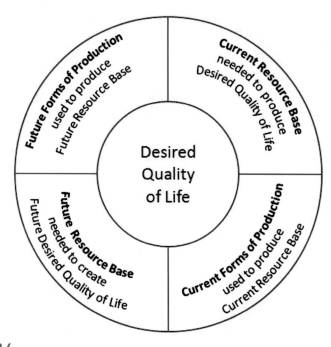

FIGURE 3.4
Simplified diagram used to begin a holistic process. This is the detail used to obtain information for the center concentric circle used in the diagram in Figure 3.2. Once these details are determined by the community, then the actions, testing questions, tools, and whole to be managed can be described. The feedback loop operates on the entire system, on all parts, and is best illustrated as a third dimension. (Dunkel, Baldes, Montagne).

(Fig. 3.4A). Second, students engage in holistic process conversation with their site mentors who are members of the community students focus on for the semester (Fig. 3.4B–D). Third, students apply the 10-course concepts (one of which is the holistic process, another is decolonizing methods) on specific issues in their community-of-focus. Fourth, students conduct participatory research with the community that requested it and maintain a mentored reflective journal of their experiences.

CONCLUSION

With over 7 billion people to support, our Earth's natural systems are more and more stressed. Reductionist (Western) science offers ways to increase capacities, but often neglects externalities and unintended consequences. We need to be able to continue to apply reductionist techniques, but within a holistic systems thinking framework like HM.

Existing TEK (Berkes et al., 2000) is again emerging to share techniques and wisdom generated and tested over eons of repetition. However, our Earth's systems are strained by human activities at higher levels of intensity and scales of area than before, when most of existing TEK was generated. Hence, we must bring TEK approaches to the table as we continue to develop both reductionist and holist technologies for management of natural and human resources. Agriculture, as one of humankind's most important activities, has a responsibility to rise to the opportunity to grow beyond straight reductionism. Components of the holistic process as expressed in the HMM and in the Savory text (Savory and Butterfield, 1999) are useful guides for this process.

HMM and TEK need to come together in classrooms beyond the few courses at Montana State University, AGSC 465R, NRSM 421, and LRES 521. We propose that HMM and TEK appreciation becomes a best practice in boardrooms, at policymaking tables, and wherever people intersect with food and agriculture. It is especially critical that HMM and TEK be directly used in working with any part of our world population that considers themselves disenfranchised.

In this book, we share how one can use holistic management, holistic process, and appreciation of TEK within communities-of-focus in West Africa (Chapter 5: Listening With Subsistence Farmers in Mali); on Native American reservations (Chapter 6: Listening With Native Americans); with nomadic communities in Asia (Chapter 7: Listening Within a Bioregion); in struggling, stereotyped communities of rural United States (Chapter 9: Listening With Students); among the diaspora of Jamaica (Chapter 9: Listening With Students); and among the disenfranchised workers in middle class United States and England (Table 3.1).

Setting the Stage: Power of the Circle as Metaphor for Learning and Research

Just sitting together in a circle gives each person an equal opportunity to be heard, and to react to what others have said. It is an effective way to start listening deeply, and in a group or organization, horizontally. The circle arrangement can promote openness, transparency, trust, and shared values. If a community group, especially if it includes an outsider, develops trust and open communication, it has a better chance of successful problem solving that incorporates multiple viewpoints and disciplines. Horizontal listening and subsequent open communication allows for more complete understanding of the ecological, social, and financial dynamics of the situation. This leads toward solutions that address the cause of the problem rather than focusing on the effects. Like peeling an onion, the outer layer "effects" can be discarded in favor of identifying and solving for cause.

I remember the circle from my earliest experiences of human community in the out of doors. Whether sitting around the campfire in the evening telling stories, or gathering around an interesting rock outcrop on a geology field trip, people tend to form a circle.

In my teaching career, the circle idea came, as mentioned earlier, from a workshop on Systems Thinking in Agriculture sponsored by Washington State University and facilitated by Dr. Kathy Wilson (Wilson and Morren, 1990). In this text, Kolb's model of the learning circle (Adapted from David A. Kolb. Experiential Learning, 1984, Fig. 22, p. 42) shows a learning process as a circle combining human tendencies to learn from personal experience through reflection, conceptualization, experimentation, and then application. The application results then cycle back into the same circle in an iterative process in which awareness, knowledge, and application become more effective. With assistance from Dr. Brian Sindelar, former range science faculty member at MSU; Roland Kroos, Holistic Management Educator; and graduate student Charles Orchard, I placed the Holistic Management Model (Savory and Butterfield, 1999, Fig. 7—1, p. 54) into the Kolb learning style circle as introduced in this chapter. Blackfoot Tribal member and MSU undergraduate student Roylene Rides at the Door and I then adapted this to the Blackfoot sacred circle, which is a model for the cosmos and how people learn and act, and then learn from their actions (published in American Indian Science and Engineering Society Journal).

Allan Savory first introduced me to an effective form of group communication and "team building." At an introductory course on Holistic Resource Management, I found myself part of a group of over 60 participants. We did not know each other and came from a variety of backgrounds, locations, and professions. Sixty seems like a large group for a circle, so instead we each found ourselves assigned to a smaller group of six sitting around a table. With six, we could easily get acquainted by sharing what made up the most important quality of life values for each of us. Then by reporting out to the whole group of 60, we established common understanding of quality of life values we all agreed on, and we then had a web within which to carry out our work. In this course developed by Allan Savory, the instructors act as content-matter practitioners and facilitators, rather than lecturers. The rhythm of discussion may start with a minicontent sharing by an instructor, but small student task-groups process in their own discussions before sharing with the entire class. In this way, students continually go back and forth from individual or small group scales to that of the larger circle, which contributes a broader diversity of background, viewpoints, and expertise.

References

Applequist, W.L., Moerman, D.E., 2011. Yarrow (*Achillea millefolium* L.): a neglected panacea? A review of ethnobotany, bioactivity, and biomedical research. Econ. Bot. 65 (2), 209—225.

Ayittey, G.B.N., 2005. Africa Unchained: The Blueprint for Africa's Future. Palgrave, Macmillan, NY, 483 pp.

Bennett, M., 2004. Becoming interculturally competent. In: Wuzel, J. (Ed.), Toward Multiculturalism: A Reader in Multicultural Education, second ed. Intercultural Resource Corp., Newton, MA, pp. 62—77.

Berkes, F., Colding, J., Folke, C., 2000. Rediscovery of traditional ecological knowledge as adaptive management. Ecol. Appl. 10 (5), 1251—1262, Ecological Society of America. Stable URL: <http://www.jstor.org/stable/2641280>.

Cajete, G., 1999. Native Science: Natural Laws of Interdependence. Clear Light Publishers, Santa Fe, NM, 339 pp.

Calderisi, R., 2006. The Trouble with Africa: Why Foreign Aid Isn't Working. Palgrave Macmillan, New York, NY, 249 pp.

Capra, F., 1982. The Turning Point. Bantam Books, New York.

Capra, F., 1996. The Web of Life. Bantam Books, New York.

Chaikin, E., Dunkel, F., Littlebear, R., 2010. Dancing Across the Gap: A Journey of Discovery. Montana State University. 56 minute documentary aired Montana PBS November 2010; Northern Cheyenne TV November 2010. http://www.montana.edu/mali/npvideos.html.

Chambers, R., Pacey, A., Thrupp, L.A. (Eds.), 1989. Farmer First: Farmer Innovation and Agricultural Research. Bootstrap Press, New York, 218 pp. (read pp. 55−59 and 77−85).

Dagget, D., 2005. Gardeners of Eden: Rediscovering Our Importance to Nature. The Thatcher Charitable Trust, Oxford.

Dunkel, F.V., Montagne, C., 2006. Discovery-based undergraduate opportunities: facilitating farmer-to-farmer teaching and learning. Second annual report for USDA CSREES higher education challenge grant program, http://www.montana.edu/mali/documents/annualreport-due14nov.htm.

Dunkel, F.V., Montagne, C., Peterson, N., 2007. D−iscovery-based undergraduate opportunities: The Collaborative Research Support Programs (CRSPs). http://www.montana.edu/mali/documents/challengefinalreport2002.pdf, 17 pp. Final qualitative report to the USDA CSREES Higher Education Challenge Grant Program.

Dunkel, F.V., Gamby, K.T., 2007. Linking biotechnology/bioengineering with Mali-based agribusiness: strengthening food and water quality for health, safety, and exports. Final report for higher education for development. http://www.montana.edu/mali/grad_annual.htm.

Dunkel, F.V., Montagne, C., 2009. New paradigm for discovery-based learning: Implement bottom-up development by listening to farmers' needs while engaging them in participatory, holistic thinking. Second annual report for USDA CSREES higher education challenge grant program, http://www.montana.edu/mali/pdfs/CRISfinalreportyear2.pdf.

Dunkel, F.V., Shams, A.N., George, C.M., 2011. Expansive collaboration: a model for transformed classrooms, community-based research, and service-learning. North Am. College Teachers Agric. J. 55(Dec), 65−74.

Dunkel, F., Giusti, A., 2012. French students collaborate with Malian villagers in their fight against malaria. In: Thomas, J. (Ed.), Etudiants Sans Frontières (Students Without Borders): Concepts and Models for Service-Learning in French. American Association of Teachers of French, Marion, IL, pp. 135−150.

Dunkel, F.V., Coulibaly, K., Montagne, C., Luong, K.P., Giusti, A., Coulibaly, H., et al., 2013. Sustainable integrated malaria management by villagers in collaboration with a transformed classroom using the holistic process: Sanambele, Mali, and Montana State University, USA. Am. Entomol. 59 (1), 45−55.

Dunkel, F., Hansen, L., Halvorson, S., Bangert, A., 2016. Women's perceptions of health, quality of life, and malaria management in Kakamega County, Western Province, Kenya. GeoJournal. 81, 25 pp, doi:10.1007/s10708-016-9701-7. Web published open source 4 May 2016.

Easterly, W., 2006. The White Man's Burden: Why the West's Efforts to Aid the Rest Have Done So Much Ill and So Little Good. Penguin Press, New York, 436pp.

Jadhav, S., Shah, R., Bhave, M., Palombo, E.A., 2013. Inhibitory activity of yarrow essential oil on *Listeria planktonic* cells and biofilms. Food Control 29, 125−130.

Kante, A., Dunkel, F., Williams, A., Magro, S., Sissako, H., Camara, A., Kieta, M., 2009. Communicating agricultural and health-related information in low literacy communities: A case study of villagers served by the Bougoula commune in Mali. Proceedings of the Annual Meetings of the Association for International Agricultural and Extension Education, San Juan, Puerto Rico, 24 May 2016. Used as a case study by USAID Feed the Future. MEAS Case Study Series on Human Resource Development in Agricultural Extension, Case Study #11, January 2013.

Kiely, R., 2005. A transformative learning model for service-learning: a longitudinal case study. Michigan J. Commun. Service Learn. 12, 5−22.

Kolb, D.A., 1984. Experiential Learning: Experience as the Source of Learning and Development. Prentice Hall, Englewood Cliffs, N.J.

Leopold, A., 1949. Sand County Almanac. Oxford University Press, Oxford.

Luong, K., Dunkel, F., Coulibaly, K., Beckage, N., 2012. Use of neem (*Azadirachta indica* A. Juss.) leaf slurry as a sustainable dry season management strategy to control the malaria vector *Anopheles gambiae* Giles s.s. (Diptera: Culicidae) in West African villages. J. Med. Entomol. 49 (6), 1361−1369. Available from: http://dx.doi.org/10.1603/ME12075.

Manning, R., 1995. Grassland: The History, Biology, Politics, and Promise of the American Prairie. Penguin Books, New York.

Makkar, H.P.S., Becker, K., 1996. Nutritional value and anti-nutritional components of whole and ethanol extracted *Moringa oleifera* leaves. Anim. Feed Sci. Technol. 63 (1/4), 211−228.

Mezirow, J., 2000. Learning to think like an adult: core concepts in transformation theory. In: Meizrow Associates, J. (Ed.), Learning as Transformation: Critical Perspectives on a Theory in Progress. Jossey-Bass, San Francisco, CA.

Norton, G.W., Rajotte, E.J., Luther, G.C., 2005. Participatory integrated pest management (PIPM) process. In: Norton, G.W., Heinrichs, E.A., Luther, G.C., Irwin, M.E. (Eds.), Globalizing Integrated Pest Management: A Participatory Research Process. Blackwell Publ, Ames, IA.

Savory, A., Butterfield, J., 1999. Holistic Management: A New Framework for Decision Making. second ed. Island Press, Washington, DC.

Tajik, H., Jalali, F.S.S., Sobhani, A., Shahbazi, Y., Zadeh, M.S., 2008. In vitro assessment of anti-microbial efficacy of alcoholic extract of *Achillea millefolium* in comparison with penicillin derivatives. J. Anim. Vet. Adv. 7, 508−511.

Turley, R. 2011. Seasonal availability of lysine and tryptophan in the Sanambelean diet. Research paper and poster submitted in partial fulfillment of AGSC 465R. <http://www.montana.edu/mali/pptsaspdfs/TurleyBeckyPoster.pdf>.

Wilson, K., Morren Jr., G.E.B., 1990. Systems Approaches for Improvement in Agriculture and Resource Management. Macmillan, New York.

Yunus, M., 2003. Banker to the Poor: Micro-lending and the Battle Against World Poverty. Public Affairs, New York, NY.

Yunus, M., (with Karl Weber), 2007. Creating a World Without Poverty: Social Business and the Future of Capitalism. Public Affairs, New York, 261 pp.

Further Reading

Ayittey, G.B.N., 1991. Indigenous African institutions. Transnational Publishers, Inc., Ardsley-on-Hudso, NY, 547 pp.

Curtin, C.G., 2015. The Science of Open Spaces: Theory and Practice for Conserving Large Complex Systems. Island Press, Washington, DC.

Grierson, R., Munro-Hay, S., 2000. The Ark of the Covenant. Phoenix, London, ISBN 0753810107.

Munro-Hay, S., 2002. Ethiopia, the Unknown Land a Cultural and Historical Guide. I.B. Tauris and Co. Ltd, New York, ISBN 1 86064 7448.

Rides at the Door, R., Montagne, C., 1996. Holistic resource management meets native culture. Winds of Change. Am. Ind. Sci. Eng. Soc. 11 (4), 136−141.

Savory, A., Butterfield, J., 2016. Holistic Management: A Commonsense Revolution to Restore Our Environment. third ed. Island Press, Washington, DC.

Sindelar, B.W., Montagne, C., Kroos, R.H., 1995. Holistic resource management: an approach to sustainable agriculture on Montana's Great Plains. J. Soil Water Conserv. 50, 45–49.

Smith, L.T., 2012. Decolonizing Methodologies: Research and Indigenous Peoples, second ed. Zed Books, New York.

Smuts, J.C., 1927. Holism and Evolution. second ed. Macmillan and Co., New York.

Wilson, E.O., 2006. The Creation: An Appeal to Save Life on Earth. W.W. Norton & Co, New York.

CHAPTER 4

Immersion

Ada Giusti[1], Florence V. Dunkel[2], Greta Robison[3], Jason Baldes[4], Meredith Tallbull[5], El Houssine Bartali[6], and Rachel Anderson[7]

[1]Department of Modern Languages and Literature, Montana State University, Bozeman, MT, United States, [2]Department of Plant Sciences and Plant Pathology, Montana State University, Bozeman, MT, United States, [3]Department of Earth Sciences, Montana State University, Bozeman, MT, United States, [4]Wind River Native Advocacy Center, Fort Washakie, WY, United States, [5]The Northern Cheyenne Nation, Lame Deer, MT, United States, [6]Department of Agricultural Engineering, Institut Agronomique et Vétérinaire Hassan II, Rabat, Morocco, [7]The Madisonian, Ennis, MT, United States

CONTENTS

Definition of Immersion 74

Case Study 1. An MSU French Professor Collaborates With an Entomology Professor on a Malian Food and Health Project 75
Introduction 75
My Naïve State 76
Growing Awareness 76
Applying Cultural Knowledge 77

Case Study 2. Arranging Successful Immersions at an Institution Level .. 80
Introduction 80
Annual Undergraduate Immersions 81
Evaluating Immersion Results 83

Case Study 3. Native Foods and Food Deserts 85
Why Is the Mini-Immersion an Essential Part of Education? 89

"*Given my personal and professional background, I, Dr. Ada Giusti, was somewhat confident that I was prepared to communicate and collaborate with Malian agricultural scientists and the villagers they were serving. I am the grandchild of subsistence farmers and have a PhD in French and Francophone Studies. Prior to my departure for Mali, a Francophone country in West Africa, I learned about Malian agriculture under the tutelage of Dr. Dunkel. I had, however, only a basic knowledge of Malian history, customs, and contemporary issues. Then I traveled to Mali; there, I began to see with my own eyes. Upon my return to the United States, I realized that while in Mali I was seeing a superficial story, one I had constructed through my own cultural lenses. Reading books authored by Malian authors about their country's history and complex culture, and making several subsequent trips to Mali was like opening many windows in a dark room.*

Understanding culture is extremely important. It is a key element of partnership building because it enables participants to develop equitable long-term relationships based on respect and trust. Both in-country and visiting faculty members and scientists focused on technical solutions to community problems may be reluctant to invest the time needed to learn how to navigate a new culture and develop personal relationships with locals. Yet, in the absence of such relationships, agricultural-based projects or any other development projects initiated and funded by outsiders are likely to fail."

In this chapter, Giusti, Robison, Tallbull, Baldes, and I, Florence Dunkel, converse about rural and urban relationships in a foreign-to-them culture. Anderson and Robison reflect on immersion as students. Jason Baldes introduces us to his immersion experiences in Africa and in the Euro-American

Incorporating Cultures' Role in the Food and Agricultural Sciences. DOI: http://dx.doi.org/10.1016/B978-0-12-803955-7.00004-3

Mini-Immersion Process........................ 90

Native Plants and Native Foods on the Northern Cheyenne Reservation.................. 92

"Let's Pick Berries" Project.......................... 95

Summary of Case Study 3................ 98

Are There Negative Consequences From Not Including Immersion in General Education? 99

Concluding Thoughts on Immersion 100

References 102

Further Reading 105

culture on the Montana State University (MSU) campus as a prelude to a more in-depth reflection in Chapter 9, Listening With Students, as a student and as a tribal member. Together we explore ideas and some fine details about what you would want to know about the background of the people in the culture you are entering as an outsider. For example, who you show up with in the village or reservation community will determine how you will be perceived initially and the communication path that follows. This chapter focuses on the particular importance of cultural immersion when you work with agriculture and food, but first, we must ask the question: Why bother with an immersion experience? And, more basic: What is cultural immersion?

The other authors and I offer three examples of cultural immersion and how each has "opened windows in a dark room" for participants by working with bottom-up approaches to understand issues of food security. In the first case study we follow a faculty member from the humanities preparing for her role on a team of agricultural scientists. In the second example, we explore a unique 4-year undergraduate immersion program that has been running for 40 years and is required of all majors in the premier agricultural-based university in Morocco. In the third example we illustrate how Native foods and Native plants can be learned about in situ, during immersion outside of a classroom, on a Native American reservation, and in the process become a platform to engage locally about food deserts and "Gardens of Eden." Ada begins our discussion with these definitions of immersion.

DEFINITION OF IMMERSION

According to the Oxford dictionary, the word immersion entered the English language in the 15th century, from late Latin immersio(n-), meaning "to dip into." Throughout the centuries, this word has kept its original meaning and gained several others. Additional definitions include: (1) deep mental involvement; (2) baptism by immersing a person bodily in water; and (3) the disappearance of a body in the shadow of or behind another (Astronomy, rare).

Various cultural immersion programs administered inside and outside academia reflect these different definitions. All are designed to engage participants in a process allowing them to "dip into" another cultural and/or linguistic experience. The depth of this immersion may vary greatly. In some cases, participants can become "deeply mentally involved" when they immerse themselves in the study of a foreign country; they may begin by investigating that country's history, then move on to reading its most influential authors, all the while keeping up with current events and learning the country's official language(s). In other cases, participants may be dipped mentally and physically in the target culture; this is akin to a "baptism" from which participants emerge more or less transformed by their experience. In rare cases, those who partake in extensive immersion programs or reside and integrate in a foreign society for many years will undergo a complete transformation. The body and mind molded by their native culture and language will "disappear into the shadow" of the host culture and language. A new person appears, fashioned by its native and host cultures and languages.

The deep mental involvement usually attained through classroom study or self-guided immersion (the latter for those who are no longer traditional students, such as policy makers and administrators) is a necessary prelude to the "baptismal" immersion. Being physically transported or "dipped" in the foreign culture is not enough. Just like in a religious or spiritual context, if the person has not completed a sufficient period of introspection and study, simply immersing oneself in holy water will not produce a transformation. The following case study illustrates the need to integrate both mental and physical processes if culture is going to be recognized and respected in the sciences. Moreover, it illustrates how cultural knowledge can build relationships and ensuing collaborations between all stakeholders in the food and agricultural sciences.

CASE STUDY 1. AN MSU FRENCH PROFESSOR COLLABORATES WITH AN ENTOMOLOGY PROFESSOR ON A MALIAN FOOD AND HEALTH PROJECT

Introduction

As a French and Francophone Studies professor, I, Ada Giusti, have always aspired to travel and work in Francophone Africa. In early 2004, Dr. Dunkel, an entomology professor from the MSU College of Agriculture, offered me the opportunity to fulfill this aspiration when she invited me to develop language and cultural immersion workshops for her students and other faculty members collaborating in a multiinstitution, government-funded project focusing on food safety in Mali. This project was part of a global program, the Integrated Pest Management-Collaborative Research Support Program (IPM-CRSP) US Agency for International Development (USAID). As far as I knew, I was the only French and Francophone Studies professor in the Mali part of the program and the only language and literature professor involved globally in the IPM CRSP in 2005.

Prior to our March 2005 departure, I attended Dr. Dunkel's seminars. There, her students and I learned about the various government agencies with whom we were to collaborate and about the agricultural practices of the Malian villagers who were to be the focus of our upcoming trip. We learned about some of the obstacles the farmers had dealt with in the past decades, including soil erosion, droughts, the prohibitive cost of pesticides, and the ensuing effect of pesticide use on the farmers' food sources. In turn, I taught French, and shared my own research on economic poverty and poverty alleviation, as well as some components of Malian history and contemporary political and cultural issues.

My Naïve State

Once we were in Mali, I felt physically, emotionally, and intellectually immersed. However, it wasn't until I returned home and began an in-depth study of this complex culture that I realized that this had been a very naïve perspective. Several factors had led me to think that I was successfully integrating into this host culture. For instance, communicating with local residents appeared to be straightforward. In Bamako, Dr. Dunkel introduced our small group to our partners at l'Institut d'Economie Rurale (IER) and the USAID. All were welcoming and forthcoming with their knowledge. Our daylong visits to several Bambara villages, accompanied by IER staff, led to many focus group discussions on the holistic goals of the various communities. (Outcomes of these discussions served as starting points for future shared work with Dr. Dunkel, her students, and villagers.) On a personal level, conversations with rural school teachers and health workers, hotel personnel, taxi drivers, and food vendors gave me a glimpse of the daily concerns of Malian citizens. To further understand current social and political events, I watched local TV news broadcasts, listened to local radio, and read newspapers. Clearly, this first visit had been very informative and personally enriching. It did not, however, consist of a deep immersion because I lacked knowledge of the complex cultural framework that underlined communication patterns and exchanges. Without this knowledge, I interpreted all interactions through my own cultural and personal lenses.

Growing Awareness

After 2 weeks in Mali, I returned to the United States equipped with a reading list compiled by a Malian historian and a sociologist, both professors at the Faculté des Lettres, Arts, et Sciences Humaines de Bamako (FLASH). These two scholars had generously provided me with a list of themes and authors best equipped to teach me about their culture (Bâ, 1972, 1992, 1994, 2000; Bâ and Hanes, 1978; Badian, 1972, 1976, 1997, 2007; Famory, 1996; Diakité, 2002). Over the next year, I focused my studies on Malian culture, exploring its history, religious beliefs, mythology, and literary canon. Through novels and sociological essays authored by Malians, I discovered the complexity of village hierarchy and family structure. From the writings of Bâ, Badian, Famory, and Diakité, I learned about shared values, attitudes, beliefs, and behaviors held by various Malian ethnic groups. I read about daily life under French colonization, the various ways polygamy impacts family life, obstacles to communication between rural and urban dwellers, and villagers' perceptions of government officials and nongovernment organizations (NGO) workers. This in-depth study led me to realize that if future collaborative relationships and sustainable partnerships were to take place, I would need to provide future teams traveling to Mali with a deeper cultural immersion.

The aforementioned Malian authors opened windows for me to truly see Malian historical and cultural landscapes. The views they offered profoundly impacted the way I perceived contemporary Malian society; they also informed my future interactions with Malians. For instance, I now understood that almost a century of French colonization continued to deeply impact Malians' perceptions of Westerners. Although we perceived ourselves as students and professors collaborating with Malian villagers in their effort to achieve their holistic goals, they inevitably perceived us first and foremost as Westerners. In their mind's eye, we belong to those who colonized them and are perhaps still intent on colonizing them. This led me to wonder how we could build relationships that would not perpetuate these past injustices and present perceptions.

Applying Cultural Knowledge

Thankfully, the same readings that had inspired such introspection also offered pathways to better communication. For instance, in their fictional and academic works, the authors had underscored the hierarchical nature of Malian village life. They illustrated how hierarchy dictated communication patterns and explained that if outsiders (be it Malian city dwellers or foreigners) wished to show respect and avoid offending villagers, they needed to follow strict communication protocols. These included greeting first the village chief and his council of elders before opening public discussions with other village dwellers. Offering kola nuts to the chief and his council, as well as engaging in proper lengthy greetings, before explaining the reasons for one's visit, is further considered culturally appropriate. In fact, only after the chief and his council have offered their opinion and support should outsiders consider that they are ready to move forward and engage in village-wide discussions.

Visiting with the chief and council of elders and consulting representatives of the village associations during the first and all subsequent visits is of utmost importance. It shows respect for the community, a basic understanding of village structure, and willingness to partake in normative behavior. It also shows humility; by that I mean a willingness to embrace the community's ways of communicating and foregoing one's own communication style. Considering that for over a century French colonizers had shown contempt for traditional village hierarchy and modes of communication (in fact most were intent on destroying these traditional ways), it is imperative that Westerners learn about culture-specific ways of communicating and adopt them. Such behavior exemplifies what it means to put the community first.

Another dimension of humility and putting the villagers first is recognizing that when outsiders—be it teams of NGO staff, professors, students, development workers, or Malian civil servants—arrive in a village, they are guests in the community. Hence, prior to a visit, it is best to consult with villagers about an appropriate date to visit. This gives villagers time to prepare; it also makes it possible for the chief, council members, and representatives of the village associations to be present. Villagers find showing up unexpectedly and expecting them to drop whatever they are doing as discourteous and unreasonable considering how busy they are "making ends meet" (meeting basic needs of one's family) which could be a barrier to building healthy relationships.

Malian authors and multiple village visits also taught me that who we bring to visit a village can impact the kind of relationships we will build in that community. As Westerners working on food and agriculture projects, we typically arrive in a village in the company of translators and in-country collaborators; the latter may include Malian agriculture extension personnel, scientists, government officials, or USAID staff. Because some of these men and women speak the local language or because they interact regularly with Malian rural inhabitants, they take on the role of intermediaries. As such, they are the ones who set the tone for the kind of communication we have with the villagers.

It is therefore best to consider what kind of interactions and relationships we wish to build with villagers and choose intermediaries accordingly. For instance in Mali, government officials can best facilitate top-down relationships. Their past legacy of spreading fear (especially during French colonization when government officials would show up in villages unexpectedly and gather able bodies for forced labor) as well as their perceived abusive taxations and inappropriate meddling in village life will erase any idea of equitable relationship building. The villagers will immediately understand that they are being asked to engage in a top-down approach to a development project they have not requested. Fearing future troubles if they do not acquiesce to our development ideas, they will politely receive us and our government intermediary, listen and verbally approve any of our proposals and send us off with much fanfare. Once we have departed, they will ask their god(s) that we should soon be forgotten and never show up again to cause further trouble and waste their precious time. They will also meet with their chief and council to discuss what they should do in the event that our team reappears in their village. In the end, the villagers may or may not engage in our agriculture development project; this will depend on whether they believe that there will be punitive consequences from the government for not participating, and if they are provided the resources needed to engage in the project.

If instead, we begin by listening to the community describe their "good life" and only then engaging with the people in their own, community-driven (bottom-up) food and agriculture projects, we are best served by intermediaries who identify with Malian villagers and interact with them in a manner that demonstrates respect and knowledge for their ways of life. If we work with translators who have always been city dwellers and show contempt for rural life, villagers will neither share their holistic goals nor agree to collaborate with our team. On the other hand, if our intermediaries are perceived by the villagers as one of them, it can open doors to more honest communication and equitable relationships. Perhaps, in addition to working as translators or agriculture extension scientists, your intermediaries grew up in a Malian village of the same ethnic group; or they may have always been city dwellers but have kept ties with their relatives residing in rural communities. Their *experiences with and appreciation of Malian rural communities can lead villagers to perceive them as people who share the same roots, values, and cultural framework.*

Clearly, there can be exceptions to the scenarios Ada has just outlined. A local government official who has earned the respect of villagers and garnered a good reputation throughout the region may be the best intermediary in a bottom-up agricultural project. Alternatively, a translator who grew up in a rural Malian community, yet speaks to villagers with contempt, may be the worst person for the job. Although cultural competencies are essential to building a team, relational competency is just as important. Hence, it is advisable to take into consideration a country's historical and cultural framework as well as the unique attributes of potential team members.

During her first trip to Mali, Ada wanted to speak directly to villagers. By reading Malian authors, however, she learned that communicating via intermediaries is part of the social norm. For instance, if a Malian child wants a parent to be swayed on an important matter, that child would not ask the parent directly but instead ask an aunt or uncle to plead his or her cause. In such case, direct face-to-face communication would be construed as inappropriate, even impudent. Similarly, coworkers or neighbors would choose intermediaries to represent them in important matters. The prevalence of this indirect communication style, practiced in Mali and throughout sub-Saharan Africa, makes it even more important to choose suitable intermediaries and refrain from engaging in our customary Westerners' direct communication style.

I also learned that personal relationships are fundamental in Malian society; they constitute the foundational building blocks to social and work interactions. Hence, for those of us who wish to engage in community-driven agriculture projects, it is imperative that we build personal relationships in the community. When our team began collaborating with Malian villagers in 2005, we understood the imperative of building relationships. During our first 10-day visit to Mali, we stayed in a hotel in Bamako and took day-trips to visit with the villagers. Within a few years, we expanded this practice complementing our day visits with overnight stays. The latter enabled us to sleep under the same roof and share meals with our hosts, meet their extended family and neighbors, and share personal stories about our daily lives and our families. These extended village stays greatly contributed to building personal and lasting relationships, and to better understanding the life of a community.

Village stays also afforded opportunities to learn about local food and agricultural practices. For example, while on solo walks through the village I discovered a variety of vegetable gardens. Gardeners who saw me walk by frequently invited me for a tour, all the while commenting on their crops and gardening practices. While walking on the outskirts of town, I met young boys eager to show me the birds they had killed with their slings. On my way back to town, I saw these children roasting their catch over a small fire. These small bits of information are extremely important for agriculture experts seeking a more complete picture of food sources and resources available to villagers. With each subsequent visit, my colleagues and I would acquire essential details about food growth, storage, and consumption practices that would have remained unknown to us had we stayed in Bamako.

Lessons I learned through my physical and mental immersions are the basis of the cultural training I offer students and faculty engaged in development work. In my workshops and classrooms, they learn that culture and history matter. They learn that communication styles differ greatly between and within countries, that understanding and respecting cultural frameworks is essential, and that taking the time to build and nourish personal relationships facilitates sustained and equitable collaborations.

These general guidelines can be applied throughout the globe. From Case Study 1 we have learned: there are no exceptions, no fast way in; we must take the time to *learn about* and *dip into* our partners' culture. Immersion, both physical and mental, is an essential component of any intercultural engagement.

Native American students and any students coming from another nation to a Euro-American university, for example, experience a physical and mental immersion. When this immersion is in a culture that was a former colonizer or aggressor, there is the tendency to adopt this new culture as one's own. Jason, in Chapter 9, Listening With Students, shares with us the advice his Elders provided to him when he began the immersion as a student at MSU. His Elders advised him *"…to take advantage of opportunities, however, to be careful to hold on to my identity as a Shoshone person, and not to become what the university wanted me to be."*

CASE STUDY 2. ARRANGING SUCCESSFUL IMMERSIONS AT AN INSTITUTION LEVEL

Introduction

In his book, *The First One Thousand Days*, Thurow (2016) shared a good reminder that welled up from his physician friend, Dr. Brent Savoie. During his work with community health systems while immersed in Guatemala, Savoie noticed, "If you don't see it or smell it, you can't get it." It is one thing to recognize the importance of immersion, but orders of magnitude more difficult to *arrange* the experience. How does an agricultural scientist, policy maker, university faculty member, or administrator create a positive immersion experience for others?

After one clearly recognizes one's own culture, as we advised here in this book in Chapter 1 entitled *Where did you come from?*, the first step in arranging a productive immersion experience is to mentally immerse one's self in books such as Helena Norberg-Hodge's (1991) *Ancient Futures: Learning From Ladakh*; Roger Thurow's *The Last Hunger Season: A Year in an African Farm Community on the Brink of Change* (2012); and William Easterly's *White Man's Burden* (2006). Step number one also includes watching 5—90 minute documentaries such as *Enoughness*; *American Outrage* (Gage and Gage, 2008); *Homeland*; *The Canary Effect*; *Broken Rainbow*; and *Wind River* (Carr and Hawes-Davis, 2000). Each of these books and films lifts readers and viewers out of their Western culture, or other culture's comfort zone to simply "live" for a few minutes in the courtyard of communal homes at 14,000 ft in the Karakoram Mountain range (Norberg-Hodge, 1991), or in a one room subsistence farmers' home in Western Province Kenya (Thurow 2012), or in several other places in sub-Saharan Africa (Easterly, 2006), or in the kitchen of two Western Shoshone sisters raising horses on rangeland in Nevada (Gage and Gage, 2008) or with Eastern Shoshone along the dry bed of the Wind River reflecting on wild fisheries vs the value of irrigated crops (Carr and Hawes-Davis, 2000).

Each book paints pictures of daily life, conversations, relationships, and decision-making processes representing the "Bottom Billion," or rather snapshots of the 2 billion people of the world living in material-resource-poor situations. Each film draws one into a community, a family with interesting externalities superimposed. Often, as one learns most intensely in *Ancient Futures* (Norberg-Hodge, 1991), these communities have a high level of cultural wealth which appears only when one delves below the surface of the physical appearances of the community. Once these mental immersions are completed through books and films, it is time to take the second step.

Annual Undergraduate Immersions

Step number two entails a comprehensive mental and physical immersion experience. Here is an example, as told to me by Dr. El Houssine Bartali, of a successful model underway for the past 40 years at l'Institut Agronomique et Vétérinaire (IAV) Hassan II, Rabat, Morocco. Houssine tells the story as a senior faculty member in Environmental (hydraulic) Engineering and also as a former IAV student in this system. This is a method of annual immersion of all undergraduates in local, rural food systems during each of the first 4 years of their university training. The entire agricultural university, the administrators and faculty, have made immersion a requirement for each student for the past four decades. Each successive year for the undergraduate at IAV Hassan II, the level of immersion deepens and builds on the learning from the previous year.

All first year students engage in the immersion activity at the same time, each in a preselected small group to insure discipline diversity (for instance a group of eight would include students planning to major in hydraulic engineering, food science, plant pathology, English, entomology, agricultural economics, and so forth). Second year students focus on a single farming community for their immersion experience with two or three other classmates. Third year students go in pairs to a single farm family to experience daily routines, eating meals with the family, as well as single household economic and other decision-making processes. Fourth year students carry out a region-based research or development project requested by the farm families or communities in a selected area.

First year students' immersion works like this. Students are asked to bring a bed roll, whatever essentials they will need, and whatever food they require. Each small group, usually eight groups of eight students, is dropped off in the community at eight points on the perimeter of a large circle with a diameter of 160 km, as is formed by the spokes of a wheel. They receive only the instructions to in 1 week meet at the center of the circle. During that week the students in each small group walk together, forming a transect or straight line through farmer's fields, small towns, orchards, and forests. Along the way, students sample the entire set of ecosystems in which the farms are nestled.

During their week students collaborate on preparing a report on the transect as they walk together. They report on the kinds of farms encountered, the plants, the soil, the rocks, the birds, the insects, etc. They walk 15 km per day for 1 week. At the end of the week, the director of IAV and the faculty meet the students at the center of the "wheel formation" for a dinner and a debriefing. The students then each prepare a written report.

In the second year students go in groups of three to one rural area, a village they encountered during their 80 km trek. Students live in this village for 4 weeks and try to learn more about rural life and food production, marketing, and consumption. Students observe life in the farm family. They eat with the farm family and visit the area souk (local market) with the farmer to observe what is sold in the marketplace. Students also observe what resources are available to the farm family. For example, what energy resources does the farm have and what are their constraints to food storage. Upon completion of their village stay, students write a group report.

In their third year, each student lives with another classmate within a single farm family. The objective of the third year immersion is for students to understand the whole of the functioning of the farm, to understand the individual economics of that particular farm. In discussions with the farmer during this week, students are to pose questions to learn from the farmer about his family's desired quality of life and the constraints in achieving that desired quality of life. The farmer lists these constraints in their perceived order of importance. This is a prelude to the year 4 immersion.

In year 4, students return to the countryside to conduct participatory, collaborative development studies at the regional level. They cover topics related to rural and *agricultural development. The research development topics and the hypotheses tested are generated by the farm families. The hypotheses are aimed at improving livelihoods and income of rural population in the region-of-focus. Typically the family chooses a topic which is close to or already does interrupt the desired quality of life of the family. As a result, the research has a heavy sense of importance for the family.*

Evaluating Immersion Results

In 1986 I, Florence Dunkel, encountered the results of this undergraduate immersion program when I was asked to travel to Rabat over a period of 6 years to provide assistance to the young faculty at IAV in building a collaborative field and laboratory research program. These new IAV faculty had all completed their undergraduate work at IAV with this series of immersions and then completed PhDs either in France or at Land Grant Universities in the United States. Now these students were back in Morocco as the new faculty at IAV, their alma mater.

My role was carefully prescribed by the director of IAV, the premier agricultural university of Morocco. His directives to me were to: (1) build an integrated, multidisciplinary team of these new IAV faculty; (2) write and submit national and international grants; (3) get the grants funded. I had just come from successfully doing exactly this with faculty at the University of Minnesota—Twin Cities campuses (Bylenga et al., 1987; Dunkel et al., 1986, 1988; Lamb and Dunkel, 1987; Lamb and Hardman, 1985, 1986; Hanegreefs et al., 1987; Edmister et al., 1986; Mestenhauser et al., 1984). I anticipated it would be much the same process at the University in Morocco—tough, but doable.

I was wrong. Building the team and preparing the team for the transdisciplinary research that directly addressed Moroccan farmer concerns went significantly more smoothly than that of preparing the University of Minnesota faculty for similar transdisciplinary research and outreach in Rwanda. Why? What made the IAV experience different?

A major difference between the two institutions is that IAV's extensive immersion program had prepared its graduates to work in multidisciplinary teams and to develop personal relationships within academia and in the rural communities. In 1986 these young IAV faculty seemed to have a deep-seated, long history of camaraderie, respect, and trust with each other. In addition, these new IAV faculty seemed to effortlessly move in and out of disciplinary boundaries. On the farms in Morocco, I was amazed at how engaging research conversations were. I derived most of my information from nonverbal cues since on-farm conversation were in Darija (Moroccan Arabic) and in many of the Berber languages. Farmers were welcoming, cooperative about setting up field research, excited about results. Technical, laboratory-based scientists who grew up in one of the Moroccan cities as well as field-based scientists seemed quite at home on these farms. There appeared to be a deep respect for the farmers and their families on the part of each faculty member.

In periods of reflection later, it seemed quite clear to me that this form of education-by-immersion was the difference between individualistic training at US Land Grant Universities and the IAV model of immersion. The new IAV faculty who I had been sent to Rabat to work with recognized this as well. In the first moments of my first meeting in Morocco with these young faculty, they turned on the slide projector (this is 1986) and proceeded to explain the details of how this immersion process worked. The process that they had experienced as undergraduates, they had noticed, was absent in their US Land Grant University experience as graduate students, and were now in charge of arranging as professors for their own undergraduate students. They respectfully, without confronting me with the difference in the two forms of education, made it clear that required immersion had advantages. These young Moroccan faculty clearly understood the value of this unusual teaching process they had experienced and were now in charge of delivering. Thirty years since that slide presentation by the young IAV faculty, this required immersion process for all IAV students is still in operation.

From IAV, over 40 years, 100% of the students in all majors at this agriculture-focused public university participated in a series of immersion experiences. Data summaries of the percent of study abroad and other immersion experiences such as farm-related immersion for students in the colleges of agriculture in the United States are difficult to obtain. At one of the Land Grant (agriculture-focused) Universities, Michigan State University, 2500 students (approximately 10% of the student body) participating in study abroad were surveyed 1999−2002. The majority were not from the College of Agriculture and Natural Resources: 62% of the students taking immersion experiences were from the colleges of arts and letters, business, communication, natural science, and social science (Ingraham and Peterson, 2004). In 2010 perceptions and aspirations of 956 College of Agriculture and Life Sciences students regarding international engagement were examined at another Land Grant University, Texas A&M (Briers et al., 2010). Seventy percent of these students recognized the positive role in enhancing their own life and improving their career competitiveness. Students surveyed preferred faculty-led programs, but no mention was made of the role their international experience would have in understanding complex global issues, including world hunger and peace.

Ross Lewin summarized the situation in 2009 that continues, "The percentage of US students studying abroad lags far behind that of most highly industrialized countries. As a percentage of all US students, study abroad participation has actually not increased significantly over the last decade." These students chose shorter study abroad periods than the previous generation (Chieffo and Griffiths, 2004). We hope, based on our observations with 131 students in the Mali Extern program (Dunkel and Montagne, 2011; Dunkel et al., 2013; Dunkel and Giusti, 2012) and the 218 additional students who completed the collaborative research during mini-immersions of AGSC 465R, that this trend continues. We also hope these periods abroad are now becoming times to live with, engage in deep discussions with host families of the visiting students. The need for students to go abroad (no mention of immersion experiences in local enclaves of Indigenous people such as in the Moroccan example) is now seen in academic mission statements, business association's best practices, and even federal legislation as Lewin observed in 2009. Since the end of World War II, high school student exchanges (homestays) between families in Germany and France have been required and have met with success in heightened appreciation of each other's cultures. Finding solutions to our complex issues including environment, immigration, poverty, health care, and food security systems depends on training armies of individuals capable of cooperating with other individuals across borders. It also depends on the training of the faculty and other instructors that lead these immersion experiences.

From Case Study 2, we have learned that it is possible to do institutional-wide, all-discipline inclusive immersion. Careful planning is essential. Logical progression of deepened immersions is suggested. The rewards are great. As Jason reminds us in Chapter 9, Listening With Students, *"Cross-cultural awareness, respect, and appreciation enhance the communities in which we live [and teach]. Building relationships with those that are not like ourselves diminishes misunderstanding and builds trust."*

CASE STUDY 3. NATIVE FOODS AND FOOD DESERTS

Three decades later, I recognized the same principles underlying the IAV education-by-immersion were leading to success in my and my colleagues' program developed on a smaller scale at MSU and at the Northern Cheyenne Tribal College, Chief Dull Knife College (CDKC) in Lame Deer, Montana. In this third example we offer a case study on how Native plants and other Native foods emerged as a rich learning experience for both Native American and non-Native students collaborating in situ, during immersions outside of classrooms, on a Native American reservation. In the process, the immersion became a platform to discover that the food desert we read about in the peer-refereed literature on Native American reservations may just be a "Garden of Eden" for the residents and a rich source of ideas for the sustainable foods movement that is growing among the students and recent graduates at our US Land Grant Universities.

The immersive process can be transformative even if it is brief and seems superficial. First, we have observed in hundreds of students the transformative power of very short exposures for those eager to learn, particularly for young students such as undergraduates. Second, international trips that require crossing an ocean are not the only way to do intercultural experiences with transformative results. Third, mini-immersions we present here in Case Study 3 may be superficial, but this initial learning can also open windows and doors. One or two of the students in an initial cohort of 15 or 20 may later participate within a reservation community in a greater degree. This involvement that would constitute an intense intercultural immersion could provide that carefully nuanced competency that Ada Giusti illustrated in Case Study 1 set in Mali and Meredith Tallbull wishes for outsiders to experience on the Northern Cheyenne Reservation (Box 4.1).

"Feet wet" mini-immersion is rare in US Land Grant Universities, even at MSU, the 1862 Land Grant Institution in a state with 7 Native American reservations and 13 tribes. Why is this mini-immersion important? Can it open windows to new ideas? Can it also be opening doors to opportunities for a more intense intercultural immersion that provides a more carefully nuanced competency?

BOX 4.1 SCIENCE PARK/BOTANICAL GARDEN, THE REZZERIA, AND IMMERSION

Photo of lake with reflection of trees: Botanical Garden/Science Park in reclamation process, located in the center of Lame Deer, Montana, capital of the Northern Cheyenne Nation.

Photo of stone oven: Stone oven made by Meredith Tallbull as the center piece of his Rezzeria, a pull up-restaurant with pizza to order during their visit to the Garden/Park.

(Continued)

BOX 4.1 (Continued)

Meredith Tallbull with a few of his Montana State University collaborating students of Dr. Dunkel during their visit to the Garden/Park on the Northern Cheyenne Reservation.

The following is Meredith Tallbull's reflection, partly written and partly offered orally. Meredith, an enrolled member of the Northern Cheyenne, has shared his thoughts on the process of non-Native American students visiting the Northern Cheyenne Reservation.

The experience with the AGSC 465R students from MSU-Bozeman was an educational experience for myself and as I am sure it was for them. The students' visit to the reservation made a long-lasting impression on me due to their concerns about the current situation on the Northern Cheyenne Reservation and its inhabitants. They made me feel like the Lame Deer, Montana Science Park/Botanical Gardens is a well-pursued endeavor. The vision I shared with the class I hope will inspire the students to go beyond their current views of the reservation.

The Lame Deer Rezzeria (a pizzeria) was an idea that I had long wanted to do, the Rezzeria along with an inspirational course I had taken at MSU-Bozeman, "Early Native America," were what gave me the idea to merge the two projects (Rezzeria and Science Park) together because of the creativity of pizza making and the long forgotten world that my ancestors had set aside during the

time of European conquest and subjugation that still to this day does not honor or recognize the first peoples and the food substances that were brought to the first thanksgiving dinner. I believe that the white population must learn to recognize and respect the first people of this continent by teaching their parents, grandparents and children of the contributions that our ancestors have made to the Turtle Island and later to the development of the "Americas".

The Science Park will be an accumulation of Native Science and Western Sciences, holistic healing for the tribe, as well as a place to admire and reconnect with nature. I would like to thank Dr. Florence Dunkel, students, site mentors for aiding other underserved people of this great country. This plot of land in the Northern Cheyenne was an eyesore. Oh my, I wish that that dump wasn't there, I would think each day as I passed by it. Respect for the place was not evident. My colleagues, mostly young folks and I spent a couple of years cleaning it up. I wanted to fully understand this place, so I waited another full year to understand the ecology before I did anything. It is really starting to take shape now. There needed to be that waiting part. The vision needs to go out to the schools and the community. Find a niche. Invite students down there to learn and to help build the place into a nature sanctuary where young people can see Native food plants and medicinal plants growing. The natural amphitheater will be the first addition. As I was clearing some brush for the ampitheater this summer, I found a mountain bluebird nest with three little white eggs. The mother is feeding the little ones now. There is some wildlife there. Second part for this summer is the foot paths. After the powwow season in early July, I will begin again to work toward this vision, my vision of a Science Park.

Students in AGSC 465R were introduced to Meredith and his dual vision for Lame Deer in its earliest stages. MSU students understood the value of the Science Park and Rezzeria and how it might serve as a model for place-based, culture-based

(Continued)

BOX 4.1 (Continued)

economic stimulus on the reservation. Students tried to respond to Meredith's suggestions for help. Some created information packets that collected in one place (a bound book) all the peer-refereed Western Science information confirming Traditional Ecological Knowledge of the Northern Cheyenne about specific plants such as yucca, yarrow, and *Echinacea*, plants that grow wild in the Science Park. One student built plant tags for the Botanical Park and linked them with plant stories told by the tribe's ethnobotanist, Linwood Tallbull, accessed through a smart phone or tablet while in the Park. Other students wrote stories for the very young children about *Echinacea* and its traditional medicinal properties introducing the Cheyenne word for the plant. During their mini-immersions AGSC 465R students participate informally, spontaneously in various conversations. They notice the desire of those there on the reservation to remain there but to obtain specific training needed on the reservation and to be able to use those new skills to earn money for technologies like cell phones, Internet access, a car, and home repairs.

When asked if the 24–96 hours students spend on the reservations are enough to call it an immersion, Meredith quickly responds, *"No. Twenty-four hours a day students 'have each other's back.' They know the exit. Even a 1–2 week 'vacation' is not an immersion. Stay for a month or two become part of the whole community. The quality of the immersion depends on who the visitors have they been talking with during the immersion. Visitors can dream, but it must be tied to reality."* For example, during their mini-immersions on the reservation, students can see the physical labor needed at the Science Park. Students also hear Meredith and other Cheyenne say *"People on the reservation are looking for work. People are wanting to feed their family."* But Meredith is trying to build his vision without government assistance, without grants or outside financial help. He understands even the work cannot be done by outsiders if the Park is to be sustainable. Ownership must be within the Cheyenne community. This requires finding Cheyenne in Lame Deer who will provide physical labor and other skilled help to, for example, build foot paths and information signs. It requires finding help from Cheyenne who are committed to just the vision without economic incentives. Meredith is keenly aware that work is available for those with new skills, even in exchange or trade for commodities needed. This is one reason Meredith initiated the Rezzeria which makes a profit by providing a sought after service, delicious stone-fired pizza made from locally produced ingredients served

conveniently in city center Lame Deer or, by special arrangement, for one's group on the shores of the beautiful lake in the center of Lame Deer at the Science Park.

Students understand this in their short visits. The Rezzeria and Science Park seem to be an ideal place for local youth to work with Meredith, learn locally useful skills, share in the profits, *and* become inspired by the vision of the Science Park and the natural environment. MSU students try their best to be creative in making suggestions, but according to Meredith, the solution to making faster progress is for outsiders to have a deeper understanding of the reservation system, a deeper understanding that is obtained by more than a mini-immersion.

To make ends meet in his effort to live off the grid in Lame Deer, Meredith took as a high school science teacher position in 2015 at Lame Deer High School. Next year he will be teaching math there. In the High School, Cheyenne make up only 10% of faculty. How was your first year teaching? Meredith says, "I need to teach another year before I can tell you. Responsibility is the tough part." Meredith is not used to this with no children of his own and being the youngest in his family. Meredith would like to provide a place for high school youth to work during the summer or after school during the academic year but school restrictions rule this out. He would like to pull youth away from their electronic devices that separate them so completely from the natural world that the youth talk about the wealth of knowledge accumulated as a child on the reservation is simply referred to as knowledge from "the olden days" and of little current use. Meredith is a "Cheetah" in his community (Ayittey, 2005). He has the ability to dream like Yunus (2007). When AGSC 465R Native and non-Native students hear about his vision and visit the Science Park and Rezzeria, it all makes sense and is inspiring... inspiring to students to want to expand their mini-immersion to a longer visit, to a "real immersion."

This deeper understanding that may come from a real immersion might be as follows. The help that could be so useful to realize the vision of the Science Park/Botanical Garden is from the teenage students from middle school and high school in Lame Deer. Teens might even share in some of the profit. The school system has disallowed this because of safety. Ninety percent of the faculty are non-Native and most all do not live on the reservation. Would the deeper understanding of what a perfect opportunity this might be to work with Meredith be recognized if faculty and administrators of the high school had more than a superficial introduction to the community.

For any veteran foreign service scientist or policy maker, this brief time living on a reservation may seem odd and out of place in a chapter on immersion. We include it here as evidence of how a small exposure such as this can bring out "push back," fear, family intervention, *and* the kind of positive cognitive dissonance that leads to transformation.

Why Is the Mini-Immersion an Essential Part of Education?

Students that come from backgrounds with limited to no experience of reservation life get a glimpse of local third world conditions when they visit the reservation, not realizing perhaps that these communities are on the interstate and in the backroads of our Western states. Submersion in a "foreign" community has transformative effects on the level of intercultural competency, especially after several visits to the community. This similarly happens to students who have been primarily raised on the reservation with limited experiences in mainstream culture. Submersion in the mainstream culture is often difficult due to differing worldviews and fundamental belief systems. Unfortunately many students who have done well academically on the reservation find the transition too difficult and return home to the reservation prior to completing studies. This emphasizes the importance of community-based and peer-based programs to create a familial atmosphere similar to home. It also emphasizes the need for mainstream culture to have some familiarity, understanding, and knowledge with respect to Native Americans, especially on a college/university campus.

Cognitive dissonance often occurs after several interactions with the reservation communities and institutions and even more so, once one begins to understand the philosophical differences between Native belief systems and that of mainstream culture. General lack of understanding about Native Americans pervades, and there is little to no educational foundation as a prerequisite for understanding treaties, sovereignty, trust responsibility, or self-determination. No other ethnic group in the United States has the unique and historic relationship as Native Americans to the Federal Government, and yet few Americans know or understand present Native Nations and their rights based in centuries of Federal Indian Law.

Few, if any, other courses offered at MSU require that students visit a reservation. Few non-Native people visit reservations on their own accord and in some instances are reluctant and have even been encouraged to not visit. These and other misconceptions exacerbate the division of Native people from being understood by the mainstream. This course is successful at mitigating those unfortunate separations of communities and fosters understanding. Even if this course only reaches a few individuals and creates opportunities for future collaboration and inclusion, then it has been successful at transcending boundaries that few universities and higher education institutions have been able to do.

Mini-Immersion Process

Here is how it worked. Each semester, the course AGSC 465R Health, Poverty, Agriculture: Concepts and Action Research is populated by a new set of students, mainly juniors and seniors with an occasional graduate student or scholarly sophomore. Within 3 weeks of the first-class meeting, all students travel 3–4 hours by car to the Northern Cheyenne and the Apsaalooke Reservations with Dunkel. To prepare for this immersion, students partake in a heavy dose of concept readings, several videos filmed in the communities-of-focus and discussions in a large group (15–20 students) and in small groups (3–4 students), and in 1:1 mentor meetings with Dunkel. Students also learn in detail the proper verbal and nonverbal behavior for greeting and departing. This includes the appropriate hierarchy of communication in that culture. At first when students enter the reservation, they notice visible and electronic differences: no billboards; no fancy homes with landscaping; limited cell phone and Internet service; few stores; one gas station; one grocery store per community and that with limited fresh produce, and a lot of boxed products high in sugar and salt with a high omega-6:omega-3 fatty acid ratio. Then we begin our visits with Elders in their homes or in homes of classmates or site mentors. Non-Native students notice: multigeneration families living in the same household; multigeneration caring for each other beyond the household; and a slow-paced feel. Non-Native American students often describe to me a feeling of peace that comes to them when they are on the reservation.

Students soon become aware of the underlying intricate set of tribal societies that dictate social interactions. This awareness comes in spontaneous conversation with Elders or with our other site mentors who have been raised traditionally, or with students themselves who are from these nations and learned the intricacies of tribal society (Chapter 6: Listening With Native Americans). For example, the ethnobotanist of the Northern Cheyenne, Linwood Tallbull, usually explains the four societies of his tribe since he belongs to one of them; Tracie or Kurrie Small explains the names of the Apsaalooke clans to which they belong and what function they serve in creating a well-balanced life; and someone usually explains the importance of "teasing" and how this is a prescribed part of the social structure. One student explained how an Apsaalooke spear is decorated for spear throwing contests. The feathers attached to the end of the spear and beautiful colors painted on the side are not simply gathered or purchased, they have to be earned from an Elder by acts of kindness, bravery, and responsibility. Students also in these visits observe the strength of family ties. Plans made in advance of our visits can change abruptly due to family emergencies. During one visit, the auntie of one of the site mentors died. Students noticed that most activities in the town shut down. Although the tribal college classes continued, class attendance was sparse.

While staying on the reservations, MSU students and faculty prepare meals cooperatively and each day bring lunch for sharing with the people with whom they meet. Sometimes it is the site mentor who makes the meal. Food is a fundamental part of social interaction for the Native Americans with whom we exchange visits. Times when there was not enough food is a fresh memory for these families. When the Native Americans were forced onto reservations, excluded from their hunting areas, and their chief food sources, bison, were massacred, food for the community was a daily struggle. During our visits, food is an appreciated way to share.

One of the student's favorite forms of engagement is taking a walk through the Botanical Garden or Science Park emerging from the abandoned wooded wetlands in the center of town, Lame Deer MT, capital of the Northern Cheyenne Nation. Preceding or succeeding the walk, site mentor, Meredith Tallbull, often prepares pizza in his Rezzeria, a homemade stone oven that sometimes is pulled behind a truck to the shore of the small lake in the Botanical Park (Box 4.1). When appropriate, students pose the holistic process questions (Savory and Butterfield, 1999) to their site mentors, such as Meredith. Students ask and listen to learn what is his: desired quality of life; current and future resources; and current and future forms of production.

Both groups coming together seem to gain from these mini-immersions. MSU students find listening sessions or plant walks inspiring with ethnobotanist, Northern Cheyenne healer and Elder, Linwood Tallbull (Box 4.2). Likewise, MSU students enjoy meeting with two Apsaalooke sisters, Kurrie and Tracie Small, who were also raised traditionally and who have taken action with their ideas gleaned from conversations with their Elders on

BOX 4.2 ETHNOBOTANY AND THE SCIENCE PARK

The Science Park is a sound concept, particularly because Meredith's vision is shared by Linwood Tallbull, an Elder who is ethnobotanist of the Northern Cheyenne people, and who also is Meredith's cousin. Linwood is along with Meredith, an important site mentor in AGSC 465R. Indeed the role of Linwood in mini-immersions for MSU-Bozeman students began during a powwow organized by the late Robert Madsen, science teacher and research entomologist at Chief Dull Knife College (CDKC). It was 2005 and the Malian site mentors had just arrived for classes at MSU. These Malians were site mentors for CDKC students when they visited subsistence farming villages in Mali. Linwood's introduction, a mini-immersion with the Cheyenne culture, can be viewed at www.montana.edu/mali in the film East Meets West, produced in 2005 by an MSU student and her professor, Dr. Dunkel.

strengthening the Elder—youth connections on the Apsaalooke reservation. Meredith often shares with us that just our interest as outsiders in his progress renovating the wetlands in the center of town gives him needed strength and confidence that his vision will become a reality. Kurrie and Tracie are also driven by their vision—meeting the desired quality of life of their Elders.

This hosting process on the reservations is reversed at the end of the semester when the Northern Cheyenne and Apsaalooke students, family, and faculty from the reservations come to the MSU Bozeman campus for the Share-the-Wealth Symposium. Many important things go on at this Symposium. It is a celebration of the traditional wealth that the outsiders (non-Native) students learned during the semester. It is a time when the tribal college students and reservation high school students learn that there is a conference room in the Plant Bioscience Building on the university campus where their knowledge is valued and there are friends whom they have worked with and learned to know over the semester and those friends are MSU students and professors.

Native Plants and Native Foods on the Northern Cheyenne Reservation

There is no substitute for the "baptismal" immersion of being physically placed in front of a Northern Cheyenne ethnobotanist to hear him speak of his relationship with plants and how he integrates the nutritive and medicinal roles of plants. There is no substitute for a fellow student or faculty member inviting you to, "come, let me show you how we cut meat for dry processing." To understand how this is tied to traditional memories of family and ancestors is another rich experience that enriches students in any area of the agriculture and health sciences, particularly for those in Sustainable Foods and Bioenergy Systems, Horticulture, Animal Science, and Agricultural Education, Pre-Med, Pre-Dent, and Nursing. To prepare for giving these students this "baptismal" opportunity, Dunkel had to develop a trusting relationship in several parts of the community, with tribal college administrators, staff, and faculty, with families, with CDKC students. We began this "mini-immersion" process in 2002 and by 2009 it was a requirement for MSU students in AGSC 456R.

Here are two specific examples of how a mini-immersion can first open windows and then may create opportunities to open doors for an in-depth immersion.

Rachel Anderson was a senior in Liberal Studies. She and the rest of her classmates in AGSC 465R were linked for the semester with Northern Cheyenne students taking a CDKC course on the History of the Northern Cheyenne. Early in the semester during the first mini-immersion, MSU students learned that the CDKC students thought these "outsiders" should learn how to prepare dry meat the Cheyenne way in order to have a deeper understanding of their history. During the next mini-immersion, Rachel and her classmates learned how to cut venison paper-thin for drying. As their MSU instructor, I had to learn how to build a drying rack and prepare dry meat stew, Northern Cheyenne style. This required considerable listening and phone calls to Elder women at CDKC. The semester following, CDKC students again challenged AGSC 465R students visiting from MSU. This time the tribal college students asked the MSU students to help them prepare an *Edible Entrails* booklet. This linked effort between MSU and CDKC students resulted in a tangible, jointly prepared product with preparation methods and stories whereas the dry meat process training just happened. In the post-project evaluation, CDKC students shocked MSU students by reminding their project partners that just the idea of a cookbook was a Western culture concept. In an oral-based culture, such as the Northern Cheyenne, recipes are not written. When you want to learn to prepare a Cheyenne dish, you go to talk with your mother, grandmother, or auntie or uncle, or an Elder.

Rachel recorded the following excerpts in her reflective journal during her mini-immersion experiences. During the first immersion, she wrote, "...*I see the broken down cars everywhere, and closed businesses. I cannot help but feel a little bit responsible. Also, I feel as if I stand out being on the reservation...Writing is helping me make sense of this learning process. Being outside, the sun is shining and there is a slight breeze. There is so much history here, you can feel it. How do I help a community that I don't understand? Do they really want my help, or is this experience meant to benefit me?*" The next morning she wrote, "*At the rise of the sun yesterday morning, I would have never guessed the way I feel at this moment... A moment of clarity... The simple life... Life discussion and building relationships with my fellow classmates. Learning about another culture and way of life is different but located so close to where I live.*" We stayed on the border between the Northern Cheyenne and the Apsaalooke Reservations.

Just prior to her second mini-immersion on the Northern Cheyenne Reservation, Rachel wrote, "*I am excited to visit the Northern Cheyenne Reservation...I need to remember the cultural barriers that will be present when we visit...the biggest challenge I face is communication with the Elders. I need to fully understand the value and tradition of the Northern Cheyenne people as they discuss the tradition of drying meat. Also, the process of doing the cutting of the meat and the drying of the meat we cut will need to be understood...I know the history of the Northern Cheyenne plays a significant role in drying meat and in its being a sustainable food for their entire community.*" Just after this second mini-immersion, she wrote, "*Our second visit to the reservation was of key importance. I met Mina*

Seminole, Kathy Beartusk, Sierra Alexander, George Nightwalker, and a lot of students from CDKC. My personal interview with Kathy was the greatest part of my day and research project. I now understand the importance of drying meat in the Northern Cheyenne culture." Kathleen Beartusk is a direct descendent of Chief Dull Knife who in the 1880s led his people in a run from Fort Robinson back to the area that is now Lame Deer, Montana, and where his descendants have remained ever since. The struggles of that breakaway and the difficult journey back to the Lame Deer area made real to Rachel the importance of being able to have food to give strength for extraordinary action that life called upon the Cheyenne People to take. Even if dry meat was not exactly part of the Fort Robinson run to freedom, Rachel could, during this second mini-immersion, begin to understand how it may have been key to survival of the Northern Cheyenne.

After the semester ended, the CDKC students and their instructor, George Nightwalker, traveled to Bozeman to participate with Rachel in the public Share-the-Wealth Symposium. It was now a mini-immersion experience for the Northern Cheyenne. Rachel wrote the following. *"I believe I have come full circle. I understand that my community-of-focus is not poor but rather rich. The strong community focus of the Northern Cheyenne people is positive. My discussions with Florence Dunkel (course professor) were the times when I realized that our [non-Native American's] understanding of the Northern Cheyenne is wrong. What we think they think is not correct. The Northern Cheyenne are not mad at us for a violent history, they are happy with their land, family, and culture. I felt that they wanted to teach me and welcome me into their community. It was very genuine. I leave this class and project growing as an individual in many directions. I feel more educated and not ignorant. It makes me want to know my own history more closely because it is a key component of who I am. I wish Western cultures were closer to that of the Northern Cheyenne, for this Indigenous culture seems so less self-gratifying and more focused on a larger picture. Overall, my interactions with everyone this semester have led me on a path of different perspectives. I will approach things in a more holistic way."* Rachel was a graduating senior majoring in Liberal Studies when she wrote this.

As a graduate student in Health Sciences, Bridget McNulty had two mini-immersions on the Northern Cheyenne Reservation focused on her research at the Indian Health Service as part of AGSC 465R and her MS degree. During a third mini-immersion, Bridget taught biochemistry at CDKC. After 1 year of medical school, Bridget shared with Dunkel, her professor and research mentor, that the use of the holistic process she learned (Savory and Butterfield, 1999) and practiced while immersed in Northern Cheyenne research with the health system put her far ahead of her medical school classmates in doing patient interviews. Bridget is now a senior in medical school at the University of Washington, Seattle. She is well positioned to be an effective health care professional with cultures similar to hers (Chippewa Cree and Western culture) and also with cultures far different from her own, particularly in working with culturally specific issues such as food and traditional plants used for strengthening and for healing.

"Let's Pick Berries" Project

The "Let's Pick Berries" project was conceived by two enrolled members of the Apsaalooke Nation while they were students at MSU. The purpose of the project was to help bring their Elders closer to their desired quality of life. Though students in AGSC 465R in different semesters, each sister was a cultural guide for the other students in the class and for the instructors. Both were former students of Clifford Montagne and with his guidance had become well-grounded in the holistic process. Kurrie and Tracie Small are middle-aged sisters, both married with four children each. Kurrie, the older, and Tracie, the youngest of the big family, were raised steeped in Apsaalooke traditions near Lodge Grass, Montana. Lodge Grass is on the Apsaalooke Reservation, 3.5 hours by car from the MSU campus. The traditional berries project emerged during holistic discussions with 36 of their Elders about the Elders' quality of life. Discussions were mainly in the Apsaalooke language, the first language of the Elders and for most of the Apsaalooke youth, as well. If the message to be shared is important, it must be spoken in Apsaalooke. To speak about one's desired quality of life, to speak from the heart about this most important topic, it must be done in one's Native language. Most of these elders do not speak English or are not comfortable speaking English. It is the colonial language.

Research questions, topics, and actions in the holistic process underlying AGSC 465R must be decided only by members of the community-of-focus. So following the discussion with the Elders, Kurrie and Tracie, members of this Apsaalooke community, then gave assignments to the students in AGSC 465R, assignments that matched student skills with what the sisters envisioned would address what was missing from the quality of life described by their Elders. The Elders wanted more connection with youth so Kurrie engaged in conversation and participatory diagramming the junior high school students in Lodge Grass and the fourth graders in a reservation border school where many Lodge Grass students also attended. Plans for managed and wild berry patches resulted. All of us learned about the vitamin C content of buffalo berries exceeding that of orange juice by 200%. Chokecherry shoots and wild plum grafts on commercial rootstock were planted, but the two sisters now both graduated from MSU and commuting to fill-time jobs at the tribal offices in the next town could not get enough water to the young shoots. The plant propagation effort failed. Not only would the young people of the community have been able to help with this, but it was those young people that the Elders wanted to see reinspired about berries. When local youth involvement seemed key, focus on the community location for the managed and wild berry patch near the powwow grounds south of town shifted to the Lodge Grass High School farm north of town high up on the bench. There, the Future Farmers of America (FFA) group and their leader, the high school biology and agricultural education teacher were getting ready to start a school farm on a section of land adjacent to the high school. Of course berries, a traditionally prized food for the Apsaalooke, were planned to be included in the demonstration farm.

A series of mini-immersions throughout this research project that began in 2012 gave an insider view to many MSU students majoring not only in Sustainable Foods and Bioenergy Systems, Horticulture, and Nutritional Sciences, but also in Earth Sciences, Architecture, Liberal Studies, and Computer Sciences when they began to make visits to the reservation on their own to work with Tracie and Kurrie Small and the FFA students with their leader, Ty Neal. MSU students found their skills and talents creatively directed. Some semesters the task was using Geographic Information Systems (GIS) to map good locations for new berry patches on the reservations or designing a recreational space based on tribal members' suggestions or planting berry shoots. Other times it was just cutting down and hauling out Russian olive trees that had invaded prime berry patch areas on the high school farm. Immersion roles reversed when Lodge Grass/high school students came to the MSU campus (about a 3.5 hour drive) to participate in an experiential class taught by the AGSC 465R students on chokecherry propagation and nutritional content of buffalo berries. The Apsaalooke high school students reciprocated by sharing traditional stories of how they used berries, such as the chokecherry pudding that is made for the toddler after their first solo steps are taken.

Greta Robison was an MSU campus leader/activist majoring in Earth Sciences and Liberal Studies (Multicultural/Global option) who had particularly honed skills in GIS mapping and writing an opinion column for the weekly campus newspaper. She participated in the reservations immersion process and during her first mini-immersion experience with AGSC 465R, accepted the request of the Apsaalooke to work on the "Let's Pick Berries" project. Although Greta was not raised in an Indigenous family, her mother's family is from a small Indigenous island community off the coast of Spain. *"In 2014 I was asked to work with the Apsaalooke tribe on the 'Let's Pick Berries' project. My GIS skills were of use as my site mentor, Tracie Small, was in the process of deciding on an appropriate site for the berry patches. Over spring break I traveled to the reservation to help Tracie survey the community of Lodge Grass to ensure the berry patch would satisfy the people's needs. During the surveying process I noticed my mentor, Dr. Dunkel, and myself were the only two white people in the room.*

Flooding into her memory at this time, Greta later shared with me were comments from the pages of the text she so revered, *Decolonizing Methodologies* by L. Twali Smith (2012). Knowing from her Native American Studies courses the history of genocide that Euro-Americans had imposed on these Apsaalooke people and other Native American tribes, she was thinking she was being perceived first and foremost as a Westerner, one of the colonizers. She felt a sudden overwhelming discomfort that she would inadvertently use colonizing language or other unacceptable action. For a moment she felt paralyzed that she would not do the right thing. As her professor, this was a planned learning moment for Greta. I also, as a co-learner with Greta, needed these experiences to be mindful of my every action and word.

This experience, along with many others on the reservation, was essential to the quality and content of my work for the tribe. It was not only the results of the survey that helped me make maps of berry patch locations, but physically shaking the hands of the Elders, listening to the conversations of the community members, and feeling and seeing the berries that held so much significance to the tribe. It is one thing to be told that groups of people are different from yourself—that they have different needs or wants. In my case I had been reviewing films, having phone calls, and researching to understand how to best meet the tribe's needs.

It was only in the site visit, though, in listening to what to the Western ear is rambling or a tangential stories, that I could understand my role in the project. It was in the discomfort of being an outsider that I learned how to be the resource the community needed instead of an imposing force. I am confident that my contributions to the project would not have been as useful or as practical for the tribe if I had not taken the time to be on the reservation without a clear agenda, without a goal other than to listen.

It is my belief that each person, regardless of discipline, should have an immersive experience in their life. These experiences can look, and will look, different for each individual. If it's working with a tribe, a different social class, traveling aboard, or a combination of these, stepping outside of one's comfort zone, or culture zone, does two things. First it transforms how the immersed individual sees the world. They do not lose their worldview, but they begin to see that their view is not the only one. This has positive impacts on a personal, academic, and societal level. Second, it vastly improves the quality of the individual's work by expanding the cultural context of proposed ideas and potential solutions. In a world with a growing population and limited resources it is truly a travesty that we, as scientists, are unwilling to look at culture. It is this separation, this block we've made, between science and culture that has made much of our hard work for naught. If we can fund and find time for new technologies in seed growth, in water management, and in pest management, then we can also find time and funds to learn about other cultures. They are truly interconnected. Understanding how to treat malnutrition in Sanambele or type 2 diabetes on the Apsaalooke reservation, things both communities want to work towards, is not possible without an understanding of culture. And what a tragedy would it be if the bright young minds of tomorrow's science create and invent, only to have no context or skills to use the technology they have discovered. How sad would it be, to have the ability to feed all 9 billion people, and fail because we do not understand the importance of culture? How sad it is that we do not feed the 7 billion now, in part for this reason."

Summary of Case Study 3

From Case Study 3 we learned that mini-immersions can lead toward a transformative experience and should be a required part of education. Even mini-immersions can lead to behavior changes in the food and agricultural sciences. My husband and I now harvest wild chokecherries which appeared over the years at our home among our ornamental bushes and make traditional Northern Cheyenne pudding. We will be planting buffalo berries in our Montana vegetable and fruit garden with the plan of abandoning the purchase of less nutritious imported oranges. We dry some venison the Cheyenne way to provide energy-free meat storage. My students and I know delicious ways to use other-than-muscle from bison, deer, antelope, and beef cattle. We know how to suggest local, traditional foods that prevent type 2 diabetes. We know how to grab yarrow to stop bleeding and to make a mosquito repellent from it. The bottom line is that we have learned to appreciate other ways by listening.

Exploring new food systems and incorporating these into our eating patterns is a visible benefit of mini-immersions, but Jason Baldes, an enrolled member of the Eastern Shoshone, sees deeper, more fundamental reasons for these mini-immersions. Jason shared this reflection on mini-immersions from his position as director of the Wind River Reservation Liaison Committee in which he serves as a mediator, peacemaker, and facilitator between the two tribes occupying this reservation (Eastern Shoshone and Northern Arapaho) and between the tribes and the County Government as well as the State of Wyoming.

Native Americans live on reservations, land reserved in perpetuity for these people. Because of that isolation Native Americans experience and lack of opportunity for interaction with the mainstream culture in the United States, this 1% of the population is faced with perpetual racism. Isolation creates misunderstanding. Mini-immersions give the mainstream population an opportunity to experience how our worldview differs from theirs (First People's Worldwide Production, 2012). Until these differences in Native American worldview, indeed the worldview of Indigenous people worldwide, are recognized, it will be difficult to build bridges of understanding.

This is why these mini-immersions, not only of university students and faculty, but also of primary school students, second grade and younger, need to take place. The State of Montana has a law entitled Indian Education for All (IEFA). All students K through 20 are required in their classrooms to learn history, food systems, and other cultural aspects of Native Americans. [Mini-immersions are not part of the requirement, but have been tried successfully in MT middle schools.] We are working toward IEFA legislation in Wyoming. Everyone is welcome on a reservation.

In the western United States, cultural differences are stronger than east of the Mississippi River. Forced assimilation in the west began relatively recently, not until after the Lewis and Clark Expedition, a while after 1804. It is a fresh memory in most families. Reservations were created out here little more than 100 years ago and became third world countries, countries within a country. Borders between reservations and the mainstream country became areas for heightened conflict, particularly between native food systems and industrial agriculture. In these border areas, distrust, law enforcement issues, and racial profiling continue to be rampant. If we could provide more opportunities for all policy makers, government workers, farmers, ranchers, and especially all school children to visit reservations, the windows of understanding will be opened. Without the opportunity to even glimpse in the windows of these other cultures with whom we share all the natural resources for food production, one loses the opportunity to open the door. Yes, mini-immersions are essential.

ARE THERE NEGATIVE CONSEQUENCES FROM NOT INCLUDING IMMERSION IN GENERAL EDUCATION?

What are the dangers or negative consequences of not including immersion in the general education of policy makers and teachers from kindergarten through graduate school? One example is the "War on Hoppers," a multinational effort in the 1980s to reduce grasshoppers and locusts in West Africa. This policy was a joint effort of the Governments of Germany, France, Great Britain, and the United States. It was a coordinated effort to remove access to grasshoppers and locusts, the traditional complete protein and most dense natural source of other nutrients (Van Huis et al., 2013) for children in West Africa. Would an immersion experience have prevented this nutritional upset? Could policy makers, entomologists, and the K−20 educators of these policy makers and scientists who organized the "war on hoppers" have made a more helpful decision and prevented the nutritional epidemic? Now, 40% of rural children of West Africa subsist on a grain-based diet. They require the nutrients in the insects (Van Huis and Dunkel, 2016), their traditional snack food, to avoid kwashiorkor, physical and cognitive stunting and to prevent micronutrient deficiency, particularly of vitamin B12 (Braget, 2016).

These specific nutritional deficiencies are now linked to building the body's own resistance to cerebral malaria (Deribew et al., 2010; Ehrhardt et al., 2006; Ferreira et al., 2015; Glinz et al., 2014, 2015; Maketa et al., 2015; Mbug et al., 2010). Yes, there are negative consequences from not including immersion in general education.

CONCLUDING THOUGHTS ON IMMERSION

In this chapter, we superimposed three very different forms of immersion onto a formal definition of immersion. In the first case study, we learned that simply "book-learning" in preparation for understanding another culture's food and agricultural system is not enough. There also needs to be an open-to-learning beyond, a book approach. Undergraduates in US colleges of agriculture recognize the value of immersion in a foreign culture, but are less likely to make room in their course load for an immersion experience in another culture, compared to their classmates in the humanities. In the second case study, we learned that immersion can be a useful learning experience for all students, particularly those focused on the food and agricultural sciences. It is possible to make immersion a required, integrating form of engagement in undergraduate education, particularly at the 70 US Land Grant Institutions. Immersion experiences, like those used in Morocco, support the Land Grant mission as defined by the US Congressional Morrill Acts of 1862, 1890, and 1994 (Brooks and Lyons, 2015). In the third case study, we were reminded that often there are other cultures existing within a university "neighborhood" that can provide ample opportunities to immerse students, faculty, administrators in enriching learning and teaching experiences. In the United States there are 564 federally recognized Native American tribes, most of which are sovereign nations themselves with their own governments, school systems, and so forth. Herein lies a rich opportunity to learn about sustainable food systems without insurmountable language barriers encountered in nations outside the contiguous United States. Simultaneously, these engagements contribute to a formal recognition of the value of these non-Euro-American food systems.

One's worldview is a complex of experiences and unconscious approaches to differences (Box 4.3). Because of the mainly subconscious nature of one's worldview, it is often difficult to change or to develop interculturally even if one wants to do so. Immersion is the first step in moving toward a transformative experience that may lead one from an ethnocentric worldview to an ethno-relative worldview. To take leadership in a food production, consumption, and resulting health care delivery system, immersion in the related food systems seems to be a necessity. The essential part of that immersion is to develop beyond an ethnocentric worldview in which the value of a crop

BOX 4.3 STUDENT VIEW OF IMMERSION

My experiences have shown that immersion does not have a goal. It does not have a destination and it is not a clear route. For example, it is the experience of immersion itself that teaches the western-born immersed individual that the linear patterns we create for the experience are unrealistic when it comes to understanding a different way of life. In reality, you do not finish immersion as much as you become so comfortable with your shortcomings and strengths through multiple cultural lenses that you are able navigate most situations. You do not resist your home culture and then move on to accept it and another culture you are immersed in as Bennett's model suggests (2004). Instead, you move from one stage to the other. There is a trend with time toward an open-mindedness, a creativity, and an acceptance of differences in similarities. But do not feel strange or wrong if, when immersed, you find yourself again returning to a frustration with your own culture or another. The process is not to get through the discomfort, but to work with it. In this way we all become better workers, researchers, and people, something I think we can all agree are results that will better our world as we face the challenges of our generation (Robison, 2015).

or food is measured only by what one's own culture holds dear to an eth-no-relative worldview (Bennett, 2004). In an ethno-relative worldview differences are noted and accepted *not* as "good, bad, or disgusting," but "just different, and that is okay." The ideal approach is an ethno-relative worldview, specifically not just to find other food systems acceptable, but to actually adapt to other cultures' view of what is food and what is not (Bennett, 2004). Social facilitation is one useful way to begin to adjust to new foods or learn to appreciate unfamiliar foods. Fairs, tasting events, whenever appetizers are served is an ideal way to make use of social facilitation to introduce new foods (Dunkel, 1996). Enjoying foods such as artichokes, fish swim bladders, sushi, sheep eyes, cow tongue, bison stomach lining and intestines, Malian tou with sauce, dry meat, lobster, oysters, algae, zucchini flowers, food insects, and even pizza are, or have historically been, cultural challenges in adaptability (Rozin et al., 2004).

Adaptability to other than one's own culture and environment requires frequent exposure and deep psychological change. Sometimes adaptability, which is a form of ethno-relativism, is acquired early in life, particularly if one has grown up immersed in two or more cultures, including each of their food systems. Studies of US college students indicate that the majority of these students do not hold an ethno-relative worldview After short-term immersion in another culture, though, college students have been documented to have at least short-term improvements in adaptability (Rexeisen et al., 2008). Adaptability strongly influences food choices, as well as policy and curriculum decisions, but adaptability flows at a much deeper level. Adaptability is like an underground river that flows silently, almost undetected affecting behavior.

When one is making policy regarding diverse food production systems or teaching sustainable cropping systems, it seems necessary to partake in an immersion, i.e., to "walk a few miles" with representatives of diverse cropping systems or food gathering systems to learn the specific meanings of the foods within the culture beyond nutrient acquisition. It is appropriate to learn their "language" and to come to understand each of their desired qualities of life and how they produce the resources they perceive are needed to maintain that quality of life.

Simply put, try it. The benefits are great. Use your imagination.

References

Ayittey, G.B.N., 2005. Africa Unchained: The Blueprint for Africa's Future. Palgrave, Macmillan, New York, NY., 483 pp. (read pp. 365–446).

Bâ, A.H., 1972. Aspects de la civilisation africaine. Présence Africaine.

Bâ, A.H., 1992. L'étrange destin de Wangrin: ou, Les roueries d'un interprète africain. Éditions 10/18.

Bâ, A.H., 1994. Oui, mon commandant! Actes Sud.

Bâ, A.H., 2000. Contes initiatiques peuls. Pocket.

Bâ, A.H., Hanes, F., 1978. (*Kaydara*). Nouvelles éditions africaines.

Badian, S., 1972. Sous l'orage. Présence Africaine.

Badian, S., 1976. Le sang des masques. Laffont.

Badian, S., 1997. Noces sacrées: les dieux du Kouroulamini: roman. Editions Présence Africaine.

Badian, S., 2007. La saison des pièges. Présence Africaine, Paris.

Bennett, M., 2004. Becoming interculturally competent. In: Wuzel, J. (Ed.), Toward Multiculturalism: A Reader in Multicultural Education, second ed. Intercultural Resource Corp., Newton, MA, pp. 62–77.

Braget, D., 2016. Micronutrient deficiencies related to the health of children in Sanambele. Report submitted in partial fulfillment of AGSC 465R Health, Poverty, Agriculture: Concepts and Action Research, Montana State University-Bozeman, MT, 27 pp.

Briers, G.E., Shinn, G.C., Nguyen, A.N., 2010. Through students' eyes: perceptions and aspirations of College of Agriculture and Life Science students regarding international educational experiences. J. Int. Agric. Extension Educ. 17 (2), 5–20.

Brooks, C.B., Lyons, L.W., 2015. The Morrill Acts of 1862 and 1890. <http://www.1890universities.org/history> (accessed 26.05.15.).

Bylenga, S.A., Clarke, S.A., Hammond, J.W., Morey, R.V., Kayumba, J., 1987. Evaluation of applying quality standards to bean and sorghum markets in Rwanda. Miscellaneous Publication 50-1987 of the Minnesota Agricultural Experiment Station, University of Minnesota, St. Paul, MN, 165 pp.

Carr, D.G., Hawes-Davis, D., 2000. Wind River. Video. 40 min. High Plains Films. Missoula, MT.

Chieffo, L., Griffiths,, L., 2004. Large-scale assessment of student attitudes after a short-term study abroad program. Interdisciplin. J. Study Abroad 10, 165–177.

Deribew, A., Alemseged, F., Tessema, F., Sena, L., Birhanu, Z., Zeynudin, A., et al., 2010. Malaria and under-nutrition: a community based study among under-five children at risk of malaria, South West Ethiopia. PLoS One 5 (5).

Diakité, D.. Les défis du multilinguisme au Mali. In: Recherches Africaines: Annales de la Faculté des Lettres, Langues, Arts et Sciences Humaines N. 00 Janvier–Juin, 2002, pp. 86–97.

Dunkel, F., 1996. Incorporating food insects into undergraduate entomology courses. Food Insects Newslett. 9, 1–4.

Dunkel, F., Montagne, C., 2011. New paradigm for discovery-based learning: Implementing bottom-up development by listening to farmers' MONE 2007-02535. 15 Sept 2007-14 Sept 2011. Final report. http://www.montana.edu/mali/pdfs/CRISfinalreportyear4.pdf.

Dunkel, F., Giusti, A., 2012. French students collaborate with Malian villagers in their fight against malaria. In: Thomas, J. (Ed.), Etudiants sans Frontières (Students Without Borders): Concepts and Models for Service-Learning in French. American Association of Teachers of French, pp. 135–150.

Dunkel, F.V., Wittenberger, T., Read, N.R., Munyarushoka, E., 1986. National storage survey of beans and sorghum in Rwanda. Miscellaneous Publication. Minnesota Experiment Station 46-1986, University of Minnesota, St. Paul, MN, 205 pp.

Dunkel, F.V., Clarke, S.A., Kayinamura, P., 1988. Storage of beans and sorghum in Rwanda: synthesis of research, recommendations and prospects for the future. Miscellaneous Publication 51-1988. Minnesota Experiment Station, University of Minnesota, St. Paul, MN, 74 pp.

Dunkel, F., Coulibaly, K., Montagne, C., Luong, K., Giusti, A., Coulibaly, H., et al., 2013. Sustainable Integrated Malaria Management by Villagers in Collaboration with a Transformed Classroom Using the Holistic Process: Sanambele, Mali and Montana State University, USA. American Entomologist 59, 15–24.

Easterly, W., 2006. The white man's burden: Why the West's efforts to aid the rest have done so much ill and so little good. The Penguin Press, New York, NY, 436pp.

Edmister, J.A., Breene W.M., Vickers, Z.M., Serugendo, A., 1986. Changes in the cookability and sensory preferences of Rwandan beans during storage. Miscellaneous Publication 47-1986 of the Minnesota Agricultural Experiment Station, University of Minnesota, St. Paul, MN, 390 pp.

Ehrhardt, S., Burchard, G., Mantel, C., Cramer, J., Kaiser, S., Kubo, M., et al., 2006. Malaria, anemia, and malnutrition in African children—defining intervention priorities. J. Infect. Dis. 194, 108–114.

Famory, F., 1996. Les poèmes de la source Saniya. Editions Donniya, Bamako.

Ferreira, E., Alexandre, M., Salinas, J., de Siqueira, A., Benzecry, S., de Lacerda, M., et al., 2015. Association between anthropometry-based nutritional status and malaria: a systematic review of observational studies. Malaria J. 14, 346.

First People's Worldwide Production, 2012. Enoughness: restoring balance to the economy. Animated presentation. 5 minutes. <https://www.youtube.com/watch?v=RxPVrr44KHI> (accessed 05.09.16.).

Gage, B., Gage, G., 2008. American outrage. First Run Features. NTSC. FRF 913829D. 56 minutes.

Glinz, D., Kamiyango, M., Phiri, K., Munthall, F., Zeder, C., Zimmermann, M., et al., 2014. The effect of timing of iron supplementation on iron absorption and haemoglobin in post-malaria anaemia: a longitudinal stable isotope study in Malawian toddlers. Malaria J. 13, 397.

Glinz, D., Hurrell, R.F., Ouattara, M., Zimmermann, M., Brittenham, G., Adiossan, L., et al., 2015. The effect of iron-fortified complementary food and intermittent preventive treatment of malaria on anaemia in 12- to 36-month-old children: a cluster-randomized controlled trial. Malaria J. 14, 347.

Greta Robison, 2015. B.S. with honors in Earth Sciences and Liberal Studies (Multicultural-Global option), Montana State University-Bozeman, MT, former student and teaching assistant in AGSC 465R.

Hanegreefs, P.R., Morey, R.V., Clarke, S.A., 1987. Alternative storage management for beans and sorghum in Rwanda. Miscellaneous Publication 52-1987 of the Minnesota Agricultural Experiment Station, University of Minnesota, St. Paul, MN, 97 pp.

Ingraham, E.C., Peterson, D.L., 2004. Assessing the impact of study abroad on student learning at Michigan State University. Interdiscipl. J. Study Abroad 10, 83−100.

Lamb, E.M., Hardman, L.L., 1985. A Catalogue of Bean Varieties Grown in Rwanda USAID-Rwanda, Local Crop Storage Cooperative Research. USAID, University of Minnesota, Minnesota Experiment Station Miscellaneous Publication 48−1985. Government of Rwanda, 35 pp.

Lamb, E., Hardman, L., 1986. Final report of the survey of bean varieties grown in Rwanda. Miscellaneous Publication 45-1986 of the Minnesota Agricultural Experiment Station, University of Minnesota, St. Paul, MN.

Lamb, E.M., Dunkel, F.V., 1987. Studies on the genetic resistance of local bean varieties to storage insects in Rwanda. Miscellaneous Publication 49-1987. Minnesota Experiment Station, University of. Minnesota, St. Paul, MN, 122 pp.

Maketa, V., Mavoko, H.M., Inocencio da Luz, R., Zanga, J., Lubiba, J., Kalonji, A., et al., 2015. The relationship between *Plasmodium* infection, anaemia and nutritional status in asymptomatic children aged under five years living in stable transmission zones in Kinshasa, Democratic Republic of Congo. Malaria J. 14, 83.

Mbug, E., Meijerink, M., Veenemans, J., Jeurink, P., McCall, M., Olomi, R., et al., 2010. Alterations in early cytokine-mediated immune response to *Plasmodium falciparum* infection in Tanzanian children with mineral element deficiencies: a cross-sectional survey. Malaria J. 9, 130.

Mestenhauser, J., Dunkel, F., Paige, M., 1984. How to pack your parachute to hit the ground running: new concepts in predeparture training for faculty. University of Minnesota-Twin Cities. 60 min.

Norberg-Hodge, H., 1991. Ancient futures: Learning from Ladakh. Sierra Club Books, San Francisco, CA, 196pp.

Rexeisen, R.J., Anderson, P.H., Lawton, L., Hubbard, A.C., 2008. Study abroad and intercultural development: a longitudinal study. Front. Interdiscipl. J. Stud. Abroad 17, 1−20.

Rozin, P., Spranca, M., Krieger, Z., Neuhaus, R., Surillo, D., Swerdlin, A., et al., 2004. Preference for natural: instrumental and ideational/moral motivations, and the contrast between foods and medicines. Appetite 43 (2), 147−154. Available from: http://dx.doi.org/10.1016/j.appet.2004.03.005.

Savory, A., Butterfield, R., 1999. Holistic Management: A New Framework for Decision Making. Island Press, Washington, DC.

Smith, L.T., 2012. Decolonizing Methodologies: Research and Indigenous Peoples. Zed Books, Ltd, New York, NY, 240 pp.

Thurow, R., 2012. The last hunger season: A year in an African farm community on the brink of change. Public Affairs, New York, NY, 294pp.

Thurow, R., 2016. The first 1,000 days: A crucial time for mothers and children—and the world. Public Affairs, New York, NY, 277pp.

Van Huis, A., Dunkel, F., 2016. Edible insects, a neglected and promising food source. In: Nadathur, S., Wanasundara, J., Scanlin, L. (Eds.), Sustainable Protein Sources. Elsevier Publ. Co., Boston, MA., pp. 341−355. (Chapter 21)

Van Huis, A., Van Itterbeeck, J., Klunder, H., Mertens, E., Halloran, A., Muir, G., et al., Edible Insects: Future Prospects for Food and Feed Security. Food and Agriculture Organization of the United Nations, Rome (Forestry Paper #171)

Yunus, M. (with Karl Weber), 2007. Creating a World Without Poverty: Social Business and the Future of Capitalism. Public Affairs. New York, NY, 261 pp.

Further Reading

Béridogo, B., 2002. « Le régime des castes et leur dynamique au Mali ». In: Recherches Africaines: Annales de la Faculté des Lettres, Langues, Arts et Sciences Humaines N. 00 Janvier–Juin, pp. 3–27.

Diabaté, M.M., 1985. L'assemblée des djinns: roman, Vol. 47. Présence africaine.

Diabaté, M.M., 2002. Le coiffeur de Kouta. Editions Hatier International.

Diabaté, M.M., 2005a. Le boucher de Kouta. Editions Hatier International.

Diabaté, M.M., 2005b. Le lieutenant de Kouta. Editions Hatier International.

Diarra, S.O., 1999. La politique linguistique en Afrique: l'enseignement des langues nationales au Mali, ouvrage collectif.Démocratisation et scolarisation en Afrique: les écoles de la IIIe République au Malis. Editions Karthala, Paris.

Doumbia, T.A. « Les relations à plaisanterie dans les sociétés mandingues ». In: Recherches Africaines: Annales de la Faculté des Lettres, Langues, Arts et Sciences Humaines N. 00 Janvier–Juin, 2002, pp. 28–42.

Dunkel, F., 2008. Discovery-based undergrad opportunities: facilitating farmer-to-farmer teaching and learning. Final Technical Report. MONE-2004-02778, Grant #2004-38411-14762.CRIS # 0200475.

First Peoples Worldwide, 2012. Indigenous peoples guidebook. http://solutions-network.org/site-fpic/files/2012/09/First-Peoples-Worldwide-FPIC-Guidebook_5.10.12.pdf.

Konaté, D., 2002. « Oralité et ecriture dans la communication usuelle au Mali: entre traditions et modernité ». In: Recherches Africaines: Annales de la Faculté des Lettres, Langues, Arts et Sciences Humaines N. 00 Janvier–Juin, 2002, pp. 43–54.

Levin, R., 2009. Transforming the study abroad experience into a collective priority. Peer Rev. 11 (4), 8–11.

N'diaye, B., 1970. Groupes ethniques au Mali. Éditions populaires.

N'diaye, B., 1997. Contribution à la connaissance des us et coutumes du Mali. Editions Jamana.

Niane, D.T., 1960. Soundjata: ou, L'épopée mandingue. Présence africaine.

Oxford Dictionary. <http://www.oxforddictionaries.com/us/definition/american_english/immersion>.

Sidibé, F.F., 2006. Une saison africaine. Présence Africaine.

Listening In and Between Communities

CHAPTER 5

Listening With Subsistence Farmers in Mali

Florence V. Dunkel[1], Ibrahima Traore[2], Hawa Coulibaly[3], and Keriba Coulibaly[4]

[1]Department of Plant Science and Plant Pathology, Montana State University, Bozeman, MT, United States, [2]Lycee El Hadj Karim Traore, Bamako, Mali, [3]The Village of Sanambele, Bamako, Mali, [4]USAID, Sikasso, Mali

CONTENTS

Introduction to Sanambele, Mali, and the Main Communicators 110

Reconciling the Role of Health in the Village's Food and Agriculture System 115

Food Security, Stunting, Amino Acid, and Micronutrient Deficiencies: What is the Culture-Smart Agriculture Answer in Villages Relying on Grain-Based Diets? 118

The Gourd Story 124

Concluding Reflections 127

References 128

Mali is a landlocked country in West Africa with an estimated population of 13.5 million people. From the formally educated families in the capital city, Bamako, to isolated rural farming and fishing villages, our team has found a joy of living, a dedication to the family, focus on children plus a reverence for the Elders, and an appreciation for music, graphic art, and design. Beneath the surface, over the decades we also have found a concern at these levels for the prevalence of disease, both infectious and diet-related, for the calling into question of certain cultural practices, and a growing interest in formal education for the children and youth, particularly the girls and young women. We see and hear the lingua franca remaining strong along with cultural pride, joy of family, and appreciation of diversity including intercultural friend relationships.

Despite many forms of cultural wealth, Mali is one of the most economically poor countries in the world. About one-third of Mali's children are stunted and over one in ten (13.3%) suffers from wasting (International Cooperation and Development, 2016). Even more stark is that given the current trend and considering population growth, the number of stunted children in Mali is predicted to be 1.11 million by 2025. Some reports deplore the further complication of locust and grasshopper infestations making the availability of nutritious food more insecure (DeCapua, 2012). Herein lies one of the problems. No mention in recent articles about the locust and grasshopper issue is recognition made of the most common snacks of rural children in Mali being roasted grasshoppers and locusts. The answer to this lack of improvement in malnutrition seems to be twofold. First, Europeans and Americans have not done their cultural homework and have, instead, advised and helped Malian farmers eliminate this best source of dense

109

Incorporating Cultures' Role in the Food and Agricultural Sciences. DOI: http://dx.doi.org/10.1016/B978-0-12-803955-7.00005-5

nutrition available physically and socially to children. Second, these same advisors have not shared nutritional biochemistry in a place-based, culture-based format for mothers with no formal education, after a trusting relationship has been developed.

The purpose of this chapter is to explore the richness in long-term relationships with a subsistence farming community and how this experience lays the groundwork for exploring and working together to solve complex health, food, and agricultural issues. This is one of the five nations in which we will explore the listening process in Part II of this book. We will also explore the details of listening in and between communities in several other venues, such as Land Grant University students, middle schools, and policy makers in Washington, DC. Here in Chapter 5, Listening With Subsistence Farmers in Mali, there will be three voices in addition to me, Florence Dunkel. Of these, the oldest relationship is with Hawa Coulibaly who we introduced in Chapter 2, Failures.

INTRODUCTION TO SANAMBELE, MALI, AND THE MAIN COMMUNICATORS

Sanambele is a small farming village composed of smallholder farmers of the Bambara ethnic group. The village has no electricity, cars, or running water, and only a few motor bikes. Almost every grand family has a bicycle. A grand family consists of the father, all of his wives (up to four), and all of their children (10−12 per wife), plus Elders of the family and other visiting relatives who have come to help or to attend school in the village. Typically, the grand family is about 40 people. There are several cell phones in the village and solar cell battery chargers at the village school and in the village health center. Daily bus service takes 8 hours to travel the 50 km to Bamako, the capital city. There is no market in Sanambele and no store, only a small storefront with a walk-up counter at which one can purchase essentials like salt and batteries. The nearest market is Bougoula, a 3 km walk, only open on Tuesday, or Dialakoroba whose weekly market day is Saturday. This market is about 15 km in the other direction.

While some of the following information may seem irrelevant in a Western culture sense, in the process of getting to know and become as much a part of a village or other community, as one can as an outsider, knowing this level of detail is essential. Hawa, unlike most women in the village, grew up in Sanambele, a Bambara smallholder farming village. Bambara villages are typically founded by men and women generally marry outside the village and go to live in their husband's village. Therefore in a Bambara village, most men are related, but not the women. Hawa's father was the village chief and it is likely he who taught Hawa to read and write. Hawa never attended a formal school and is one of two women her age in the village who can read and write. The first time I met Hawa she impressed me as a leader. She presented me with a pen and lined notebook with a matrix she had created in the notebook for village visitors to sign in with their pertinent information. This was the first time I had been given a guest book in a Malian village. I later learned that Hawa had been elected the President of the Women's Association of Sanambele. These details are important to note since in a Malian village, it is the leader of the women with whom cooperation with outsiders must be solidified, particularly if the outsiders are interested in anything having to do with children or food.

Hawa is the first wife of Mr. Samake, a member of the village Elder Council, the highest decision-making body in the village. Whereas this relationship gives her some status in the community, Hawa herself, as an individual, seems to command a separate respect within the community. Hawa is an excellent farmer and farm manager. She maintains about 25 head of cattle in addition to her several innovations in long-term storage of cowpeas and in shea butter production as we discussed in Chapter 2, Failures. In her role as Women's Association President, she holds weekly teaching sessions for village women. She also stepped right up when we began to sample anopheline larvae around the village and in the partially dry bed river near the village. Hawa made her own sampling tools, participated in the sampling, and made the neem leaf slurry to prevent larval development. There were many other roles she took related to the village's successful elimination of deaths from cerebral malaria (Luong et al., 2012). Clearly Hawa is an early adopter and she is an important village leader in Sanambele, concerned about all of the women and children of Sanambele, a model farmer, and an innovator.

Governance and social organization in this small village is complex; the main decision-making groups in Sanambele are the exclusively male Council of Elders comprised of noble men and the members of the Sanambelean Women's Association, which includes all noble women and women in the artisan guilds.[1] Leaders of these two organizations are elected on the basis of their leadership qualities, and are termed the Chief and the President, respectively (Dunkel et al., 2013). The Elder Council is ultimately responsible for

---

[1]Castes, guilds, and noble families are defined in this Bambara village of Sanambele as follows: noble families who farm and provide governance are assisted by the guilds and castes or *nyamakala*. We are using the terms guilds and castes interchangeably. *Nyamakala* are composed of the following castes: blacksmiths, weavers, potters, woodworkers, leatherworkers, and *griots*. *Griots* are divided into three groups: musicians who preserve ancient music and compose new melodies; ambassadors specializing in conflict resolution between great families; and genealogists, historians, or poets who preserve oral culture (Ba, 1981; Beridogo, 2002). Another group sporadically part of the Sanambele community are nomadic Fulani herders and their families. Nomads are neither nobles nor *nyamakala* (Dunkel et al., 2013).

all decisions made in the village. Specifically, they are responsible for new housing or allocation of land; for whether or not to receive visitors; for instituting, abandoning, or enforcing cultural practices such as FGM (female circumcision) and the gender ownership of cashew trees and nuts. All new information, supplies, and equipment given from outside the village are required to first be presented formally to the village Chief and the Council of Elders. If decisions made by the Women's Association would affect the village as a whole, the Association seeks approval from the Council of Elders before implementing the decision. There also is a young men's group, whose leader sits in the Council of Elders.

The decision-making process is consensus-based. The organizational location in which the decision is made may vary, the fundamental process does not. Wherever the decision is being made in the village government, the process is consensus. In meetings with the village Chief, whoever wants to make observations speaks first before the Chief speaks. The Chief analyzes all that has been said by the Elders and visitors, and then pronounces the conclusion that may be discussed again if any Elders are not comfortable with the decision. Each group can make suggestions to the other groups. Sometimes differences in perceptions by Elders and younger groups make resolution complex. Solutions to the problem are proposed by each group to the others and discussed until a consensus on the issue is reached by every party involved (Dunkel et al., 2013). Again, these details of village governance seem excessive, but they are examples of what is necessary to know for an outsider who wishes to learn from the community and exchange information with the community, be it about health, food, agriculture, or another area.

This decision-making process sounds complex, but because we are using the holistic process, the issues we are working with are *from* the decision-making process of the village. We merely provide requested information that the Women's Association and the Elder Council use in making decisions of which we are informed. There are three keys to this process—the interpreters and their deep understanding of the holistic process; their dedication to their people, the rural Malians, and appreciation for village life; and respect for traditional ecological knowledge of the villages. Without those qualities in Keriba Coulibaly and Ibrahima Traore, the productive listening process these past 10 years in Sanambele would have been impossible. Now we will introduce both Keriba Coulibaly and Ibrahima Traore, two outsiders like me, but who hold a strong caring and deep respect for the village of Sanambele and for rural communities in general.

Keriba Coulibaly is Bambara and he grew up in a small village like Sanambele. His university degree was in agronomy and he has worked professionally as an agronomist with l'Institut d'Economie Rurale (IER) in

Sikasso, Mali, his entire career. Keriba became involved with Montana State University (MSU) when he won a national competition to receive English language and graduate training in plant sciences and entomology at MSU for 2 years. We faculty had received a Higher Education for Development grant for graduate training of mid-career scientists working for the government (IER) and the national agricultural university l'Institut Polytechnique Rurale / Institut Formation Rurale Applique (IPR/IFRA) in Mali. Keriba had won the Malian national competition for one of these seven fellowships that MSU offered.

The two years that Keriba spent at MSU was an essential part of the success in communicating with the village of Sanambele. During the 2 years Keriba was on campus, he became a member of our families, he worked with our students, and actually helped to choose the texts that formed the core readings of the course AGSC 465R. Keriba became a mentor for the MSU students and the Montana tribal college students going to work in Mali (Dunkel and McCartney, 2005), and he took the course from Clifford Montagne on Holistic Thought and Management. In 2007 when he returned to live in Mali, I took him with me to Sanambele to meet the villagers. Because Keriba so well understood the holistic process by that time, it was quite natural for him to work with my students and me there. Keriba quickly became a welcomed visitor to the community and assumed the role of cultural advisor, translator, holistic process guide, and partner with my students and me for the next 7 years (Chaikin et al., 2010). In 2012 Keriba accepted an assignment with United States Agency for International Development (USAID) based in Sikasso, and went on leave from IER. Although Ibrahima Traore is now able to make more frequent visits to Sanambele than Keriba, Keriba remains an important cultural interpreter. His special combination of knowledge is valuable. His understanding of US institutions such as MSU and USAID, of crop science in Mali, village life in Mali, the people of Sanambele specifically and their farming system, as well as his experience as a chicken farmer in Sikasso, selling eggs in that large urban area in southern Mali.

In September 2014, we met Ibrahima Traore and Salifou Bengali. They had been selected to be Teaching Excellence and Achievement (TEA) participants and were assigned to Bozeman, Montana. TEA is a customized set of seminars on curriculum development, lesson planning, instructional technology, and new teaching methodologies. Ibrahima was a mid-career educator/ administrator trained in international law and the principal of Lycee El Hadj Karim Traore, a private secondary school founded in memory of his grandfather in the peri-urban area of Bamako, Mali. Salifou was a language trainer in both the Bambara and Senafou languages for Peace Corps trainees in Mali. Ibrahima and Salifou brought cultural authenticity to our discussions and skills workshops in the classroom. This enabled students to see different points of view within the mix of Malian cultures for although they were both Malian, Ibrahima was Bambara and Salifou was of the Senafou ethnic group. Both men brought a whole new dimension into our learning community—the formal educational process—for Malian young people (high school students) and for US young people (Peace Corps).

In Chapter 3, Decolonization and the Holistic Process, we met Carly Grimm and her struggles with identifying the crops high in tryptophan and lysine in the village of Sanambele. She was majoring in Sustainable Foods and Bio-Energy Systems and enrolled in AGSC 465R the same semester in which Ibrahima and Salifou were serving as in-class site mentors in Bozeman, and in addition to shared class time, the TEA participants met outside of their classes with Carly and other students working on cricket farming in a Malian rural household, ultimately advising on how roasted crickets might work into the Malian village cuisine. They consulted with their wives back home on how to prepare some of these dishes. Soon a set of additional crops were developed for Carly to determine lysine and tryptophan content and consider including in her wordless, numberless, quantitative nutritional food choice chart for the mothers to use in Sanambele.

During the semester, Ibrahima learned that the village of Sanambele was located near his own family farm. He offered to carry the chart back with him and bring it to Sanambele for the women to give comments. He brought the chart to the women and shared their responses with Carly who made the corrections and sent back the diagram via Ibrahima to the women in Sanambele (Fig. 5.1). This research on cropping systems and food preferences and the follow-on participatory research in this small rural farming village would not have happened had the holistic process not been under way resulting in an ever deepening level of trust, respect, and appreciation. Our Western culture approach to solving problems evolved into an approach that was more free of self-oriented thinking and accepting of the uncertain and unexpected (Cavanaugh, 2017). Carly's participatory, quantitative chart and the role of Ibrahima in delivering it provided a way to continue the back-and-forth learning and sharing about nutritional biochemistry and the farming system in Sanambele. This marked a new phase in our relationship with Sanambele.

To understand the new phase focused on nutritional health, it is important first to go back in time to understand the order of actions that took place as requested by the village. Requests of an Indigenous community to a person or group who represents their colonizers is a bold step and in my experience, it usually does not take place at all. Most often this is because of a failure in the listening process, either in a failure to create a safe place to share digressions or intrusions into the community's resources that create their desired

FIGURE 5.1
Sanambele villagers reviewing draft of Carly Grimm's participatory diagram of high lysing high tryptophan foods grown/wild collected by villagers November 2014 (photo by Ibrahima Traore).

quality of life or in a refusal to provide assistance once that disruption in a community's resources is shared. Using the holistic process helped us establish a trusting relationship in which this Indigenous community seemed to feel safe requesting what they actually wanted to do. Once the holistic process was clearly under way, the Sanambeleans gently took control of what it was that we accomplished together.

RECONCILING THE ROLE OF HEALTH IN THE VILLAGE'S FOOD AND AGRICULTURE SYSTEM

The "live chicken award" had been given to us in 2005 (Chapter 2: Failures) because we outsiders had listened long enough and carefully enough to hear the real concerns of the village: malaria first; hunger (malnutrition) second. Briefly, now, we will take the reader back to this time to capture the thread of the malaria elimination story. In doing so, we hope the reader will be able to visualize how a seeming digression may actually be a clear path to gain an understanding of why the village must be in control of the action. Although malaria is not directly about food or agriculture, it was the villagers' most pressing problem and it had to be dealt with first.

In 2005 we had discovered children's health in this village was central to the villagers' desired quality of life and not, at that time, crop yield or crop pest management or economic gain (Kante et al., 2009). Specifically, 4.2% of the village children died from malaria *each year* in this small village during the wet season (June through September) (Dunkel et al., 2013) and 23% of the children had or were at-risk for the stunting effects, both physical and mental, from kwashiorkor (protein-energy malnutrition). Indeed, women farmers were experiencing about 40% mortality of their children from birth to 5 years. After a series of holistic discussions in that village, villagers entered into a trusting and a "teaching-us-at-MSU" relationship,[2] *and* the villages developed their own plan to locally eliminate malaria (Luong et al., 2012; Dunkel and Giusti, 2012).

With some advice from us, the villagers developed an integrated management plan to eliminate deaths from malaria. This included learning the life cycle stories of the protozoa causing malaria, particularly *Plasmodium falciparum*, and the vector, primarily the mosquito *Anopheles gambiae* Giles s.s. (Diptera: Culicidae); physical removal of small larval rearing pools in the village; and larvicide treatment with neem leaf slurry produced by villagers for larger rearing pools including pools in the Zangolo River, a seasonally dry bed river channel bordering village gardens and grain fields (Fig. 5.2). Also part of the villagers' action was the development of a cottage industry controlled by the women farmers, a handicraft enterprise (www.mmama.net) providing ready cash for mothers to seek transportation to the capital to buy medication; and an art design-based community awareness program that village children in 7th to 9th grades brought back to their grand

---

[2]The villagers, particularly Hawa, seem to have a clear curriculum for us faculty and our students. We learned from these experiences, year after year, to be good students, we also learned that Sanambeleans were exceedingly proud of their village and its culture. The curriculum that the villagers of Sanambele seemed to have developed for us was as follows:

- First and most important was to learn Bambara, particularly the greetings and the departure phrases.
- Secondly, was to learn all the first and last names of everyone with whom we interacted.
- Thirdly was to learn the blessings one said to the host after eating and the thankful phrases one said to the wife who prepared the meal.
- One year Hawa prepared a tray with small seed samples wrapped in plastic of all the 10−12 main crops grown in Sanambele. She expected the students to learn all of these.
- Another main teaching goal that Hawa seemed to have was for us faculty and students was to learn the medicinal plants that grew near the village and were used for managing malaria.

Knowing the traditional stories of the Bambara people was also an important teaching goal that the villagers had for us outsiders.

FIGURE 5.2
Hawa Coulibaly and Florence Dunkel in Hawa's cattle pasture, Sanambele Mali, 2009 by her fields.

family in the village. We did laboratory studies and preliminary field bioassays with neem leaf slurry and found the 72-hour LC_{50} with A. *gambiae* was 8825 ppm (Luong et al., 2012). We translated this laboratory bioassay data into real actions suggested to villagers and observed how they organized the neem leaf slurry production themselves. In a trial run June 2011, we observed 32 men, women, and children collecting leaves and producing enough slurry in 2.25 hours to treat the dry bed river pool. Prior to the use of these neem tree leaves, my student and I did a village survey, we found 81 neem trees of sufficient size to sustain harvesting leaves to treat the seasonally dry bed river channel pool every 2 weeks during May and June.

This story now becomes an example of how solving complex problems, which some call the "wicked problems" of the world, can work. In wicked problems, when one issue is solved, another emerges. So too in Sanambele. By November 2016, the village had grown to 1250 people from 750 in 2005 when the malaria elimination process began. There were 7 years of wet seasons without deaths from malaria and one season with two deaths attributed to cerebral malaria during the national malaria epidemic of 2015 (1 in 10 children 0–5 years old died nationally); but in Sanambele, this rate was c. 1 in 175 children 0–5 years of age. Of those two young children who died, one was thought to have visited a neighbor village (that had not eliminated cerebral malaria) with their family that summer and the other child had kwashiorkor (stunting) and therefore a compromised adaptive immune system.[3]

Villagers are now asking us to help the village understand how to sustain this result of local malaria elimination. We hold weekly Skype meetings now to ensure mortality from cerebral malaria is eliminated sustainably from Sanambele. When it was clear that we had answered their first request, we began, 7 years later, to move on to their second most important village issue—hunger, malnutrition in the children. Eventually, sheer size of the community may outstrip the ability of the land to produce.

FOOD SECURITY, STUNTING, AMINO ACID, AND MICRONUTRIENT DEFICIENCIES: WHAT IS THE CULTURE-SMART AGRICULTURE ANSWER IN VILLAGES RELYING ON GRAIN-BASED DIETS?

The second issue these African villagers recognized as a deterrent to their quality of life was their children's nutrition, specifically, my students and I discovered, was the supply of complete proteins in their diet. We discovered this is a complex cultural issue. All the plant and animal products needed for normal growth and development of the children, it seems, are grown or could be grown in the village in sufficient quantities. We collected these data informally over a 10-year period by living with farming families in Sanambele and eating all meals with them during our stays.

In Sanambele, as in most of rural West Africa, the farming communities subsist on a grain-based diet. In Mali in 1999, cereals accounted for 73% of the energy and plant foods for 74% of the protein (Barikmo et al., 2007). The constraint to avoiding kwashiorkor, the most common cause of "hunger" or malnutrition in Mali, and in Sub-Saharan Africa, we found in our collaborative research, was *understanding* the nutritional biochemistry concept of essential amino acids and micronutrients in building human body proteins for ideal physical and mental growth and development. Sanambele is typical of many villages in Mali, although their measured kwashiorkor rate was half of

---

[3]Stunting is a debilitating condition that usually occurs around the time of weaning. In Africa, it is often called kwashiorkor, a Swahili term meaning, "the next child comes too soon." The first 36 months after birth, the times of most rapid growth in humans, are the most vulnerable to nutrient deficiencies and therefore this age is the most likely time for stunting to occur. If the mother had during her pregnancy a nutritionally compromised diet, stunting is more severe (Thurow, 2016). To prevent stunting a child needs an adequate supply of all 10 of the essential amino acids. Often the deficiency of essential amino acids is confounded with a deficiency of one or more of the micronutrients, such as calcium or vitamin B12. Stunting is reversible, but if not addressed aggressively, the child will die. Further complications with stunting are the effect it and micronutrient deficiencies have on the proper functioning of the immune system.

the reported average for Mali (40%). As of June 2011, 99.5% of the adult women in Sanambele had no formal education and did not read or write. Since the village school (K–9) was built, most residents under 20 years seem to be able to read and write, but we have not taken this data specifically. Since 2008, Hawa Coulibaly has been holding literacy classes that focus on health and nutrition for adult women.

Informal analysis of children's food intake (ages 2–4 years) in Sanambele indicated that their diet was low in the essential amino acids, lysine and tryptophan (Grimm et al., 2014; Stein et al., 2012), which cause stunting (both physical and cognitive) and malfunctioning of the adaptive immune system (Richards et al., 2016). Following laboratory estimation of what 2- to 4-year-olds were consuming in Sanambele (Turley et al., 2011), there was a flurry of research, by students who followed her, examining the amino acid composition of specific foods grown or wild-collected by villagers. Bambara groundnuts, cashews (Howe and Coulibaly, 2013), dry meat processing of beef cattle (Gambill, 2011), goat meat and milk, and grasshoppers (Fejes, 2009) were some of these products explored. There was, we discovered, a wide diversity of foods in Sanambele rich in the essential amino acids that were missing in the grain-based diet, essentially toh, made from corn or millet or sorghum (Stone et al., 2015) (Table 5.1).

The "backstory" that is often overlooked by early childhood nutritional scientists is that many of these foods, although grown locally, are not available to the very young children because of various cultural practices. We discovered at weaning, the intense nutritional needs of the young child and of the mother during pregnancy and lactation are not specifically understood at the village level. There are no stories, at least we have not yet found any, that are still available in the traditional ecological knowledge which identify certain

Table 5.1 Growth-Limiting Essential Amino Acids Missing in Sanambelean Young Child's Diet (Stone et al., 2015)

Type of Toh	Total Protein g/500 g[b] Toh	Available (g)		Missing (g) for 10 kg Child[a]		Missing (g) for 20 kg Child[a]	
		Lysine g/500 g Tou	Tryptophan g/500 g Tou	Lysine	Tryptophan	Lysine	Tryptophan
Maize	47.50	1.27	0.335	3.23	0.265	7.73	0.865
Millet	48.50	1.66	0.945	2.84	−0.345	7.34	0.233
Sorghum	50.05	1.02	0.615	3.48	−0.150	7.98	0.585

[a]Missing on a daily basis.
[b]Mean daily portion of tou for Sanambelean child.

nutritionally dense foods to use especially during these times of life. As the villagers began to get to know Keriba and me, we shared stories with the Sanambeleans of how he and I cooked and baked with local insects in Bozeman, Montana. This generated a number of fragments of stories from the women who individually came to us and shared what they remembered about insects other than grasshoppers that their family ate when these adults were young children, but none seemed to be tied specifically to the intense needs of the "first 1000 days"—from pregnancy through lactation 12 months beyond, 36 months after the child is born.

The specific cultural practices related to food we found in use in Sanambele were age-based and gender-based. Cashews are a man's crop, the trees, the fruits, and the nuts. That is cashews are for the men to harvest and not the women. They are wild-harvested and in recent years the men have chosen to sell them outside the village now for roasting and export, rather than roasting them in the village and sharing them. The children, I found, though, frequently helped themselves to the cashew fruit that they take from the grove of cashew trees. The Chief and Elder Council shared with us that one or more generations ago, these were distributed among the people of Sanambele for consumption within the village (Howe and Coulibaly, 2013).[4] During our conversations in the Elder Council, the leaders now understand that the cashews could supply some of the missing nutrients that cause stunting in their children.

Chicken meat is reserved for the men and visitors. Chicken bones, especially the knuckles and the broth, are reserved for the women and children. Children seemed to "own" the chicken intestine and this they wound around a stick and roasted it on their father's charcoal stove. Shea nuts, on the other hand, belong to the women, although the trees belong to the men. Women made shea butter from the nuts but when this was used in cooking, it was shared through all genders and age groups. Grasshoppers and locusts are widely known to belong to the children. They are the traditional children's snack food. Goat meat and mutton are reserved for feast days. The annual day for having goat or mutton is Tabaski, celebrating the end of the Hadj. Beef is reserved for a wedding celebration or a funeral.

---

[4]The cashew tree blossoms in February and the nuts are picked from late April to June, just at the beginning of the hunger season. These trees have a wide precipitation range from 87.5 to 375 cm but they must have a long dry spell in between for the tree to recover from its previous crop (Aremu et al., 2007). This tree is well known to be a drought tolerant tree in Mali (Coulibaly, K. personal communication). Cashews analyzed by Aremu et al. in Nigeria contained 17 g protein per 100 g cashew nuts. This protein contains all the essential amino acids. This would mean that a child would need to eat 200 g per day if cashews were their only protein source. Certainly just the cashew would contribute to eliminating stunting in Sanambele if it were made available to the children.

Culturally appropriate sources of tryptophan for children in Sanambele are peanuts (Sousa et al., 2011), amaranth leaves (for traditional sauce), Bambara groundnuts, fish, frogs, milk, and grasshoppers (Ramos-Elorduy, 2012). By culturally appropriate we mean are appropriate or culturally allowed children's food. In this cultural system, some foods are only for children (grasshoppers and locusts). Some foods are for men first and then women and children if there is enough left over (chicken meat). All of these are seasonally available in the fresh state in Sanambele, but with proper storage, many could be available year-round.

This completed the first step in the holistic process in Sanambele. The Sanambeleans defined their desired quality of life: after elimination of deaths from malaria, they wanted no more stunting or micronutrient deficiencies in their children. Then in a series of living and eating together, garden walks, informal discussions, the Sanambeleans defined for us the resources they had within their village to achieve this desired quality of life. They made it clear that our role at MSU was to do the library and laboratory research to figure out how those resources they had in their village, given their gender and age and other cultural constraints, could be used to eliminate stunting and micronutrient deficiencies.

The next step was taken by Carly (Grimm et al., 2014) (Figure 5.3). Two examples of how Carly's wordless, quantitative diagram described in Chapter 3, Decolonization and the Holistic Process, that led the Sanambeleans closer to their desired quality of life are the story of amaranth leaf sauce in the school lunch program and the story of the egg. In January 2016, Hawa announced to Ibrahima and me that there was now a school lunch program at the preschool through 9th grade school in the village. The menus had been developed by a nongovernment organization (NGO), PLAN Mali, and then prepared by mothers in the village. The meal plan seemed somewhat strange to Hawa. It included spaghetti, for example. Hawa wanted us to nutritionally analyze their school lunches. The nutritional contribution of the main grain base, toh, had already been analyzed by the students (Stein et al., 2012; Stone et al., 2015), so when Amy Bordeau, another Sustainable Foods major, joined the class, she took on the role of analyzing the sauces (Bordeau et al., 2016). Of the sauces chosen by PLAN Mali, the highest level of the missing lysine and tryptophan were found in amaranth leaves. We passed this information along to Hawa in another of the expanding series of quantitative wordless diagrams. During a summer Skype session with Hawa and Ibrahima, I learned that Hawa shared this diagram with the other women and they noticed that when the rains began in June (which signals the beginning of the malaria season), there were amaranth plants springing up throughout the village. Hawa shared the suggestion with the women that they should try to serve amaranth leaf sauce with millet toh at

FIGURE 5.3

Gathered in Dr. Dunkel's office at MSU, recent MSU Sustainable Foods graduate Carly Grimm (right) returns from her home in Wisconsin to meet with new Health Sciences graduate student, Jordan Richards (left) to explain how the revised lysine/tryptophan chart for village mothers works so that Jordan can carry on the participatory diagramming process with the women to make the connection between the malaria season, the seasonable availability of these crops, and the importance of these nutrients to strengthen the immune system to build the body's defenses against malaria.

least three times a week. Besides that, she shared with us that the children really like the amaranth leaf sauce! The intricate, nonlinear holistic process had worked. Current resources in the village plus new resources requested from outside (nutritional information in an understandable form from their partners at MSU) had brought the village a step closer to their desired quality of life, nearly a daily supply of the essential amino acids to reduce stunting and strengthen the adaptive immune system just as the malaria season was beginning.

The egg story is more complex. For several years, my students and I had been investigating the nutrients in eggs and how difficult it would be for Sanambelean women to raise laying hens, and what the cultural barriers there would be for rural mothers to introduce this odd product to their children. Eggs were just beginning to be a commodity available in the city but not in the countryside. Our research team discovered the absence of vitamin B12 and the lack of adequate calcium in the current diet of young children (Braget et al., 2016) and the quantitative opportunity to supply those nutrients in chicken eggs (Anderson et al., 2016) (Table 5.2). For these nutrients,

Table 5.2 Essential Amino Acid Needs for Children Ages 1–8 Years Old and Percentage of These Essential Amino Acids for One to Two Chicken Eggs (Anderson et al., 2016)

Amino Acid	Recommendations (1–3 Years Old) (mg/kg/day)	Recommendations (4–8 Years Old) (mg/kg/day)	Daily Amount Needed for 12 kg (26.4 lb) Child (mg)	One Medium Egg (mg)	%/Day Met for 12 kg (26.4 lb) 1- to 3-Year-Old Child	
					One Egg (%)	Two Eggs (%)
Tryptophan	6	5	72	73	101	203
Threonine	24	19	288	245	85	170
Isoleucine	22	18	264	295	112	223
Leucine	48	40	576	478	83	166
Lysine	45	37	540	401	72	149
Methionine	22	18	264	167	63	127
Phenylalanine	41	33	492	299	61	122
Tyrosine	41	33	492	220	45	89
Valine	28	23	336	378	113	225
Histidine	6	13	192	136	71	142

vitamin B12, and calcium, we are working on both eggs for children and crickets as feed for the laying hens that is part of the individual farm system of each mother in the village. The participatory diagrams for egg nutrients (Braget et al., 2016; Anderson et al., 2016) answered the question: How many do you need for a family of 12 so everyone, especially the children, can have two eggs per day, the minimum to prevent stunting.

We (villagers, research students, and I) found the most rapid way to return the micronutrients and essential amino acids lacking in their diets with food grown in the village was to provide information on seasonal, smart choices. We developed a way to provide that information in wordless, numberless quantifiable drawings (Grimm et al., 2014; Braget et al., 2016; Anderson et al., 2016). This year we began to see several important results. These were: (1) tripling of the frequency of using amaranth leaf sauce for their main grain dish (tou) at the beginning of the hunger/malaria season; first ever introduction of eggs (guinea fowl eggs followed by local laying hens) into children's diets; and exploration of cricket farming, *Acheta domesticus*, to provide high-quality chicken feed for women's hens and eventually to replace nutrients missing when children's grasshopper snacks had to be abandoned (2009) due to pesticides used in hopper hunting fields.

The significance of the crickets is that women do not have access to high-quality chicken feed and so have an impossible situation with raising laying hens. Men are able to leave the household more easily and so feed their chickens termites. Termites are not found in the nearby fields surrounding the village and so it is necessary to walk quite a distance (c. 2 km or more) to obtain his high-quality chicken feed. They collect the termites that cling to a stick put directly into the termite mound. Those chickens are sold for cash by men and this does not necessarily improve children's diets. Crickets, women villagers discovered with us this year, can easily be raised in their household. Not only that but eventually, kids will discover that these house crickets that their mothers and grandmothers raise can be roasted like they used to do grasshoppers and are similarly delicious (crickets are an adequate source of micronutrients, such as vitamin B12 and calcium, as well as the two essential amino acids, lysine and tryptophan, missing or in low supply in the diet of children and women).

To support this solution to food security in West Africa, we have now initiated the following laboratory research:

- Continue bioassays of growth, development, and nutrient content of food/feed crickets, *A. domesticus*, comparing those raised on cracked millet (current diet village women use) vs squash (also available in the village) and hand-milled squash seeds.
- Analyze omega-3 fatty acids and other essential polyunsaturated fatty acids at key life stages in crickets and on the human village diet combinations being evaluated.
- Further analyze omega-3 content of eggs produced by chickens fed the highest level of omega-3 crickets compared with appropriate controls.
- Continue in-depth interviews with villagers to determine seemingly minute dietary changes, but-in-reality-big nutrient changes; address suggestions with our unique system of wordless, numberless quantitative posters.
- Make quantitative measurements of stunting in collaboration with the Sanambele village Women's Association and the clinic staff.

This is opposite from the usual pattern of food and agricultural scientists developing a new method or new crop and then searching for a group of people who seem to need the specific research result that this result addresses. The following story is an example of "reverse technology," a flow of technology in the opposite direction of how colonizers think technology should be transferred.

THE GOURD STORY

After the Sanambele mothers announced that grasshopper hunting fields were off limits to the children because of the pesticides used on the cotton grown there, we thought immediately of suggesting our main commercial food insect in the USA and Canada, the house cricket, *A. domesticus*, that we knew also lives in the village homes in Sanambele. Whether the villagers used the crickets for a new children's snack to replace the grasshoppers or used the crickets for high-quality feed for the chickens didn't matter. It seemed either way the children would get those essential nutrients, amino acids (lysine and tryptophan), calcium, and vitamin B12, that were missing in their diet now that grasshoppers were not an option. The first thing my students and I thought of was how would the Sanambeleans be able to contain the crickets and then what would they feed the crickets. It is a subsistence farming system and so nothing is wasted. There was nothing we could see or think of that was left over from food preparation or eating, that didn't have some other animal depending on as a source for nourishment.

Five students over 5 years in my laboratory undertook extensive cricket growing projects to discover what the crickets could be fed and how to protect them from being eaten too soon by the chickens or just simply escaping (Fraser et al., 2013; Stein et al., 2012; Kashian et al., 2017; Stokhof de Jong and Dunkel., 2014; Lebel et al., 2012). Students tried all sorts of local materials for containers—mud or clay pots—sticks to climb into places to lay their eggs and cloth crocheted by the women to put over the top to prevent crickets from escaping. We experimented with crickets surviving on millet (a highly valued grain for human consumption in Mali), on corn husks, and orange peels, and many other local Sanambele materials. Once a week or so, my students and I would report our results to Ibrahima and Keriba who were now back continuing their work in Mali. Then Ibrahima and Keriba would translate the results from my students for Hawa.

One day, in a discussion with Hawa and Ibrahima, we learned that Hawa had meanwhile done her own experiments. She had used a callabasse, a big spherical gourd grown in the village and used for serving bowls and spoons[5] (Fig. 5.4C) turned upside down and hid underneath the callabasse grains of millet (Fig. 5.4B). Each day her house crickets would scurry under the callabasse inside

[5]The calabash is a vine grown for its fruit, which can either be harvested young and used as a vegetable or harvested mature, dried, and used as a bottle, utensil, or pipe (Fig. 5.4A–C). For this reason, one of the calabash subspecies is known as the bottle gourd. The fresh fruit has a light green smooth skin and a white flesh and was one of the first cultivated plants in the world, grown not for food but as a container. Hollowed out and dried calabashes are a typical utensil in households across West Africa (Fig. 5.4C). They are used to clean rice, carry water, and also just as a food container. Smaller sizes are used as bowls to drink palm-wine. Calabashes are used by some musicians in making the kora (a harp-lute), xalam (a lute), and the goje (a traditional fiddle). They also serve as resonators on the balafon (West African marimba). The calabash is also used in making the Shegureh (Women's Rattle) and Balangi (a Sierra Leonean type of balafon) musical instruments. Sometimes, large calabashes are simply hollowed, dried, and used as percussion instruments, especially by Fulani, Songhai, Gur-speaking, and Hausa people.

(A)   (B)

(C)

FIGURE 5.4

A large bottle gourd (callabasse): (A) in the edible phase, useful for cricket food; (B) dried callabasse after harvest cut in half to form a sphere similar to the callabasse used by Hawa Coulibaly in Sanambele to capture house crickets (used inverted, opening on ground) for feeding to her chickens and thereby significantly improving the growth and development of her free-ranging chickens; (C) homegrown kitchen utensils made from callabasse.[6]

her home to protect themselves from foraging chickens and eat the millet. At the end of the day, Hawa would lift up the gourd and let her chickens collect the crickets in her home. She continued this for about six weeks. The chickens grew well with the "new" food. Chickens there are free-ranging and I'm sure had occasionally found crickets when scavenging around inside the mud hut homes and

[6]Category:Gourds—Wikimedia Commons (WWW Document), 2017a. Commons.wikimedia.org. <https://commons.wikimedia.org/wiki/Category:Gourds#/media/File:Gourd-1.jpg> (accessed 03.06.17.).

in the courtyards so crickets were not a new food for the chickens of Sanambele. This way of deliberately collecting the crickets and then feeding them to the chickens, however, was new and apparently effective. A feast day came soon after the experiment ended and Hawa was happy to announce to the Montana students and me that she had had sufficient chickens for the feast day for everyone because of the crickets.

The importance of this story is that Hawa had been nudged by the students' efforts to find local housing for the potential food crickets. The success happened because of the creativity of both the students and Hawa and the partnership with Hawa who provided familiarity with local materials and willingness to experiment with them. We all had completely forgotten about the bottle gourds as a potential shelter for the crickets and other squash as food and a water source for the crickets. Hawa had solved a serious bottleneck in the nutrition of the children of Sanambele. It is important to note that this nudging and partnered creativity was going on only electronically, sharing photos, emails, posters and written reports via Ibrahima, and Skype. The effect on the chickens was clear and visible to the other women in the village. This research impact bridged an ocean and a cultural barrier.

CONCLUDING REFLECTIONS

In many respects, Sanambele became and continues to serve as a field school for the MSU and Northern Cheyenne students and faculty as well as for mid-career Malian scientists and educators.[7] Hawa Coulibaly and the Women's Association of Sanambele have twice been recognized by two presidents of MSU

---

[7]The Sanambeleans together with the Northern Cheyenne students at Chief Dull Knife College helped me understand the meaning of indigenousness. Meanwhile they shared advice for each other as Indigenous people and creating a professional film for others (Chaikin et al., 2010). Another example of how the Sanambeleans became a field school for us and for Malians is that Faculty at MSU and at the University of St. Thomas, St. Paul, MN, helped develop the Mali AgriBusiness Center in 2007. The Center, composed of the seven mid-career Malians who had received the fellowships in 2005, operated cooperatively, for several years, mainly supporting the work of teaching, being cultural interpreters, and holistic process liaisons within the villages, including Sanambele, where our two universities and other universities we collaborated with. Nine villages was another NGO nonprofit (www.ninevillages.org) that emerged in 2007 in a strong collaboration between a retired MSU finance professor, Dr. Dean Drenk, on our initial team and one of the seven fellowship awardees, Dr. Sidy Ba, an environmental engineering professor from IPR/IFRA, the agricultural university of Mali. This NGO used the model we had developed in our interactions with Sanambele and continues today based among rural Fulani ethnic group villages between Mopti and Timbuktu and Sidy continues to work toward starting a school for engineering based on the holistic process principles we all learned together, mainly in our interactions with the Village of Sanambele.

for their work in educating MSU students and faculty as well as their own community.[8] The humanities came together with the food and agricultural sciences in the form of a set of mentors and instructors, most of whom had never been to a formal school, who used best practices to provide an experiential learning environment that has positively changed many professors and Millennial Generation students' lives. Perhaps this community will, with other communities highlighted in this book, lead the way to a new global appreciation of other ways of knowing, of an appreciation of diversity, of cultural wealth, and of a way to incorporate cultures' role in the food and agricultural sciences.

References

Anderson, T., F. Dunkel, I. Traore, K. Coulibaly. Egg potential. Poster submitted in partial fulfillment of AGSC 465R, May 2016. Montana State University-Bozeman. 2016. Poster on website. < http://www.montana.edu/mali/pptsaspdfs/Anderson%20Taylor%20AGSC%20465R%20Poster%20Egg%20Potential.pdf > .

Aremu, N.O., Ogunlade, I., Olonesakin, A., 2007. Fatty acid and amino acid composition of protein concentrate from cashew nut grown in Nasarawa State, Nigeria.. Pakistan J. Nutr. 6, 419–423.

Ba, A.H., 1981. The Living Tradition. Vol. 1. General History of Africa: Methodology and African Prehistory. United Nations Educational, Scientific, and Cultural Organization, Paris.

Barikmo, I., Ouattara, F., Oshaug, A., 2007. Differences in micronutrients content found in cereals from various parts of Mali. J. Food Compos. Anal. 20 (8), 681–687.

Bordeau, A., Dunkel, F., Traore, I., Coulibaly, H., 2016. A first look into the school lunch program in Sanambele, Mali and its nutritional content. < http://www.montana.edu/mali/pptsaspdfs/BordeauAmyPosterSchoolLunchSanambele.pdf > .

Braget, D., Dunkel, F., Traore, I., Coulibaly, H., Finke, M., 2016. Micronutrient deficiencies related to the health of Sanambelean children. < http://www.montana.edu/mali/pptsaspdfs/Braget%20Danielle%20Poster%20Micronutrient%20Deficiencies.pdf > .

Beridogo, G., 2002. Le regime des castes et leur dynamique au Mali. Recherches Africaines: Annales de la Faculte des lettres, langues, arts et sciences humaines de Bamako.

Cavanaugh, J.C., 2017. Point of view: You talkin' to me? Chronicle Higher Educ. A48.

Chaikin, E., Dunkel, F., Littlebear, R., 2010. Dancing Across the Gap: A Journey of Discovery. Montana State University. 56 minute documentary aired Montana PBS, Northern Cheyenne TV. <http://www.montana.edu/mali/npvideos.html>.

DeCapua, J., 2012. FAO warns of locust threat in Mali and Niger. July 17, 2012, 9:24 AM. < https://www.voanews.com/a/mali-niger-locusts-17jul12/1418085.html > .

Dunkel, F., Giusti, A., 2012. French students collaborate with Malian villagers in their fight against malaria. In: Thomas, J. (Ed.), Etudiants sans Frontières (Students without

[8]In 2009 and 2010, Montana University's presidents (President Geoffrey Gamble in 2009 followed by President Waded Cruzado in 2010) awarded the MSU President's Service-Learning Community Engagement Award to Hawa Coulibaly and the Women's Association of Sanambele. The 2009 award recognized the work of the Sanambeleans in teaching Dr. Dunkel and her agricultural science and cell biology students and in 2010 the award recognized the Sanambele women's work in teaching Dr. Giusti and her French language and literature students.

Borders): Concepts and Models for Service-Learning in French. Amer. Assoc. of Teachers of French, pp. 135–150.

Dunkel, F., McCartney, H., 2005. East meets West: the Northern Cheyenne of eastern Montana and the Malians of West Africa. <http://www.montana.edu/mali/npvideos.html>.

Dunkel, F., Coulibaly, K., Montagne, C., Luong, K., Giusti, A., Coulibaly, H., et al., 2013. Sustainable integrated malaria management by villagers in collaboration with a transformed classroom Using the Holistic Process: Sanambele, Mali and Montana State University, USA. Am. Entomol. 59, 15–24.

Fejes, D., 2009. Kwahiorkor, cotton, and grasshoppers: how the use of pesticides in cotton production impacts the diet of children in Sanambele by contaminating a valuable source of protein. Research paper submitted in partial fulfillment of AGSC 465R. Montana State University-Bozeman, 24 pp.

Fraser, H., Dunkel, F., Coulibaly, K., 2013. Cricket farming to present kwashiorkor in Sanambele, Mali (West Africa). <http://www.montana.edu/mali/pptsaspdfs/FraserHannahPoster SanambeleInsectFarming.pdf>.

Gambill, K., 2011. The power of protein. Research Paper Presented in Partial Fulfillment of AGSC 465R. December 2011. Montana State University-Bozeman, 36 pp. <http://www. montana.edu/mali/kwashiorkor/pdf/GambillKelsiKwashiorkorProteinResearchPaper.pdf>.

Grimm, C., Dunkel, F., Ibrahima, I., Bengaly, S., Keriba Coulibaly l., 2014. Participatory diagramming to eliminate malnutrition in Sanambele, Mali.Research paper submitted in partial fulfillment of AGSC 465R December 2014. Montana State University-Bozeman. 24pp.

Howe, J.C., Coulibaly, K., 2013. The cashew, *Anacardum occidentale*: stopping kwashiorkor in its tracks in Sanambele, Mali. Research paper submitted in partial fulfillment of AGSC 465R. Montana State University-Bozeman.

International Cooperation and Development, 2016. Mali Nutrition Country Fiche—Child stunting trends. European Commission. <http://ec.europa.eu/europeaid/mali-nutrition-country-fiche-and-child-stunting-trends_en>.

Kante, A., Dunkel, F., Williams, A., Magro, S., Sissako, H., Camara, A., et al. 2009. Communicating agricultural and health-related information in low literacy communities: a case study of villagers served by the Bougoula commune in Mali. Proceedings of the Annual Meetings of the Association for International Agricultural and Extension Education. San Juan, Puerto Rico, May 24, 2016. Used as a case study by USAID Feed the Future. MEAS Case Study Series on Human Resource Development in Agricultural Extension, Case Study #11 January 2013.

Kashian, E., Dunkel, F., Coulibaly, H., 2017. Mass rearing *Acheta domesticus* in rural Mali. <http://www.montana.edu/mali/pptsaspdfs/KashianEmmaPoster.pdf>.

Lebel, J., Dunkel, F., Kone, Y., 2012. Ideal physical and environmental conditions for oviposition, embryogenesis, hatching of Malian food cricket, *Brachytrupes membranaceus* eggs. <http://www.montana.edu/mali/pptsaspdfs/LebelJacquiPosterMalianCrickets.pdf>.

Luong, K., Dunkel, F., Coulibaly, K., Beckage, N., 2012. Use of neem (*Azadirachta indica* A. Juss.) leaf slurry as a sustainable dry season management strategy to control the malaria vector *Anopheles gambiae* Giles s.s. (Diptera: Culicidae) in West African villages. J. Med. Ento. 49 (6), 1361–1369, http://dx.doi.org/10.1603/ME12075.

Richards, J., Dunkel, F., Traore, I., 2016. Research paper submitted in partial fulfillment of M.S. in Health Sciences at Montana State University.

Ramos-Elorduy, J., Pino Moreno, V., Camacho, M., 2012. Could grasshoppers be a nutritive meal? Food Nutr. Sci. 3, 164–175. Available from: http://dx.doi.org/10.4236/fns.2012.32025.

Richards, J., Dunkel, F., Traore, I., 2016. Research paper submitted in partial fulfillment of M.S. in Health Sciences at Montana State University.

Sousa, A., Canuto, D., Fernandes, A., Medeiros Alves, J., Borgesde Freitas, M., Veloso Naves. 2011. Nutritional quality and protein value of exotic almonds and nut from the Brazilian Savanna compared to peanut. Food Research International. 44(7):2319–2325.

Stone, M., Dunkel, F., Traore, I., 2015. Meal plans to aid in providing children with adequate levels of tryptophan and lysine. Research paper submitted in partial fulfillment of AGSC 465R. Montana State University-Bozeman.

Sousa, A., Canuto Fernandes, D., Medeiros Alves, A., Borgesde Freitas, J., Veloso Naves, M., 2011. Nutritional quality and protein value of exotic almonds and nut from the Brazilian Savanna compared to peanut. Food Res. Int. 44 (7), 2319–2325.

Stein, C., Dunkel, F., Kone, Y., Coulibaly, K., Jaronski, S., 2012. Potential approach to regulate and monitor moisture for *Brachytrupes membranaceus* eggs. < http://www.montana.edu/mali/pptsaspdfs/SteinrCarissaCricketsinSanambelePoster.pdf > .

Stokhof de Jong, S., Dunkel, F., 2014. Can house crickets, *Acheta domesticus*, be farmed in Sanambele, Mali? < http://www.montana.edu/mali/pptsaspdfs/StokhofdeJongSebastian Poster-CanCricketsBeFarmedInSanambele.pdf > .

Thurow, R., 2016. The First 1,000 Days: A Crucial Time for Mothers and Children—and the World. Public Affairs, New York, NY, 277 pp.

Turley, R., Dunkel, F., Coulibaly. K., 2011. Seasonal availability of lysine and tryptophan in a Sanambelean diet. Paper submitted in partial fulfillment of AGSC 465R, December 2014, Montana State University-Bozeman. 22 pages.

USDA. National Nutrient Database for standard reference, Release 28; Software v.2.3.8. The National Agricultural Library.

Category:Gourds—Wikimedia Commons (WWW Document), 2017b. Commons.wikimedia.org. <https://commons.wikimedia.org/wiki/Category:Gourds#/media/File:Calabash_bowls_-_2_ hand_processing_millet_flour,set_aside_dry_flour.jpg> (accessed 03.06.17.).

Category:Gourds—Wikimedia Commons (WWW Document), 2017c. Commons.wikimedia.org. <https://commons.wikimedia.org/wiki/Category:Gourds#/media/File:Calabash_bowls_and_ spoons.jpg> (accessed 03.06.17.).

CHAPTER 6

Listening With Native Americans

Florence V. Dunkel[1], Linwood Tallbull[2], Richard Littlebear[3], Tracie Small[4], Kurrie Small[4], Jason Baldes[5], and Meredith Tallbull[6]

[1]Department of Plant Sciences and Plant Pathology, Montana State University, Bozeman, MT, United States, [2]The Northern Cheyenne Nation, Busby, MT, United States, [3]Chief Dull Knife College, The Northern Cheyenne Nation, Lame Deer, MT, United States, [4]The Apsaalooke Nation, Baaxuuwaashe, MT, United States, [5]Wind River Native Advocacy Center, Washakie, WY, United States, [6]The Northern Cheyenne Nation, Lame Deer, MT, United States

CONTENTS

Phase 1: Awakening 132

Phase 2: Indigenous Teaching and Learning 136
Beginning a Relationship With the Apsaalooke. 141
"Let's Pick Berries" Project 143

Phase 3: Linked Courses, Shared Curricula, and Classrooms 146

Closing Reflection 155

References 156

Further Reading 157

Some of the best advice that I received from my Native American colleagues came from my student, Tracie Small, in a kind, compassionate, but direct way. Tracie is an enrolled member of the Apsaalooke. She was raised traditionally, on the reservation in a family replete with degrees in higher education. Tracie and her husband are also raising their four children traditionally. After a successful semester of research using the holistic process with her people, Tracie was standing at the whiteboard during a small group mentor session. She and her classmates were designing their poster to share in our university Share-the-Wealth Symposium. Tracie turned to me and said, "Here is what we Apsaalooke hold dear and how we preserve it." She moved her arm over the entire center of the poster where there were statements in concentric circles (the holistic diagram filled in). "Here is what we think of information from the University, from other students, from you and other professors." Tracie pointed to several small circles at the far edges of the poster. It was a moment of kindness, caring, and loving concern for her professor and the two non-Native students who were her site-mentorees for the semester. It was also a moment of stark truth:

Outsider's knowledge is respected and appreciated but has a low value compared to that of one's tribe.

In listening with Native Americans, this is the fundamental first step, to recognize, understand, and accept the low value of information that we, non-Native Americans, bring to the meeting. In this chapter, we take you to the nations of

Incorporating Cultures' Role in the Food and Agricultural Sciences. DOI: http://dx.doi.org/10.1016/B978-0-12-803955-7.00006-7

the Northern Cheyenne and the Apsaalooke. They live side-by-side on two reservations a "stone's throw" from my office and laboratory at Montana State University-Bozeman. I, Florence Dunkel, will be your guide as we one-by-one put together experiences over two decades to create a learning environment for students, faculty, and policy makers that engages us all in the food and agricultural sciences with health as the bottom line. The whole emerges with experiences intertwined like the ultra-thin strands of each leaf of sweet grass do in the fragrant braid used in purification rituals of these two nations.

PHASE 1: AWAKENING

I first encountered the Native Americans in Montana as a newly arrived natural product toxicologist/entomologist, a professor whose underlying principle in my research was to work only with plants or other materials that had survived the test of time in the local Indigenous culture. By the time I took the headship of Entomology at Montana State University, I had already spent a decade with cultures in South China and East Africa and knew that plants with effective antibiotic properties often held potential as pest management materials. After one of my first seminars, Robyn Klein, a Montana herbalist, introduced herself and strongly urged me to talk with an Elder of the Apsaalooke, an ethnobotanist, Alma Hogan Snell. Robyn introduced us. Alma made suggestions of Native Montana plants that I should consider for my research and I followed up on her suggestions (Weaver et al., 1995; Dunkel and Sears, 1998). The result was a fruitful back-and-forth and a deep appreciation for Alma's knowledge and world view (Snell, 2006, 2000). Alma came to my laboratory, used my microscope to observe the glandular trichomes that held the healing and bioactive compounds produced by these plants so revered by the Apsaalooke. Alma also gave standing-room-only talks in the MSU Student Union Building. Years went by with Alma and Robyn suggesting that I should to go to the reservation and listen. Finally I did. After getting lost a few times near St. Xavier, MT, I arrived at Alma and her husband's place, a cozy trailer in the plains near the mountains with no other homes around. During the visit, Alma's love and reverence for plants and the amazing health benefits she found in them overflowed. The breadth and depth of her own knowledge and the gratefulness of what she had learned about them from her grandmother, Pretty Shield was clear (Snell, 2000, 2006; Linderman, 2003). Just before we departed, Alma surprised me with a gift of a blanket. The signal was clear. We must go and listen. When invited, we must go to the reservation, to the homes of the Elders and listen. In the next decade, Alma passed and was posthumously awarded an honorary doctorate by Montana State University. The blanket remains as part of my course AGSC 465R either under the meal for the evening or as a gathering place on the floor for break-out groups. It is a weekly reminder during class that it is essential to visit the Native American Elders to understand their important message to us.

We must go to the land of the Native American Peoples and listen.

It was many years before I took my own advice again. The reservations seemed far away, but in actuality they were only a 3- or 4-hour car ride from my office at the University. In 2000 I began to intertwine my natural product research, primarily based now in West Africa, in Mali, with my teaching responsibilities and launched the Mali Extern program (Chapter 5: Listening With Subsistence Farmers in Mali). Working for me in my laboratory at that time was Michelle Madsen, an MSU student who grew up in Hardin, MT, just over the northern edge of the Apsaalooke reservation. She also worked as a peer counselor with the Montana Apprenticeship Program (MAP) that brought high school students to the MSU campus to engage in research in laboratories and build science skills as well as to demystify the University. Michelle suggested as did the Northern Cheyenne and Apsaalooke students who were working on projects in my laboratory that the Native American students from the reservations would also like to participate in this program in Mali. Michelle is the daughter of the late Bob Madsen, who was an entomologist and science instructor at the Northern Cheyenne tribal college, Chief Dull Knife College (CDKC). She introduced us to each other and we found much common ground, particularly in our interest in learning from other cultures. Bob was not Native American, but a much-appreciated instructor at the CDKC and he cared deeply about inspiring his students to find their "good life." When Bob heard about the Mali extern program that I was initiating at MSU and requesting USDA National Institute of Food and Agriculture (NIFA) funding to support, he suggested that might be just the thing for infusing a new way of learning science for his CDKC students. It was, and it grew in ways we could not predict.

With the USDA NIFA Higher Education Challenge grant in 2002, CDKC launched its first international study program, and one of the first tribal college international, service-learning courses in Montana, maybe also in the United States. CDKC is the smallest of the seven tribal colleges in Montana. This College serves mainly the Northern Cheyenne reservation, the smallest reservation in Montana. It was an unlikely place to take such a bold step. During the planning visit that Bob made with me to Mali so that I could introduce him to my colleagues and the villagers that I had listened with and worked with for 8 years, he shared with me that he had been a Peace Corps volunteer in Morocco. Now the unpredicted success of this project at this small tribal college made sense.

The key was a faculty member who had had a significant immersion experience.

Bob Madsen was a scientist with seemingly boundless energy. His students were already engaged with doing science: research on how honey bees detect pollution, and on water quality on the reservation; collaborating with scientists at the University of Montana and at MSU; and giving national presentations. The international dimension with other Indigenous people was new. Bob was a perfect guide. His earlier experiences in Morocco had prepared him to be a successful listener with the Northern Cheyenne. Now he was making use of his earlier Africa experience to provide a rich learning experience for his Northern Cheyenne students and, eventually, for the whole reservation community.

CDKC's first extern was Janelle Beartusk. She was a direct descendent, on her mother's side, of Chief Dull Knife himself (CDKC), who had courageously organized a break from the prison in Fort Robinson, Nebraska. With his small band of Northern Cheyenne families, he fled the prison and led their run back to their chosen land along the Powder River in what later became the State of Montana. Encouraged by her mother, Janelle traveled as far as France with her and then the rest of the way with Bob Madsen and an African colleague, a social scientist, who worked with the faculty and administration at CDKC. I met them in Mali. Janelle was the first of her people to go to Africa, we believe. None of us non-Native Americans were comprehending the enormity of what Janelle had done. President of CDKC, Dr. Richard Littlebear, later explained. Even stepping out of the reservation for a short time was equal to losing one's Northern Cheyenne-ness, one's cultural identity. Even Richard, fluent in the Cheyenne language, was taunted as being "not Cheyenne" when he returned to the reservation after his formal Western-culture education. Likewise, Janelle was risking her Cheyenne identity by making this journey across the ocean.

For a Native American to step out of their own reservation for any significant period of time, compromises their identity in the perception of fellow tribal members.

Thanks to the courage of Janelle to take that first step, the program succeeded, and a total of 11 students at CDKC became Mali externs during the series of three projects funded by the USDA NIFA Higher Education Challenge grant program. By 2005 the Mali extern program was thriving at both MSU-Bozeman and the CDKC campus as well as at other US university campuses. Thanks to Bob Madsen's foresight and enthusiasm to travel by car across Montana, Mali externs from the two campuses were able to engage in peer learning and teaching with each other. Bob frequently reminded me, though, that this was not enough and that I get out of my office and laboratory and *must go to the lands of the Native American peoples and listen*.

This chapter focuses on the land and the development of relationships with the Northern Cheyenne and the Apsaalooke, because that is where I have focused my time and energy. It is important that we pause for a moment to review the basics we hope to convey with these specifics. First, Native Americans, and most Indigenous people, have an intimate connection with the land. They feel themselves part of it and so to know them is to know their land and to be there. Second the richness of the culture and the wealth of knowledge about food and Native Science are learned by an outsider over many moments of coming together. Just the building of trust to have those moments of learning takes time. Hence, in one's lifetime, this level of sharing can happen with only a few tribes, truly—with only a few indigenous groups worldwide. Third, it is essential to recognize the unique nature of each tribe, their history, their path of change going forward, and their approach to food—its gathering, its production, its social ownership, and its role in celebrations.

In 2005 a grant parallel to the Mali extern program, but from the US Agency for International Development via Higher Education for Development (Dunkel, Montagne 2004) provided funding for six Malian mid-career scientists and an engineer to come to work on graduate degrees there. These were the only Malians in Bozeman, and likely the only Malians in the state of Montana at that time. That meant there were now site mentors for the Mali externs right in Bozeman so it was they who began to share responsibility with me for predeparture preparation of the externs. I noticed their communication with the Northern Cheyenne was rich, even though English skills of these Malians at this time was minimal. For full immersion of the Malians in the United States, I knew I must take them to the reservation. All these years I had never set foot in the Northern Cheyenne reservation. I thought I was too busy to travel so far. We departed for the reservation in October with a former Mali extern, Heather McCartney, who had also worked on the reservations. This was a lucky choice since she provided us with some tribal history and geography of the area as we made the 4-hour car trip together. To our surprise when we arrived, there was a room full of Cheyenne and a formal seating for the three Malians, Heather, and me. Soon there were formal welcomes by the former CDKC externs and Cheyenne dancers appeared in their regalia. Narrating was Linwood Tallbull, the ethnobotanist of the Northern Cheyenne. Drummers, singers, and dancers, with Linwood explaining to all of us their traditional societies within the tribe and the origin of the dances. Then all the visitors were invited to participate. Drumming and dancing are, I noticed, over the years, part of any celebratory gathering in Mali. Indeed, the Malians seemed to feel more "at home" with this celebration on the reservation than the few Euro-Americans in this group of about 40 people. Malians danced and drummed and all of us made short speeches and everyone got up out of their chairs and moved in a circle to the beat of the drum and the trills of the women singers. Dinner followed—a typical boiled meat and vegetables—prepared by one of the students and her mother. It was clear that the Northern Cheyenne were very proud of their heritage, their culture, their music, regalia, food, and societal organization. There was no question about this (http://www.montana.edu/mali/npvideos.html).

Our understanding of this traditional knowledge they shared with us and our appreciating the pride of these peoples were essential to our scientific collegial relationship.

What fascinated me most about the surprise powwow was that these three Malians whose English, at that time, was quite minimal, sensed a kinship with the Cheyenne that was not being communicated solely in words. I pondered this for several years. In 2009 with the third USDA NIFA Higher Education grant focused in part on capturing this story for a PBS audience, we set out for Mali. From the previous 11 externs, Janelle Beartusk, Stacy Bearcomesout, and Shari Ewing were chosen to tell this story. Josette Woodenlegs was selected as the videographer from the Northern Cheyenne.

PHASE 2: INDIGENOUS TEACHING AND LEARNING

The film, *Dancing Across the Gap* (Chaikin et al., 2010), touched us on a deeper level than even food and agricultural sciences and the related health issues. On-camera and particularly off-camera, there was an intense feeling of loss emerging in the Northern Cheyenne women when they realized these people in the village of Sanambele, Mali were culturally richer than they were though they seemed to be at an incomprehensibly lower level of economic poverty than that which the women saw around them all their lives on the reservation. The ability of everyone in the village to be fluent in their Native language (Bambara) was at the same time moving and sad when the Cheyenne women comprehended what they had lost during their forced assimilation by Euro-Americans (Weist, 2003). Likewise impressive to the Northern Cheyenne women was the daily hard work of each member of the family from the very young children of 5 or 6 years of age, and the energy and pride with which every member of the community carried out their role. While living in the village of Sanambele, Janelle, Stacy, Shari, and Josette were impressed with the happiness of the elders and parents surrounded by their children. This awareness is helpful in understanding responses to curriculum, pedagogy, and policy discussions back in the United States.

Outsiders should be aware of these fundamental feelings of loss and yearning and the origins of those feelings whether the topic of discussion is the food and agricultural sciences or any other topic.

During the making of the film, both in Lame Deer, Montana and living together in the village of Sanambele, Mali and when spending weeks interacting with the Northern Cheyenne on their reservation, I realized that this immersion was a rich experience worth sharing with my AGSC 465R students. Many times in the making of the film, I said to myself "I wish my AGSC 465R students were with me." These were ideal teaching moments. I often found myself role playing. "What would the students be learning now, if they were standing in my shoes?" Learning was taking place on many levels simultaneously.

Thanks to the Northern Cheyenne women, I was beginning to comprehend the importance of stories for a culture. Because the Cheyenne were with me, the topic of stories came up for the first time in discussions with the women in Sanambele. The Sanambelean women shared with the Cheyenne women how their children were having less and less time to learn the traditional stories of their culture. Storytelling was declining because women were more busy and now the village had a television. Although the television was locked in a cabinet whose key was kept by one of the elders, occasionally it would be brought out in the evening to play movies rather than have a storytelling session. Compassionately, the young Cheyenne women advised the Sanambelean women to be careful and not to let the stories be forgotten like their generation and their parents' generation had on the Northern Cheyenne reservation in Montana.

The next year, AGSC 465R students picked up on this conversation from the film. They brought supplies for the fifth and sixth graders to illustrate each part of one of the Sanambelean children's favorite stories, Masake ni Faritalenw (English translation of Bambara: The King and the Orphans). It is now a children's book about drought and appreciating diversity written in Bambara, French, and English, used in the public school system in Sanambele as a primer for children in fourth grade (national Malian rule is French must begin in fourth) and as a primer for seventh graders (national rule is English training must begin in seventh) (Dunkel et al., 2011). This experience reminds us of the importance of storytelling among the Northern Cheyenne as a communication pattern and its fundamental usefulness in communicating information in the food and agricultural sciences in other cultures. We are also *reminded of the special gift Indigenous peoples have in communicating with each other*, often overcoming language barriers.

Storytelling is an important tool in communicating information in the food and agricultural sciences.

Making the film marked the beginning of an entirely new level of interaction with the Northern Cheyenne. The film recognized basic feelings of loss following the forced assimilation by Euro-Americans. *Dancing Across the Gap* also provided a venue for the Cheyenne women to reach beyond these intense feelings of loss to help another Indigenous culture half way around the world hold firm to the parts of their culture that they treasure. Now the film was being shared nationally and AGSC 465R had a new community of focus, Lame Deer, Montana and CDKC. *Dancing Across the Gap* premiered at the Charging Horse Casino. This may not have seemed the appropriate place to air a USDA NIFA-funded documentary produced by CDKC and Montana State University and students from both institutions, but it was the largest place near Lame Deer (Capital of the Northern Cheyenne Nation) that could seat 250 people for dinner (the bingo room) on the reservation and could also show the film theater style. In the afternoon before the premier event began, a few of the CDKC staff held an impromptu miniworkshop on frybread. There we were, six non-Native Americans: three students in AGSC 465R; my husband, Robert Diggs; one of the Mali site mentors, Abdoulaye Camara who had been a guest at the "surprise powwow" in 2005, and had come all the way back from Mali for this event; and me. We huddled around the table in the teaching kitchen of CDKC watching Larissa Spang mix the flour, water, milk powder, and baking soda while she told us the story of how "Indian frybread" came to be. It is *not* a story of which we Euro-Americans can be proud.

The oil began to make bubbly sounds in the big pots on the stove as we each took a stout rolling pin and as we began the process of making the triangles to drop into the hot oil, Larissa began the story. In the late 1800s, Euro-Americans slaughtered almost all of the bison, the main source of food and shelter for the Plains Native Americans. The Native American tribes were herded onto specific, small parcels of land called reservations, sometimes just one tribe was put on a reservation, like the Apsaalooke. Sometimes two warring tribes, like the Eastern Shoshone and the Northern Arapahoe, were put together on the same reservation in hopes that they would annihilate each other. Other tribal enemies were put in side-by-side reservations like the Northern Cheyenne and the Apsaalooke in hopes that they would make war also. Without bison now or the big lands where they could roam and hunt, or the intricate trading system across the United States for food and other commodities, the Plains Native Americans were living in a food desert.

The US Government was and still is their trustee, so the Government was responsible for providing them with food. At first the government shared unused military commodities, usually large quantities of not very useful materials or rather deteriorated materials. Sometimes the flour was discarded and the cloth flour bags used to make clothing. Then the US federal commodity program was developed. This provided: flour, sugar, oil, baking soda, lard, corn meal, alcohol, and tobacco (Bass and Wakefield, 1974). Without berries and meat, women were hard pressed to develop other "comfort" foods.

"Frybread" was invented by these women as a make-do treat from the box of commodities that the US Government provided. The message we took away from that afternoon was: know the stories behind "traditional" food. It may not be a positive story. When frybread is put into a health context, it is easy to see how this contributed to the rise of Type 2 diabetes, and the resulting renal problems as well as other associated issues related to the human circulatory system.

Frybread was an invention of creative mothers and grandmothers during the difficult days adjusting to the reservation period. It is an embarrassing and grievous reminder that leaders in the food and health care system in the United States would provide commodities, at first unknowingly, that would cause a diabetes Type 2 epidemic on these reservations, which in turn, would lead, in part, to a 20-year difference in the mean life expectancy of women on reservations versus non-Native American women in the United States (Lee, 2016). Seldom, if any, powwows now are held without there being a booth that sells frybread, a grim reminder of how the US food policy on reservations led, in part, to a health crisis among this group of Native Americans.

Know the food history of each Native American tribe with whom you are listening and collaborating.

Learn the story of the origin of favorite foods and survival foods during the past history of your community of focus. For example, for us it was essential that we understood the reservation period, the nomadic phase, and the agricultural period of the Native Americans who later became the Plains Native Americans. Especially important is to hear these stories embellished with personal experiences from members of that specific tribe.

Later that spring, there was a premier in Bozeman, with the "tables turned." Elders from Lame Deer and two of the Cheyenne women from the film traveled to Bozeman. Elders gave the blessing before the meal. The Cheyenne women led a question and answer session with the audience following the film. An Apsaalooke dancer and an Eastern Shoshone singer and drummer, Jason Baldes, along with two African drumming/dancing companies from Bozeman, all non-Africans celebrated the Indigenous peoples communication documented in the film, *Dancing Across the Gap*. About 350 people from 1 to 90 years of age streamed into the Emerson Cultural Center, on Main Street in Bozeman, Montana.

In May of that year, 2010, the producer of the film and I traveled to Washington, DC, Waterfront Center, home of USDA NIFA, to share the film with our program officer, Gregory Smith, and his colleagues who had supported construction of the film, both financially and through their unwavering personal encouragement over many years. It was shown to two small groups, first the program officers and Center Directors, such as Hiram Larew. Then it was shown to visiting grant recipients from Native Alaskan and Hawaiian USDA NIFA projects. Tears flowed. We recognized without numbers, charts, tables of data that the cultural part of the food and agricultural sciences and the background of Indigenous people and what they hold dear, calls us to pay attention—to pay attention in our research, teaching, outreach, and policy making.

When non-Native American faculty ask what is the most important knowledge we need to know to make their mentoring of students from the tribal colleges in Montana successful, the answer from each Native American to whom I pose the question is about diversity. Understand there is no generic Native American. Understand the diversity of history and other parts of their cultures, including their varied food systems.

Learn that all Native American tribes are different, have a different history, and each has a history to appreciate. Listen to those from each tribe you work with tell you their own people's story.

This listening process, I noticed years later, changed my own behavior, especially the visit to the off-grid home of Sierra Alexander's family and the stalwart stance of nonexploitation of the Northern Cheyenne with respect to their coal reserves under the reservation. One story of "appreciative history" that I like to tell and retell is how this small tribe of Native Americans, women, men, and young children, who broke out of prison at Fort Robinson January 9, 1879 and literally ran back to the land they had chosen, all the while during the run, dodging the US military that pursued them on horseback with rifles (Nebraska State Historical Site, 2004). Decades later, these Northern Cheyenne find that the land that they had chosen, and that finally the US Government ceded to them in a treaty, the Northern Cheyenne Reservation, was actually sitting on top of a fortune of natural resources. Under the reservation was the highest quality coal. This was coal with the lowest sulfur content in the world situated in such a way near the earth's surface that it could easily be strip-mined. The land that constitutes the Northern Cheyenne reservation is held cooperatively and equally by each member of the tribe. Selling this fortune of nonrenewable natural resources would easily make a multimillionaire of each member of the tribe, woman, man, and child, if they voted to allow their reserves to be mined (Grossman and Thomas, 2005), but the Northern Cheyenne are standing steady amidst strong overt and subtle pressure to sell their mineral rights. They are truly a symbol of stewardship for other Native Americans and certainly for Euro-Americans.

Long-term relationships with other cultures often results in some permanent changes in one's own life style. I can see how this listening/watching process on Northern Cheyenne and other reservations has changed my own approach to energy, to obtaining food, and to be respectful in using drinking-quality water for anything but drinking water. We compost all kitchen waste, except meat, to fertilize our subsistence vegetable, fruit, and herb gardens. We use the wild chokecherries that have invaded our gardens for food and gifts. We save and gift seeds.

Beginning a Relationship With the Apsaalooke

In 2009 we had our first Apsaalooke students in AGSC 465R. Francesca Pine and Jade Three Irons grew up in different parts of the Apsaalooke reservation. Jade spoke Apsaalooke as a first language and Francesca did not. By the end of the semester when we had the Share-the-Wealth Symposium with a poster session, formal PowerPoint-illustrated speeches by the students, and then a home-made dinner for all who come to the Symposium, we knew the Apsaalooke had brought a new community of focus to our course. As we gathered in the conference room for the presentations there were 10 Apsaalooke in the audience. Friends and family had come from the reservation to support and encourage Francesca and Jade. The Three Irons family brought berry pudding and many of us had our first taste of this traditional treat.

Only in later semesters did we learn about the role of berry pudding in rites of passage within an Apsaalooke family. Children are celebrated in the traditional Apsaalooke family (as in many other Native American tribes). Berries are part of this recognition of the child, especially to celebrate the growth and accomplishment of toddlers. When the little baby takes its very first steps, berry pudding is cooked and fed to the child as a treat to celebrate their achievement. Berries are also important resource to use in sharing among the tribe and of honoring one's elders. In Alaska, the Tlingit have times when they come together as a tribe at the end of the summer just to compare the berry harvest that the individual families made. There the main berry is the salmon berry, *Rubus spectabilis*, one of the brambles in the rose family, native to the west coast of North America from west central California to Alaska. This practice is reminiscent of the epidectic flights of birds that nest in colonies. There are specific times during the year when the birds suddenly take flight as a group. The flight or swam is a way of signaling to the community of birds before the mating season begins if, for example, their population might outstrip the space for expansion of the bird colony on their rock outcropping. The Tlingit might use their "epidectic" gathering to observe the total berry harvest to gain some knowledge of climate fluctuations. This would be a cultural practice that leads to information gathering in a Native Science process. We will explore Native Science process in greater detail in Chapter 11, Couples Counseling: Native Science and Western Science.

In the summer of 2010, an entirely new level of interaction began with the Apsaalooke. Francesca had graduated and had taken a job on the reservation as greenhouse manager at the campus community garden at Little Big Horn Tribal Community College (LBHC). She asked for help from our class in generating ideas for reaching out to her community with gardening and healthy eating. Francesca wanted to see more vegetable and fruit items in the diets of her people. Her mission was to engage students in AGSC 465R with the Apsaalooke *while* providing ways for these non-Native students to help her achieve her goals of encouraging the Apsaalooke on the Tribal College campus and in the community of Crow Agency, Montana in eating *and* growing fresh fruits and vegetables.

That summer of 2010, Francesca and her supervisor, Birgit Graf, a German citizen and horticultural specialist, arrived at our home in Bozeman for a look at our method of producing all the vegetables we use in a year and a lot of the fruit and herbs we use each year. Birgit was the director of the agricultural program and Ag Experiment Station at LBHC. Before we walked very far into the meandering garden, we made a few calls to locate an Apsaalooke Elder Bill Yellowtail who lived in Bozeman. We invited him to come to lunch with us at our farm/garden and join in the conversation. Francesca helped me carefully harvest the spinach and strawberries for the salad for our lunch along with a few grasshoppers. In the kitchen, we had a great time talking about her new job and how my class was going and her vision for student involvement with her work in the campus garden with the community. Then, in the midst of our preparations, Francesca looked over at me as I put the hoppers in the melted butter in the skillet, and said, "I know I should like the beautiful spinach and strawberry salad, even with sautéed grasshoppers from your garden, but it is just not the food I grew up with. I would much rather have frybread, French fries, or a big Mac hamburger." She went on, "We struggle with obesity and diabetes on the Rez, but changing our eating habits which are now so comfortable, is really difficult."

For the next 2 years, AGSC 465R students worked with Francesca surveying perceptions and interests of clients at the Crow Mercantile (only grocery store in the capital of the Apsaalooke Nation). Other students conducted a nutritional analysis of recipes made from vegetables produced in the LBHC campus garden and tested the recipes with 14 Apsaalooke volunteers (ages 10−60) in the MSU classroom kitchens. When the grasshoppers, *Melanoplus differentialis*, became a nuisance in the campus garden, one of the students worked on developing a repellent spray in my MSU laboratory made from vegetables the grasshoppers avoided. We also developed an easy-to-harvest process for collecting the grasshoppers for food from this organic garden. In-depth interviews indicated the Apsaalooke were more interested in the small enterprise opportunities for the harvested grasshoppers than for a nutritious snack as their friends to the south, of the Eastern Shoshone (J. Baldes, personal communication, 2015). Another student worked with Francesca to create a management plan for the greenhouse with an integrated diagram to share with visitors all the ways the greenhouse could serve the community. Across the street from the campus garden and greenhouse is an elementary school. One of the AGSC 465R students conducted in-depth interviews with the school nurse, the school counselor, and some of the teachers to determine the level of interest in incorporating visits of the students to the garden and greenhouse or in actually growing plants.

The next fall, 2011, Francesca took a position at Lame Deer High School as the school counselor for students focused on transitioning into college. Several Apsaalooke students at the LBHC filled the void to manage the garden and greenhouse for the College. When AGSC 465R students who were Apsaalooke from Lodge Grass, Montana asked to do their community-based research project with their own people in Lodge Grass, it sounded like a good mutual learning collaboration both the Apsaalooke and the other students in AGSC 465R.

"Let's Pick Berries" Project

"Let's Pick Berries" is a good example of how a holistic process develops and evolves a new morphology, but in unpredictable ways. By definition, community action based on the holistic process is unique to that moment in time at that location with the specific people involved. It is always different and not predictable. The "Let's Pick Berries" was designed by the community itself to addresses Apsaalooke needs for improved nutrition, community engagement, and youth-Elder connections through the holistic process in a community-based, service-learning, action research course. The intertwining of health, nutrition, and food production research with traditional ecological knowledge, cultural practices, and economic opportunities from women-based cottage industry has the potential to be exemplified in the "Let's Pick Berries" project.

The project did not begin with berries. Not until the second semester of the involvement of the Small sisters, Kurrie and Tracie did traditional berries surface as a resource to achieve the missing parts of the desired quality of life of their community. When Kurrie graduated, the project continued with her sister, Tracie Small. Actually, Tracie told us the first day of class that she was there because her sister Kurrie said that she must take the course. Tracie was the youngest sister. "In an Apsaalooke family," Tracie explained, "when your older sister tells you to do something, you do it without questioning why." Now it was a team working. Kurrie had returned to the reservation with her family to live and Tracie was still living with her young family in Bozeman until she graduated.

The first big step happened when Kurrie and Tracie received approval to have a holistic discussion with the 36th Meridian Committee in Crow Agency, capital of the Apsaalooke Nation. This meeting of about 35—40 Elders was conducted in Apsaalooke, which worked since both Kurrie and Tracie had been raised traditionally and knew the language. When Kurrie and Tracie summarized the meeting, they said it was the berry finding, picking, and sharing that the Elders felt were an important resource that would improve their quality of life.

In the midst of the Elders' discussion, the berry picking of the Hutterites was brought up and the room became electric. As the Elders reminded each other that the Hutterites, a German-derived community that lived and farmed on the Apsaalooke reservation, were trespassing on their tribal berry patches and picking their berries, the mood in the room instantly changed. Not only were they picking the berries, but they were selling Apsaalooke berries back to the Apsaalooke. "This needed to stop. We, Apsaalooke should be picking our own berries and practicing all the good traditions that go along with the berry picking."

The next big step was the similar holistic process discussion that Kurrie undertook with fourth graders in Hardin, MT (all Apsaalooke students), and high school students at Lodge Grass High School. It was clear. From fourth graders to high schoolers, Apsaalooke young people echoed their Elders' decision that importance of going berry picking contributed or would contribute strongly to their desired quality of life.

The Small sisters began the revival of community berry picking in earnest, then, on several levels. First they conceived of a combination of a managed patch near Lodge Grass where families could easily walk to located in a place where there already were wild berries growing. They found a site, managed to lease the land from the Tribal Council. The next semester, Tracie tested the results of the holistic process with the high schoolers by offering a morning with an Elder to go out to pick berries. The response was enthusiastic. Native berries were a resource to address the desired quality of life of Apsaalooke Lodge Grass high schoolers (Small et al., 2013).

Soon Tracie had engaged most of the students in AGSC 465R with her ideas. Projects with the students were: greenhouse-based berry propagation projects at MSU (Setzer et al., 2014); construction of a website was constructed to provide a learning framework for Lodge Grass high schoolers to contribute to (www.apsaalookeberries.com); and a charrette conducted with 36 Apsaalooke giving an architecture student suggestions to design the family gathering place beside the berry patch in Lodge Grass (Killian et al., 2014). Tracie and Kurrie brought AGSC 465R students to the homes of their Elders to continue the conversation about berries: buffalo berries, June berries, chokecherries, and wild plum. A Geographic Information Systems (GIS) analysis of the chosen site was undertaken to find additional good berry sites. Set how climate change, a drier climate, would affect berry production on the reservation (Robison et al., 2014). Two MSU students took internships that brought them to return to the Lodge Grass berry site to transplant cuttings and to provide a GIS workshop. Both sisters continued after their MSU graduation to work on the project, and serve as site mentors for the research portion of the project with the AGSC 465R course at MSU-Bozeman. But, as can happen with projects of community activists, there simply was not enough person power to do what needed to be done, like hauling water from the river to the young tree transplants every day or planting the riverbank with plants to prevent the fast moving erosion. In 2015, at the suggestion of Tracie and Kurrie, we began a listening process with the agricultural education teacher at Lodge Grass High School, Ty Neal. Ty was not Native American, but he had grown up on a farm/ranch in the Lodge Grass area. He was also advisor for the Lodge Grass FFA (Future Farmers of America). The timing was good. The high school had just been given a section of land for the school farm. A berry patch was planned for the north end of the farm that received drainage from the nearby water tower. The only obstacle to establishing the school berry patch was the stand of 14 mature Russian olive trees that were making use of the water drainage. Russian olive trees are invasive along riparian areas on this reservation and no one questioned the need for the trees to be gone. Cutting the 14 massive trees was not directly about berries, but in the pure practice of the holistic process, whatever the community perceives to obstruct their desired quality of life, is what we focus our expertise on. Not all Apsaalooke now are brought up steeped in traditional ways. This space-clearing activity also served another purpose: youth sharing information about traditional uses of berries, and what it is like going to school at a large university off the reservation.

During their visit several weeks later to our university class in Bozeman, the Apsaalooke high school students shared a few of their berry traditions (Bagoly et al., 2016). For example, when a child is beginning to walk and has just taken their first solo steps, the mom or grandmother quickly brings out their frozen or dried chokecherries and cooks chokecherry pudding. This takes several hours to cook. Then the little one is given chokecherry pudding. It is a celebration with chokecherries at the heart of it. Traditional "pudding" is a not-artificially sweetened berry sauce made with water and a thickening agent like cornstarch, but no milk. AGSC 465R students shared surprising information about the nutritional content of the berries: one half cup of buffalo berries is equal in vitamin C content to the juice of two oranges (Loucy et al., 2016). This place-based fruit thrives in riparian areas, stabilizes stream banks, but does not require irrigation, this is a great alternative for fruit growers in Montana (Vaczy et al., 2016).

All in all the entire project resulted in a fair share of learning. Steady, sustainable indicators that both the community-based berry patch and the school patch will continue are tentative. Not all beginnings like this with the holistic process will thrive, but the chances are much better than if the project were a top down idea from the "outsiders" as Tracie described to us.

"Let's Pick Berries" was one of a number of research projects that emerged, welling up from indigenous communities as a result of using the holistic process. The result is a community-based rather than discipline-based research, teaching, and outreach program. Technical information is used when needed but the aim of the project is determined by the community and directly linked to their expressed quality of life. A word of caution is important to remember in projects using the holistic process. Over-commitment is not a culture-specific attribute. Be sensitive to particularly busy schedules of your collaborators. An outsider must wait patiently until schedules free up again. Sometimes the community leader will suggest an alternative. Sometimes a project will be redirected to produce resources in a different way. When founded on the holistic process, there is the continual use of the feedback loop and realignment in the path to the desired quality of life.

And so the course AGSC 465R then moved forward with a listening process on two reservations, located side-by-side in south central Montana. These two Native American nations were historically arch enemies, fought on opposite sides of the Battle of the Little Big Horn, and have a markedly different history of how they both came to settle in the Northern Great Plains area of the United States, the territory which later became part of the state of Montana.

PHASE 3: LINKED COURSES, SHARED CURRICULA, AND CLASSROOMS

In October 2013 Dr. Richard Littlebear, President of CDKC, was on the MSU campus for a few days to participate in a Montana Native American poetry reading. Sierra and I were in my kitchen rapidly teaching ourselves to make chokecherry pudding for class that day. After a series of phone calls, Sierra negotiated a visit of Dr. Littlebear to our classroom that same afternoon.

It was the day of the Northern Cheyenne role play. We use role plays in AGSC 465R to help us understand how a colleague or a group might respond to an issue. It helps us understand their viewpoint. On this day, students who had chosen the Northern Cheyenne reservation as their community of focus presented their role play. Stewardship of natural resources on the reservation was the issue. Students had studied and reflected on the people portrayed and how they might respond when certain issues were discussed. Today, the issue was: Why preserve the coal deposits under the reservation from not just exploitation, but even exploration. Quietly Dr. Littlebear watched the characters respond in an extemporaneous process. At the end, he simply commented that he would like frank and open discussions like this on his campus at CDKC. Others in the class suggested how this might work and a new level of listening resulted.

In Spring 2015 I received a surprise call from George Nightwalker, one of the few Northern Cheyenne instructors at CDKC. Ninety percent of the instructors at this Tribal College are non-Native American. George teaches: History of the Cheyenne People; Native American Art History; and several other courses. He wanted to know how I did this listening and action research. After describing the diversity of his students, in age, background, academic level, George challenged me with the ability of this action research and holistic method to engage and excite his students. I accepted the challenge. The linked courses began Fall 2015. We were catapulted into a new way of listening.

The rules George and I had decided on before physically bring the two classes together was that his students would decide on the research topic we would work on together and my students would follow through on the hypothesis testing. Sierra Alexander was our in-residence teaching assistant, and she announced that the students had chosen "dry meat." At first, I was puzzled about how this fit into the History of the Northern Cheyenne and could not imagine how we might develop a hypothesis and test it with this topic. By the end of the semester, the wisdom of this choice was clear and we at MSU all realized it had been a perfect choice.

Together, MSU students, teaching assistants, and instructors worked creatively to uncover the details. We explored and hypothesized how dry meat processing might be related to the history of Northern Cheyenne survival, particularly after the hunting and gathering phase in the forests around the Great Lakes of Northern Wisconsin and Minnesota followed by the sedentary agricultural period prior to the nomadic plains years. We investigated safety and nutritional factors.

At the beginning of the semester, Mr. Nightwalker gave us some of his venison dry meat to take back to Bozeman, and we experimented with making dry meat stew. This traditional knowledge was not a "look up" on the Internet or in peer refereed journals, so the research process involved interviewing several Elders at CDKC. Each time we made it, the Cheyenne said it was improved markedly. Toward the end of the semester, we had a final workshop on the CDKC campus. We served dry meat stew to several Elders including Dr. Richard Littlebear, President of CDKC. We learned what we had to work on to improve our stew preparations.

For the students, the informal one-on-one conversation with the Elders and students following dinner was a highpoint of the semester. Following dinner, we retired to a public kitchen where we were each given a special knife and a slab of raw venison meat. We learned the paper-thin cutting process from a professional dry meat processer, a young mother who was a student in our linked course at CDKC. We built a meat drying rack and had the dry meat on display beside the research posters during the Share-the-Wealth Symposium.

"Proof of the pudding" was by the end of the poster session where there were plenty of Western-culture snacks available, the dry meat display entirely disappeared into the mouths of the young Northern Cheyenne who had come with their parents to the Symposium. Not until the next semester when we sat down for long conversations with Linwood Tallbull did we truly understand the importance of dry meat and how it saved the life of the tribe many times in the past couple hundred years, including as they were pursued by the US Cavalry.

That following semester Mr. Nightwalker and I decided the learning on both of our linked courses was significant: history of the Northern Cheyenne People on his side; and health, poverty, agriculture on my side. We continued this process of linking courses and the result was even more surprising. "Entrails" was the challenge topic posed by the CDKC students to my students and me this semester.

Three MSU students, each majoring in Sustainable Foods and Bio-Energy Systems, accepted the challenge. Each student went in different directions. One dove again into the history, another into nutritional profiles of entrails, and the third MSU student studied the effect on the CDKC students' interest in traditional foods following our intense discovery process with them on one of their people's unusual food groups. CDKC students further challenged MSU students to create an actual collection of traditional Northern Cheyenne recipes. After completing this tough assignment, Cheyenne students shared with a bit of humor, "We Cheyenne don't really use cookbooks, if we want to know how to make a traditional dish, we just ask our mom, our auntie, or grandma."

Elder Linwood Tallbull, Ethnobotanist for the Northern Cheyenne, reminds us not only that food is health and health is food but also that health is actually a way of life, Linwood thrives on students eager to learn about Cheyenne ways. When students sought his advice with the challenge from their linked classmates at CDKC, he was their best ally during the "Edible Entrails Semester." Toward the end of the semester, students asked Linwood for a personal interview specifically on this topic. Linwood agreed. The students met Linwood at his family farm near Busby, Montana. When they arrived, he already had lunch ready for them, a hearty plate of cow's hooves. These are highlights of Linwood's comments from the conversation that ensued.

> We use different animals and different things for the gifts that they give, like antelope and their speed. Deer and elk, all of those, like a mule deer, women have to be very careful with them because it is believed that they can turn into a man. White tail, men have to really watch them because they turn into a beautiful woman. They have scent glands that will get you sick on them. Their scent glands are where [for] their little ones [so they] can find them.

My students already understood the commodities that became part of the reservation period, but Linwood's cooking details made this period of Northern Cheyenne history far more understandable for the students.

> Probably the most common food was the gravy, made out of flour. You just get a skillet, put a little oil in there or tallow, and then you pour flour in it. You keep mixing it up until the flour turns light brown. Then you just pour water in it. That is what most of the people lived on, just that gravy. If you were fortunate enough, you could break meat up and put it in there. Each person only got so many ounces of flour, so they had to make it stretch.

Linwood's personal history of the hardships, though, made historical food choices vivid.

> During that period of time, most of our people went to prison. Most of the young men went to prison for poaching, cause they were starving to death. We had 70 ranches on our reservation and around and there was no more deer, no more wildlife, nothing. So they got to be good poachers. They [*the poachers*] would see where the cows went into the pond (?) bush. [*Bushes grow around the drinking water sources, the ponds. When the cow was drinking water with its body in the bushes and the hind legs exposed,—easy to cut the ham strings when the cow was in that position.*] They would have riders looking who would rope this cow and then go in and hamstring it. They would cut its hamstrings in the back and that cow couldn't get up. Then they'd leave it. They would come back at night and hit it in the head with an ax and butcher it. They were pretty resourceful.

With Linwood's storytelling, the historical development of Om Sta Nah, Road Bread, became clear. Its origin was about as dramatic as the development of matzo for the Hebrews as they fled Egypt. Oppression, slavery, threatened slaughter of one's first born son were strong selection pressures to give up temporarily a traditional food, leavened bread. Now, 3000 years later, matzo is still eaten as a remembrance of oppression, escape, and survival.

> Many of them [*the Northern Cheyenne*] got sent to prison and they were just trying to survive. And most of them just lived on gravy and fried dough. They make a dough that just uses baking powder and flour and salt. You mix it with cold water and make a dough out of it. Then you just put a little bit of fat in your skillet and you make these things called Ogiman (?) Bread. It isn't a fry bread. You make it flat and you cook one side and wait for bubbles. It's almost like a tortilla. But they're called 'Om Sta Nah' (?). That means it's called 'Road Bread'. You made it real fast. Cooked it in a skillet like that. It makes a real fast bread when you're traveling.

Even meat preparations had a dramatic origin for the Northern Cheyenne during their nomadic period, often pursued by the US cavalry. It was with the following comment that finally made the historic significance of dry meat for the Northern Cheyenne became real to the students and to me, their instructor. In contrast to the Hebrews, Northern Cheyenne still use dry meat not only because of the remembrance of overcoming oppression that it invokes, but also, it seems, because it is a nutritionally and environmentally "smart food." Dry meat has no additives, not even salt, and it is entirely portable (light, occupies a small space, and requires no energy for long-term preservation).

> In their earliest times, they [*the Northern Cheyenne*] were pursued by an enemy, like the soldiers when they left Oklahoma. The soldiers were always right behind them. So the warriors would go up ahead and kill one of their skiddy [*easily spooked*] horses that they could afford to lose, and they would butcher it. The warriors would put that meat on their saddle and on their arm and they would go among the people and they would hand meat to them. The warriors would make it into dry meat. They'd cut it thin. Then they'd go up on big tufts of grass and they would throw that meat over that grass and they would take their rifle and that powder would ignite that grass and they would cook their meat. They got back on their horses and they would hand the meat out. All of the men would cut meat as they rode along. The flavored part was the black grass that was in it, like that blackened bark you get when you barbeque stuff. It's like the same thing. They'd flip the meat right over the buffalo grass, ignite it, put it over their arm, get another one.

Another Elder, Mina Seminole, the curator of history at the Cultural Center on the CDKC campus, reminds us often when we visit her that the Fort Robinson Run is the most important recent story of the Northern Cheyenne.

Knowing the story of this escape from the US military prison in Fort Robinson and the struggles survivors had once returning to the land in what later became Montana, helps us understand the origins of the present food systems in the Northern Cheyenne Reservation. Understanding the whole life system is essential to understanding the part related to food and agriculture. Most of Linwood's comments were from the nomadic period of the Northern Cheyenne and the beginning of the reservation period, not the sedentary, agricultural period about 300 years ago.

> The soldiers, *Linwood continued*, 'could never catch stuff [*wild game*] cause they [*warriors*] didn't have to stop for lunch breaks. See, there's no such thing as breakfast, lunch, or supper. It was always the same meal. Boiled buffalo ribs—that's what they had for breakfast. The thing about the Indian diet is that there wasn't a hell of a lot of greens. Most of their stuff came from organ meats—a lot of their vitamins and stuff. Bile, the intestines, things like that. The other day I was talking about afterbirth—the buttons on afterbirth. If you ever get to see one sometime look inside the afterbirth and you'll see these things that look like little donuts. That is pure veal. But nobody [*else*] eats it. We always ate it. You just fry them and if you've got two or three, for some reason a whole car would pull up with old folks. I don't know how in the hell they knew. We usually just got the afterbirth from cattle. We had a calving ground just over the hill and that's where we always use to go—about April [*each year*].

The CDKC students had asked the MSU students to work with them to jointly prepare a cookbook, *The Edible Entrails Cookbook*. As the students listened to Linwood discuss the different foods, they specifically asked Linwood about recipes.

> As far as recipes go, the only things they really cooked with were wild onions and wild garlic that we have here. Those are about the only spices they really had. The rest of their stuff came from marrow bones. I think a lot of vitamins came from marrow bones. Marrow bones and berries...They [*the Northern Cheyenne*] always had salt. Whenever we ran out of salt we would walk across the calving grounds and the ranchers would take their ax and cut a big chunk off of their salt blocks. We would take it home, pound it up real fine, and fill our salt shakers.

Needing information about entrails as food, the MSU students paid particular attention to the following comments from Linwood who seemed to like knowing that MSU students wanted to know more about the entrails and the story of the stomach and its individual sections in ruminant animals.

> You wash them out with a hose or [*at*] the creek. You really have to wash them well if you're going to dry them. Then you hang them where you have a

fire going. You put box elder leaves in there and get a fire going. Or chokecherry. It kind of coats it and gives it a different flavor. It also keeps the flies away from it. As soon as it was dry they would cut it up and put it in their bags. It dries really fast. It would last forever.

Other important culture-specific parts about understanding a food system is knowing food taboos, food hierarchies, behavior codes when eating, and cooking utensils available to a culture or community, as we described in Chapter 5, Listening With Subsistence Farmers in Mali, with the Bambara subsistence farmers in Mali. Part of understanding the food systems of Native Americans in understanding the extensive and far-reaching trading system they were part of before the Europeans arrived. The Cheyenne, Linwood reminded his students who had come to lunch, used to trade with Mexico. It is from the Mexican traders that clay pots and iron utensils were obtained for preparing meals. Although the Tribal College students' challenge was about entrails, Linwood took the opportunity to also share information on Northern Cheyenne dishes prepared from other parts of the food animal's body.

What nobody uses anymore is deer tongue. They use to take the deer tongue out and then slice it down the middle. They would keep folding it and slicing it like dry meat. The tongue was also dried. If a guy pledged a Sundance, he had to have at least 125 deer tongues (all dried). People will take the skin off of the deer tongue before drying it, but they use to dry it with the skin on. After you boiled the tongue, you took off the skin. All the organ meats were basically consumed raw. My uncle use to throw a whole liver into the campfire and the ashes. And then he would cut off the black part. Deer liver is really sweet. We ate porcupine, muskrat; the only one we didn't eat was weasel because they have a really strong scent gland. Basically they (muskrat, porcupine, etc.) were skinned and boiled in a soup pot. Beaver tail was roasted on a fire. They loved beaver tail because it has a lot of fat in it.

In their traditional nomadic diet, fat was a valued component of the diet of the nomadic Northern Cheyenne, and also in the early reservation period when it was a rare commodity.

Skunk used to be the greatest thing. We would harvest skunks in the fall, when they went into hibernation and were really fat. You had to have that fat—prosperity. If you had that fat in your lodge door, no sickness could come in (famine/sickness).

The MSU students, therefore, had a hearty diet, figuratively, of Northern Cheyenne nomadic diets and diets during warfare against the Euro-Americans. These MSU students, literally, had a steady diet of dry meat, simple, boiled dinners with venison, root crops, and wild berries. A few of the students were actually able to sample cow hoof stew served with generous helpings of Northern Cheyenne stories told by Elder and the tribe's Ethnobotanist, Linwood Tallbull.

Northern Cheyenne students reported in a student survey their interest in Northern Cheyenne traditional foods increased. At the end of the academic year, their instructor in this CDKC course linked to AGSC 465R, George Nightwalker, commented in a summative evaluation that "Cheyenne don't like to take credit for anything." He continued to once again remind us that "It is important to recognize the challenges the two student groups are facing. . .making common ground. . .is really important. . .work outside what they have been conditioned to know. . .need to walk in each other's shoes. It is a challenge what they are facing. . . .It is in their subconscious." Over the academic year, without lengthy discussions about pedagogy, or hours of collaborative lesson plans with me, the tribal college instructor engaged with MSU students in substantive way. These non-Native MSU students shared with me during mentor meetings as well as in their reflective journals that they had learned about the food system of the Northern Cheyenne in a way they are likely to never forget. I watched the MSU students listen intently and comment appreciatively on the reservation, and then I read and heard the results of their research after they returned from their reservation visits and dove into the literature.

After their times on the reservation, each of the MSU students seemed compelled to make sense of food choices from history of the tribe; to understand the nutritional content; and to find examples of other food systems that used similar foods. As a professor watching this take place, it seemed to me to be a good example of a learning process example where Native and Western Science mixed together in a way that seemed at first haphazard, but in the end facilitated deep learning. We were firmly warned at the beginning of the course linking by George Nightwalker, the CDKC instructor that the tribal college students would be "stand-off-ish." His warning came to pass. Communication was difficult for at least a couple of reasons. I was reminded of Tracie Small's advice to us, "Outsider information" is little valued. Cell phone service, if one has such a piece of technology, is nonexistent on much of the reservation. Internet is available on the CDKC campus, but likely not in most homes. E-mail is not a typical communication path for these students. Euro-Americans, these tribal college students learned as they grew up, Mr. Nightwalker reminded us, were going to be prejudiced against Native Americans. CDKC students and their own Northern Cheyenne people had strong reasons to be "stand-off-ish." Their ancestors faced annihilation at the hands of Euro-Americans. The action of watching the MSU students try to make historical, ecological, and nutritional sense of Northern Cheyenne food choices helped dissolve this cultural distance.

In addition of using food to dissolve diversity issues, Mr. Nightwalker and AGSC 465R students improved communication pathways by holding weekly informal discussions by speaker phone with Mr. Nightwalker each Thursday at 3:30 p.m., just between George's final class of the day and the start of our class. In addition to meeting with the students three times in person, in the beginning and middle on the reservation, and in Bozeman at the end of the semester, students passed surveys from AGSC 465R students and answers back from CDKC students via Mr. Nightwalker. George suggested in his evaluative comments that interacting on social media such as Facebook might be a better way than E-mail to "break the ice" between these two groups of students. All in all, he considered it a successful two semesters and so did I. We share these details with the readers who likely will have more ideas for using technology to bridge the gap.

In evaluating the entire year of linked courses, Richard Littlebear, whose idea it was initially, observed that it was interesting that now, the Millennial generation of Northern Cheyenne have to be paid with cash prizes to try these staple survival foods of their ancestors. During Native American Week 2016, one of the events held on the CDKC campus was a traditional food tasting. Cash prizes were offered for students who tasted each dish. The Elders, Richard observed, just sat down at the table and ate, never minding the tasting contest that was underway with the students. There are many reasons that we need to pay attention to survival foods. Humans all over the world are looking at a drier environment, higher water marks, sharing the land with more people, scarce potable water, growing food, and consuming locally. We may never know the real reason these food challenges were chosen by the students and instructor at CDKC, but we do know that it is useful knowledge, useful knowledge that may come in handy someday.

One important result of all of these student projects and collaborations is that a dialogue has begun between MSU students and faculty and tribal college students and faculty with a set of shared experiences. The newest challenge presented to faculty and students at MSU is from the Northern Cheyenne: How to inspiring high school students in "credit recovery," that is students who have failed courses which are jeopardizing their high school graduation. Reservation-based public schools in Montana usually have the lowest graduation rates in the state. Meredith Tallbull, a Lame Deer High School teacher, challenged MSU. "Can you at MSU,... you AGSC 465R students, teaching assistants, and MSU faculty create a 24-hour visit of inspiration, engagement, and Native American plant research for that would restart my high school students?...A visit that will jump start the excitement about learning that is hidden away inside these high school students?" It worked. Lame Deer High School students arrived, were warmly welcomed AGSC students with dinner and a film produced in part by Northern Cheyenne about plants solving complex problems. They stayed overnight with faculty and the next day talked, listened, and walked with scientists in Native plant research in Laboratories of MSU and at the MSU Horticulture Farm. Watching closely, one could see that *while* this inspiration of Northern Cheyenne young people is going on, MSU Plant Sciences faculty were becoming more inspired about Native Montana food plants and medicinal plants, and more specifically interested in engaging with more Native American students. Sierra Alexander, former MSU Indian Princess (2013–2014) and AGSC 465R TA and Northern Cheyenne site mentor for many semesters as this course linkage emerged and took place, shared her wisdom with us as she departed for her next life challenge:

"Be ready to learn in not the usual, formal learning places from people who are not the designated teachers and professors." — Sierra Alexander

For students in all of the seven tribal colleges in Montana, USDA NIFA Education Literacy Initiative has now funded an extern program for engaging tribal college students with their own community-based research. Support from MSU STEM faculty and our colleagues in Washington DC will, it is hoped, bring together the vast wisdom of traditional ecological knowledge and useful technology and skills of Western-culture scientists (Hunts et al., 2016). Thankfully from these projects described and others not mentioned, the new scientists involved with this grant can easily meet the advice of Fredericka Lefthand, Vice President of Little Big Horn College, to

Be sure that all the faculty know the basics before they begin interacting with the tribal colleges in Montana: how many tribes there are in MT, Which tribe is on what reservation, what their tribal names are and where the tribal colleges are. The bottom line of this set of details is to understand that there is no generalized Native American. They are all different and want to maintain their separate sovereignty and identity.

CLOSING REFLECTION

To learn about and from these different-from-Euro-American food systems requires leaps in old patterns of Euro-American formal learning. Why bother? This answer comes, in part, from USDA NIFA itself. In September 2016 at the International Congress of Entomology in Orlando, Florida, Sonny Ramaswamy, Director of USDA NIFA shared the answer with us in each of his two keynote addresses to separate assemblages of the 6000 attendees. NIFA's goal, he explained is to reduce the ecological footprint of food and agriculture by at least 50% in next 15–20 years. Food and agriculture currently play significant roles in several footprints, such as fresh water, greenhouse gas emissions, ammonia production, and nonrenewable energy resources. Being open to answers from other cultures may in the end save our lives. Listening is essential.

Broader than even the ecological footprint are the lessons learned through listening. It is these lessons that impact the quality of one's whole life and the personal sense of belonging, of integration, and of responsibility.

References

Bass, M., Wakefield, L., 1974. Nutrient intake and food patterns of Indians on Standing Rock Reservation. J. Am. Diet. Assoc. 64 (1), 36–41.

Chaikin, E., Dunkel, F., Littlebear, R., 2010. *Dancing Across the Gap*: a journey of discovery. Montana State University. 56 minute documentary aired Montana PBS November 2010; Northern Cheyenne TV November 2010. <http://www.montana.edu/mali/npvideos.html>.

Dunkel, F., Montagne, C., 2004. Linking biotechnology and bioengineering with Mali-based agribusiness: strengthening food and water quality for health, safety, and exports, 2004–2008. U.S. Agency for International Development via the Association Liaison Office for University Cooperation in Development (later: Higher Education for Development).

Dunkel, F., Sears, L., 1998. Fumigant properties of big mountain sagebrush, *Artemisia tridentata* Nutt. ssp. *vaseyana* (Rydb.) Beetle for stored grain insects. J. Stored Prod. Res. 34, 307–321.

Dunkel, F., Sullivan, M., Heneveld, T., Samake, C., Coulibaly, K., Giusti, A., et al., Le Roi et les Orphelins: Masake ni Faritalenw: The King and the Orphans. Ecko House Publishing, Sandy, UT, 28 pp.

Grossman, R., Thomas, L.B., 2005. Homeland: four portraits of native action. The Katahdin Foundation. 88 minutes. DVD ISBN: 1-59458-267-X.

Hunts, H., Dratz, E., Sands, D., Dunkel, F., 2016. Research and experiential learning for undergraduates in agriculture, food, and nutrition: from lab to table. USDA NIFA Undergraduate Research and Extension Experiential Learning Fellowships. Proposal Number: 2016-06429.

Killian, C., Small, T., Dunkel, F., Robison, G., Nyman, T., Setzer, D., Ostrovsky, Z., 2014. Let's pick berries: an Apsaalooke community-based project. <http://www.montana.edu/mali/pdfs/KillianCaleb>, 2014. ApsaalookeLetsPickBerries.pdf <http://www.montana.edu/mali/pdfs/KillianCalebApsaalookeLetsPickBerries.pdfs>.

Lee, T.H., 2016. Indigenous wisdom and academic learning converge at Native Nutrition Conference. <http://indiancountrymedianetwork.com/2016/10/06/indigenous-wisdom-and-academic-learning-converge-native-nutrition-conference-165990> (accessed 06.09.16).

Linderman, F.B., 2003. Pretty-Shield: Medicine Woman of the Crows. University of Nebraska Press, Lincoln, 148 pp. (original publ. date: 1932 by Harper Collins Publ., Inc.).

Nebraska State Historical Site, 2004. Cheyenne outbreak. Official Nebraska Government Website. <http://nebraskahistory.org/sites/fortrob/outbreak.htm> (accessed 19.11.16).

Snell, A.H., 2000. In: Mathews, B. (Ed.), Grandmother's Grandchild: My Crow Indian Life. University of Nebraska, Lincoln, N212 pp.

Snell, A.H., 2006. A Taste of Heritage: Crow Indian Recipes and Herbal Medicines. University of Nebraska Press, Lincoln, NB, 192 pp.

Vaczy, D., Dunkel, F., Scrantom, I., Small, K., Neal, T., 2016. Continuing "Let's Pick Berries" project: 'Outsiders' interest in native berries history/culture may spark the interest of lodge grass high school students. In: Paper and Poster Presented in Partial Fulfillment of AGSC 465R, Montana State University-Bozeman.

Weaver, D., Phillips, T., Dunkel, F., Nance, E., Grubb, R., 1995. Dried leaves from Rocky Mountain plants decrease infestation by stored product beetles. J. Chem. Ecol. 21, 127–142.

Weist, T., 2003. A History of the Cheyenne People. Council for Indian Education, Billings, MT, 227 pp.

Further Reading

Chief Dull Knife College, 2008. We, the Northern Cheyenne People: our land, our history, our culture. Red Bird Publishing, Inc., Bozeman, MT, 183 pp.

Little Big Horn College, No date listed. Absaroka: History of the Crow (Apsaalooke) People. Materials provided for course at Little Big Horn College. Acquired 2012. HR5-1.

Smith, H., Old Coyote, L.G.M., 1995. Flag and Emblem of the Apsaalooka Nation. Apsaalooka Heritage Series. Bethel Park Printing, Inc, Bethel Park, PA.

Straus, A., 1977. Northern Cheyenne ethno-psychology. Ethos 5, 326−357.

Posters (Archival) from the Apsaalooke *Let's Pick Berries* Project are available at http://www.montana.edu/mali/nppartnerscoursesMSU.html:

1. Summary poster presented at North American College Teachers of Agriculture (NACTA) annual meeting June 28, 2014, Bozeman, MT, by T. Small, F. Dunkel, G. Robison, T. Nyman, D. Setzer, C. Killian, Z. Ostrovsky. **"Let's Pick Berries":** Addressing Apsaalooke Needs for Nutrition, Plant Propagation, Community Engagement,and Youth-Elder Connections through the Holistic Processin a Community-Based, Service-Learning Course. http://www.montana.edu/mali/nppartnerscoursesMSU.html.
2. First poster in series, "Revitalizing berry picking in the Apsaalooke community: Preserving traditions and improving community health" by T. Small, C. Dauw, A. Berg, F. Dunkel. http://www.montana.edu/mali/pptsaspdfs/BergAndrewCrowTeamPoster.pdf.
3. "The cultural and nutritional significance of traditional berries to the Apsaalooke People: Using GIS as a tool to encourage community involvement and cross-cultural communication." By G. Robison, T. Small, F. Dunkel, C. Killian, T. Nyman, D. Setzer. <http://www.montana.edu/mali/pptsaspdfs/Ap%20-%20Robison%20Greta%20Poster%20-%20Cultural%20Nutritional%20Significance%20Traditional%20Berries%20Apsaalooke%20People.pdf>.
4. Architectural design of outdoor space for "Let's Pick Berries': An Apsaalooke community-based project" by T. Small, F. Dunkel, G. Robison, T. Nyman, D. Setzer, C. Killian, Z. Ostrovsky. http://www.montana.edu/mali/pdfs/KillianCaleb, 2014. ApsaalookeLetsPickBerries.pdf. http://www.montana.edu/mali/pdfs/KillianCalebApsaalookeLetsPickBerries.pdfs.
5. "Propagation of culturally significant berry plants for the Apsaalooke" by D. Setzer, T. Nyman, T. Small. <http://www.montana.edu/mali/pptsaspdfs/Ap%20-%20Setzer%20Nyman%20Propagation%20Significant%20Berry%20Plants%20Poster.pdf>.
6. "A guide to transplanting native berries: Preserving culture and learning new skills" by M. Logatto, J. Hoy, T. Small, F. Dunkel. http://www.montana.edu/mali/pptsaspdfs/HoyJeradWebSiteLetsPickBerriesPoster.pdf.
7. "Development of a website for the 'Let's Pick Berries' project" by J. Hoy, T. Small, F. Dunkel. http://www.montana.edu/mali/pptsaspdfs/HoyJeradWebSiteLetsPickBerriesPoster.pdf.

Northern Cheyenne

Babb, C., F. Dunkel, G. Nightwalker, L. Tallbull, 2016. Preserving Culture by Memorializing Traditional Foods: Edible Entrails Booklet. http://www.montana.edu/mali/pptsaspdfs/BabbCoryPosterEntrailsNCheyenne.pdf.

Duchin, J., Tallbull, L., Reusch, L., 2016. Interactive Children's Books to Preserve Culture, Knowledge, and Language. http://www.montana.edu/mali/pptsaspdfs/DuchinJackPosterInteractiveChildrensBooksToPreserveNCheyenneCulture.pdf.

Fletcher, K., Alexander, S., Dunkel, F., 2014. The Effects of Coal Mining on the Northern Cheyenne Quality of Life. Research paper submitted in partial fulfillment of AGSC 465R. Montana State University-Bozeman. http://www.montana.edu/mali/docsaspdfs/FletcherKatherineEffectsOfCoalMiningNorthernCheyenne.pdf.

Howard, S., Nightwalker, G., Tallbull, L., Dunkel, F., 2016. Preserving Traditional Food Knowledge in the Northern Cheyenne Community. http://www.montana.edu/mali/pptsaspdfs/HowardSydneyPosterPreservingKnowledgeOfTraditionalFoodsNCheyenne.pdf.

Murphy, K., Dunkel, F., Nightwalker, G., Tallbull, L., 2016. Edible Entrails: Preserving Traditional Foods in the Northern Cheyenne Community. http://www.montana.edu/mali/pptsaspdfs/MurphyKellyNCheyennePoster.pdf.

Apsaalooke

Bagoly, A., Dunkel, F., Small, K., Neal, T., 2016. Strengthening Youth-Elder Connections With Native Berries and the Aid of Film. http://www.montana.edu/mali/docsaspdfs/FletcherKatherineEffectsOfCoalMiningNorthernCheyenne.pdf.

Eltzroth, J., Dunkel, F., Baldes, J., 2015. Nutritional comparison of common traditional Apsaalooke foods versus common modern Apsaalooke foods. < http://www.montana.edu/mali/AGSC465RSpring2015Posters/AGSC465RPosterELTZROTHJULIA.pdf > .

Logatto, M., Hoy, J., Small, T., Dunkel, F., 2014. A Guide to Transplanting Woody Shrubs/Native Berries for the Apsaalooke Nation: Preserving Culture, Learning New Skills. http://www.montana.edu/mali/nppartnerscoursesMSU.html.

Loucy, M., Dunkel, F., Neal, T., Small, K., 2016. Inspiring the Youth: Apsaalooke Youth and Their Interest in Traditional Berry Nutrition. http://www.montana.edu/mali/pptsaspdfs/LoucyMaddieBerrysAsTraditionalNutritioPoster.pdf.

McAvoy, T., Neal, T., Dunkel, F., 2016. Rehabilitation of a Landscape With Native Berries on the Absaalooke (sic.) Reservation. http://www.montana.edu/mali/pptsaspdfs/McAvoyTannerNativeBerriesPoster.pdf.

McGill, M., Dunkel, F., Baldes, J., Small, T., 2015. Stream bank restoration on Lodge Grass Creek. < http://www.montana.edu/mali/AGSC465RSpring2015Posters/AGSC465RPosterMCGILLMEGAN.pdf > .

Myers, J., 2012. A Feasibility Study: Developing a *Melanoplus differentialis* (Orthoptera: Acrididae) Rearing Colony at the Little Big Horn College Greenhouse Project. Research paper submitted in partial fulfillment of AGSC 465R. Montana State University-Bozeman. http://www.montana.edu/mali/docsaspdfs/MyersJasonRearingAColonyOfGrassHoppersLittleBigHornCollegeGreenhouse.pdf.

Myers, J., F. Dunkel, F. Pine, 2012. The Community Garden at Little Big Horn College. <http://www.montana.edu/mali/studentsposterspdfs/MyersJasonLBHCCommunityGarden.pdf>.

Robison, G., Small, T., Killian, C., Nyman, T., Setzet, D., Dunkel, F., 2014. Spatial thinking, holistic action, and native berries. <http://www.montana.edu/mali/pptsaspdfs/Ap%20-%20Robison%20Greta%20Poster%20-%20Cultural%20Nutritional%20Significance%20Traditional%20Berries%20Apsaalooke%20People.pdf>.

Setzer, D., Nyman, T., Small, T., 2014. Propagation of Culturally Significant Berry Plants for the Apsaalooke. <http://www.montana.edu/mali/pptsaspdfs/Ap%20-%20Setzer%20Nyman%20Propogation%20Significant%20Berry%20Plants%20Poster.pdf>.

Small, T., Dauw, C., Berg, A., Dunkel, F., 2013. Revitalizing Berry Picking in the Apsaaloooke Community: Preserving Traditions and Improving Community Health. http://www.montana.edu/mali/pptsaspdfs/BergAndrewCrowTeamPoster.pdf.

Vaczy, D., Dunkel, F., Scrantom, I., Small K., Neal, T., 2016. Continuing "Let's Pick Berries" Project: 'Outsiders' Interest in Native Berries History/Culture May Spark the Interest of Lodge Grass High School students. http://www.montana.edu/mali/pptsaspdfs/VaczyDeaApsaalookeLetsPickBerriesPoster.pdf.

CHAPTER 7

Listening Within a Bioregion

Clifford Montagne[1], Badamgarav Dovchin[2], and Florence V. Dunkel[3]

[1]Department of Land Resources and Environmental Sciences (retired), Montana State University, Bozeman, MT, United States, [2]Renchinlhumbe, Darhad Valley, Mongolia, [3]Department of Plant Sciences and Plant Pathology, Montana State University, Bozeman, MT, United States

CONTENTS

Phase 1: Formation and Evolution of BioRegions 161
Initial Visits to Mongolia 161
Learning Through Crisis 164

Phase 2: BioRegions Program Matures to Include Annual Work Visits and Fund Raising 168
Education 168
Environment 169
Health 171
Boiling Duration and Water Quality for Tea 172
The Arts and Traditional Knowledge 172
Festival of the Darhad Blue Valley 173
Whole Community and Business 174

Phase 3: Deepened Listening Exchanges 176
Reflections 176
Mongolian and Native American Students Working Together 177

In this chapter we will take you to Mongolia. Your guide will be Clifford Montagne who will introduce you to a diverse cast of characters in two geographically similar BioRegions but culturally vastly different. These characters come together in the story in such a way over two decades as to create a teaching and learning environment that engages participants in the food and agricultural sciences, together with health. This whole becomes thoroughly tangled and pressed together with culture like a piece of felt.

Imagine yourself suddenly entering onto a lush green plain nestled between two blue and gray mountain ranges under a brilliant blue sky, cloudless in stillness. Seemingly out of nowhere two horseback riders come together on this vast grassland. Immediately they dismount and squat on their heels, using their boots as a short stool. Facing each other, close to the ground, Mongolian style, they share the latest news. Again in Mongolia's remote Darhad Valley for the 18th year, I, Clifford Montagne, a professor of soil science, sit on the grass in a similar way with a group of American students and Darhad herder families to discuss how the growing season is going, how the herds are doing, and how we can reduce overgrazing. We sit in a circle. No one is "boss," but conversations are animated, frequently interrupted Mongolian style with several conversations building on top of each other. We share tea and cheese along with fresh butter and sourdough bread. We all feel "at home and heard" by our colleagues, who are also becoming our friends.

In the martial art of Aikido, people are participants or players, not competitors. The most skillful players start by giving way, by giving something up, before returning to the opponent (or player) with even more power and advantage. In that circle on the grassy plain in Mongolia, success will blossom if the outsiders (the BioRegions team of students, faculty, and visiting community members, in this case) give up the first action, like Aikido players. We give up the

Incorporating Cultures' Role in the Food and Agricultural Sciences. DOI: http://dx.doi.org/10.1016/B978-0-12-803955-7.00007-9

Including Health in Food and Agriculture *179*
The Yellowstone Connection *181*

Outsiders and the Importance of Listening 183

Conclusion 184

References 184

Further Reading 184

"first word." It is a new situation, we must first listen. The art and the act of listening often softens sharp edges of separation between viewpoints. It may dissolve "lines in the sand" and lead to back-and-forth dialogue. Then, such "players" can find synergies way beyond the "sum of the parts." This is the art of interplay between people which often starts with a give-and-take conversation. Those most skilled at this art know when to listen and when to talk. Physical settings, the landscape, can promote or discourage conversation, especially in group situations. Simply how people are arranged when talking makes a big difference in conversation synergy. Are participants seated or standing? Are they formed in a circle on chairs with tables in front of them? Are participants seated on the ground with no chairs or tables? The power and importance of the circle in effective learning is closely related to concepts of Native Science, Traditional Ecological Knowledge, and Indigenous Research Methodology. The circle is also an important concept in the newly emerging Western thought pattern called systems thinking and a new mode of research called systems research (Meadows, 2008). Interesting that this parallels ancient ways. These approaches, circle seating and systems thinking, as described in Chapter 3, Decolonization and the Holistic Process, all help us create the critically needed pathways to connect culture, which includes connection with the Natural World (including humans and their communities), with food and agriculture.

These approaches that we collectively call the holistic process were instrumental in each of the three phases in the story that Cliff reveals in this chapter. In each phase of the story, there were mistakes made that Cliff and his team learned from. Cliff has highlighted these in the text and labeled these learning-from-failure moments as "RED FLAG WARNINGs to learn from."

In 1990 Montana State University (MSU) adopted the semester system. I used this time of change to convert a course titled Soil Conservation into a course titled Holistic Thought and Management. The course was, and remains to this day, highly experiential with minimal lecturing and a great deal of participation involving close listening and sharing of differing viewpoints dealing with natural and human resource management. The instructors act as both content-matter practitioners and facilitators, rather than lecturers. The rhythm of discussion may start with a mini-content sharing by an instructor, but small student task-groups process ideas in their own discussions before sharing with the entire class. In this way, students continually go back and forth from individual or small group scales to that of the larger circle, which contributes a broader diversity of background, viewpoints, and expertise.

We start many of these classes by taking time to rearrange the theater-style chair arrangement (lines of chairs, each line behind the other) into a circle. This takes extra time and energy, but also allows for everyone to participate in creating the new arrangement. There we are on the first day of class,

40–60 strangers all equally positioned, everyone visible to every other person, no single person in charge. We give everyone an equal chance to introduce themselves, often starting by suggesting they first talk with a nearest neighbor and then share something about their neighbor with the entire group first, providing abundant "airtime" for everyone. [*Phase 1 of Cliff's story begins with just that, the holistic process and a new course offering at MSU.*]

PHASE 1: FORMATION AND EVOLUTION OF BIOREGIONS[1]

Initial Visits to Mongolia

We had navigated the teaching of the concepts of Holistic Management on campus, but now we wanted to connect the concepts to real issues of food security, health, environment, and culture. BioRegions began with a 1996 visit to MSU by Kent Madin of Boojum Expeditions. Kent and his wife Linda Svendsen both attended Prescott College in Arizona and worked to promote outdoor experiential education before moving to Bozeman, Montana, and starting an outdoor adventure company, Boojum Expeditions (http://www.boojum.com/history.html), based in Bozeman. He gave a presentation to the Mountain Research Center about travel with his company in northern Mongolia. I sat in the midst of the other local mountain enthusiasts that evening and trembled with excitement over his photos of vast grasslands and mountains peopled with nomads living in gers (yurts) and traveling with their herds. The ger is a Mongolian term for a circular, collapsible wooden frame covered with thick felt, a tent used as portable homes for nomadic herder families. The whole structure can be carried by one camel. Yurt is the Russian language equivalent.

After the Berlin Wall came down and the Soviet Block crumbled, so did many of the pro-Russian regimes associated with the USSR but not a part of the USSR. Mongolia was one of those autonomous regions. Their "revolution" was largely peaceful. One of the activists, Batchuluun, was invited by the US State Department to tour the United States in the early 1990s. Coming through Bozeman, probably because of the proximity to Yellowstone National Park, Batchuluun met Kent, who promptly took him on a ski trip. Consequently they formed a partnership to take international tourists to Mongolia's remote lands for horseback expeditions. Kent had come to MSU that evening to recruit and stimulate interest from faculty in Mongolia because he knew Mongolia needed more than just more tourists in the struggle to reform its economy and government.

---

[1]BioRegions refers to the BioRegions Programs, consisting of (1) MSU BioRegions Program (an affiliation of faculty and students) and (2) BioRegions International, a nonprofit 501 c 3 organization.

In 1996 I found myself in Ulaanbaatar on a Boojum Expeditions tourist trip to the remote Darhad Valley. After many days of driving rough and seemingly nonexistent roads through grassy valleys and up and over mountain passes in the back of an open former Russian Army truck, I found myself camped in the middle of the Darhad Valley. There in 1996 I witnessed one of the first free national elections. I was transfixed as Darhad residents rode into the voting place, a small cabin, by horse, some having come 50 km in their traditional deel clothing, women done up in finery, men in bright sashes, tall black leather boots and a fantastic variety of hats ranging from cowboy to dapper Eastern European. The voting ended in a rousing volley ball game, and I saw that beneath the traditional robe clothing (deel), almost everyone had an Adidas, NIKE, Inc., or other brand warm-up suit. I was feeling more and more "at home."

Kent had asked Batchuluun, really a city guy, to find a partner in a remote place as far away from city life as possible. On the trip he kept talking about this partner whose name is Mishig. And there came Mishig on his motorcycle to join our expedition at the voting cabin. A short and very outgoing man, trained as a sheep veterinarian in Mongolia and Kazakistan, and a native of the Darhad Valley and former local administrator (*Daraq* or Governor in Mongolian), Mishig would prove to be pivotal to BioRegion's success. I enjoyed the trip and was captivated by the environment and the culture, which was just emerging in a more free way after 70 years of socialism. On the way up the pass to leave the Darhad Valley, I sat alongside Mishig, now driving the Russian Army truck along with a reporter from the Minneapolis Star Tribune. Mishig told both of us "You will never understand the Darhad way of life until you are here in the winter."

Cliff is well known in the Rocky Mountain area as an adventurer, particularly with winter sports. Yes, he did follow through on Mishig's advice by immersing himself in the real Darhad way of life by participating in a late autumn migration. This is documented in a National Geographic article (Hodges, 2003). When fall comes to the Darhad Valley, hundreds of families load up their oxen and move with their sheep, goats, and cattle to winter camps which are out of the cold air inversion of the valley floor and the grass is long enough to get their herds through until spring. A wall of 10,000 foot peaks stands between the 1300-square-mile Darhad Valley and the winter camps. Surrounded by mountains, the Darhad Valley has severely cold winters. The small amounts of snowfall persist overwinter but don't provide sufficient insulation against the severely cold winters. The summers are never warm enough to thaw the soil so permafrost limits the growing season. Families living in the valley can predict the carrying capacity of the land and know they must migrate south for the winter to avoid over-exhaustion of the land. The Mongolian government has tried to encourage permanent grazing, but seminomadic way of life is more suitable for this type of environment. These herding families are seminomadic.

They tend to migrate in similar patterns back to the same camps year after year unless weather conditions force them to adjust. The fall migration occurs in two directions, to the east over the high mountains into the Lake Hovsgul area and to the west to higher elevations which are above the cold air inversion and have more available forage and wood. The severely cold winters with relatively thin snowpack make the soils susceptible to freezing. Cool summers retard melting, and the cold soils and short growing season limits vegetation (and largely eliminates potential to grow crops).

Afterward, back at MSU I tried to discover a way to return to Mongolia and start the dream of applying lessons from Holistic Management to improve the environment as well as the lives of the people whose lives and economies had been interrupted by the change from socialism and a top-down economy to democracy and the free market economy. I wanted to protect the environment and help the people by telling them *what they should do.* **RED FLAG WARNING #1 to learn from: Listening and learning first is imperative. Going in with an intention, however good, should never preclude learning from a group first. Learn first, teach afterward.**

In 1998 Tim Swanson and I co-taught an MSU Honors course titled The World Citizen in Sustainable Communities: Exploring the Geographies of Hope. We attracted a number of geography majors among others. I had been taking students to Japan to learn about environment and culture. The MSU Office of International Programs said "Why not take students to Mongolia?" That concept, frankly, was beyond my paradigm at the time. But I thought "Why not?" We decided to go to Mongolia that summer. In March 1998, Boojum brought Mishig to Bozeman as part of their program to acquaint their Mongolian employees with realities of American life. Mishig needed a translator, so a young Mongolian woman with impeccable English, Teki Tsagaan, came too. Kent and I had met Teki after our trip in 1996 as we searched around Ulaanbaatar for local contacts.

Thus, a series of opportunities presented themselves to Cliff. Along with these opportunities was his quiet, but intense curiosity of how other cultures managed their food production and the land resources they used to produce the food. The vision of setting out to discover this food and agriculture system mixed with this far different culture than Euro-Americans farming and ranching in similar landscapes in Montana put Cliff on a path least traveled.

My wife Joan Montagne has been a participant in many of my professional activities, for I have always felt the need to be doing more things together than apart, and I have time and again benefitted from her counsel and enthusiasm. In May 1998, right after a meeting of the Greater Yellowstone Coalition, which was the mother of the BioRegions idea for me, Joan and I left for Mongolia with 13 students. Hearing that the Darhad people needed help with bridges across their rivers, the students had already staged a fund-raising concert in Bozeman and raised $700 for a "bridge fund." At the Ulaanbaatar airport we were met by Teki's smiling face, reassured by her words of welcome in "our language," and began the first of many field trips to the Darhad Valley. I hoped that we would have impactful learning but had no idea how we could have a positive influence with the Darhad people beyond the small amount of additional seed money income they might receive from us or the curiosity of seeing people from the country they had recently been told were their enemies (at the governmental level).

And so a multidecade relationship began intertwining culture with the food and agricultural sciences in the Darhad Valley of Mongolia.

The Darhad Valley is a big landscape. The Darhad people are a subethnic group within Mongolian culture and have adapted to the isolation and severe winter climate much like other landscapes and cultures in Mongolia. Globalization processes affect them as well with outside control of the prices of cashmere wool prices, loss of ability to market meat and hides to Russia, rising commodity prices, etc. Coming out of the Socialist Period, they found themselves for the most part without jobs, outside income, and services they had come to expect. So they have had to quickly adapt. The valley has abundant forage for livestock along with a low background level of wildlife including argali sheep, ibex, brown bear, musk deer, elk, wolves, and wolverines. The valley is also known for its wetlands and migratory waterfowl (Nishida and Jamsran, 2009). The estimated 8000 people living in the four Darhad soums (equivalent to counties) are primarily meat and dairy herders. There are very small family gardens and most herders cut, dry, and stack some hay by hand to tide their animals through the challenging early spring birthing season. In the early 2000s, gold mining by hand grew out of control until the local people petitioned the Mongolian government to establish preserves and eliminate the mining. Now there are three National Parks and Special Protected Areas in the valley and local people in the "buffer zones" are pushing back against new regulations designed to stop mining and control hunting and forest products harvest. There is confusion and uncertainty with tension between enforcement and exploitation.

Learning Through Crisis

Much of our learning happened through mistakes or crisis. One student became very sick with stomach and influenza-like symptoms. We happened across a local physician's assistant who instructed us to give her salty tea, the traditional Mongolian drink (usually with milk). This really helped, probably because our student was hot and dehydrated. She was stabilized and better, but in no condition to travel. Our entire group was about to take a 3-day pack trip into the mountains to visit the Dukha people, who come from

Tuva (in adjoining Russia). The Duka people represent the furthest south of the reindeer herding cultures. I was distraught about what to do, but Mishig and Teki prevailed with "Just leave her here with a local family and she can rest and get well." We made some vocabulary and conversation cards for the student and the family and guiltily left our sick student, only to come back 3 days later to hear her exclaim "This was the best experience of my life." Her sickness, and her willingness to stay with a family, had opened the door to mutual understanding and sharing. Meanwhile the rest of our group rode horses with pack camels overnight to the Dukha camp. Nearing the camp, we sent Mishig ahead to ask the elders for permission to visit. They agreed and we joined their camp. I spent a long time speaking with the leaders about strategies for them to use the Holistic Management process to improve their lives, which seemed bleak to us because of monetary poverty, alcoholism, and minority discrimination. Looking back on this it seemed futile to engage the elders in rosy Holistic Management talk with the first day of meeting them, and then to leave, with no solid obligation for follow-through. **RED FLAG WARNING #2 to learn from: Let the conversation flow naturally. The holistic process has specific unfamiliar language. As in Akido, let the others, in this case, the Elders, set the topic, especially in the first meeting. The holistic process can begin at almost any point in the process**. Now, 16 years later, in 2016, Mishig again wants BioRegions to focus on how to help these people with issues of both reindeer and human inbreeding due to the regionally closed border with Russia.

BioRegions now has much more experience in facilitating change in rural communities, change, that is sustainable rather than top down and short term. Although the holistic process is based on formal, linear logic, in actual practice, BioRegions participants have learned the process is informal, and nonlinear. BioRegions, as well as the Mali Extern Program (Chapter 5: Listening With Subsistence Farmers in Mali), and AGSC 465R (Chapter 5: Listening With Subsistence Farmers in Mali and Chapter 6: Listening With Native Americans) have all independently learned that the holistic process is not a lockstep process, but a set of categories of action rather than steps followed in a sequence (Dunkel et al., 2013).

Setting up an immersion process that is smooth and detailed, yet allows for the spontaneity of interactions with an Indigenous community is an important skill in incorporating culture into the food and agricultural sciences. Kent Madin, his wife Linda Svendsen, and the Boojum staff helped greatly during the first years. With their help, and employee Mishig's assistance, I had confidence in the logistical arrangements of safely bringing students into a remote place in a different culture. Teki Tsagaan's ability to bridge between languages and cultures provided invaluable leadership during these immersions of faculty and student from the United States.

Enriching the immersion experience for university students by including peers, students from the host country in the work group and the conversations is also a way to deepen the immersion experience. With Kent's guidance, we arranged to take along on our expedition three Mongolian university students from the Ecology Department of the National University, Ulaanbaatar, Mongolia. They improved their English skills and got the experience of a multidisciplinary field expedition and learned a new-to-Western-cultures way of interacting with rural Indigenous people, the holistic process. Subsequently BioRegions has involved Mongolian university students coming from ecology, agriculture, and health each year. Like many American students, they have faced challenges of engaging in disciplines beyond their own, and in most cases have enjoyed and appreciated these opportunities to broaden their awareness.

Getting information back to the local communities is part of the Holistic Management Feedback Loop. Remembering to document activities and leaving results in the community is a decolonizing methodology and important for maintaining trust. Our first work trip in 1998 with this student team linking US and Mongolian students resulted in an inventory of information drafted before the US part of the team departed Mongolia. This was assembled as the Darhad Valley Bioregional Management Plan (unpublished report, 1998) and remained there for use in the community Mongolia. Mongolians have a high literacy rate, well over 90%, and devour written products, so all the products we provide back to the communities are in both English and Mongolian. Another example of responding to the desire of the community to celebrate their cultural wealth is the books by the children of the valley sharing their experiences with the land and their food production in words and in drawings. Visions of the Blue Valley book (Poulsen, 2008) and The Written River book (Prince, 2010) are the best examples of providing products back to the source communities. BioRegions needs to do a more complete job of providing yearly written reports back to our host communities. There is always something to improve on.

In recognizing culture and in using decolonizing methodologies, it is necessary to emphasize the leaving of originals with the people of the village or community and giving back of the reports in a form that can be understood and used. This is not easy when working in a nomadic community. Nomadic life requires people to carry everything they need. Wind generators and 12-V car batteries are the main source of electricity, but recently the small county-center settlements have access to power line electricity.

These student-centered "work trips" have occurred every summer (except 2003) since 1998. My own view, as a field-centered soil scientist also interested in community, was to continue to offer chances for students, and then other faculty, professionals, and citizens, to participate in our work trips. I felt that expanding personal horizons through learning about life and environment in the Darhad was sufficient reason to continue. Although I was frustrated about lack of time for thorough connections and follow-through, I stubbornly continued to offer and organize student-funded annual trips, realizing that for the most part, we were not conducting validated research that could result in peer-reviewed publications. I was also somewhat frustrated that I did not have time or energy to pursue soil science in the Darhad. However, several students stepped into this vacuum as we repeated our annual visits.

In these decades (1990—2010), US professors in Colleges of Agriculture at Land Grant Universities were discouraged from doing this kind of in-depth, get-to-know the community kind of research. What was encouraged was a well-controlled, scientific field experiment bringing in the latest technology from the US agricultural scientific community. Many professors consequently had to forego the promotion and salary advantages of a more scientifically rigorous productivity in lieu of this transdisciplinary research using a holistic involvement process.

Initially we entered the Darhad to learn about the environment and culture. But soon we were being asked to assist with projects ranging from building fish hatcheries to funding school activities and supporting hospital needs for equipment and supplies. Mishig remained our Darhad Valley "local coordinator and mentor." It was a lucky choice because Mishig and his wife, both veterinarians, were well known by everyone in the community. Government employees in the area had lost their jobs and therefore their income was severed. When this happened, Mishig and his wife stepped up to the plate and began to bring basic supplies into the valley. Through these two veterinarians, we learned about basic needs in the Darhad Valley. We now had a good idea of local needs to improve quality of life, one of the main premises of Holistic Management.

In the first phase of activity that began in 1996, BioRegions took students each year to learn and do some documentation, and to get acquainted with the landscape, its environment, and culture. During Phase 1, we learned that food and agriculture are at a crossroads in the Darhad Valley, with the Siberian cold and dryness largely prohibiting crop agriculture and large-scale gardening. The traditional meat and dairy diet is now influenced by the increasing popularity of processed and packaged food coming from the outside. There are suspicions about food purity from outside-sourced food. Sugar consumption is quite high compared to the traditional diet among these nomadic herders. Domestic grazing animals take a large portion of available forage which may limit wildlife populations, especially during the critical winter and spring months.

PHASE 2: BIOREGIONS PROGRAM MATURES TO INCLUDE ANNUAL WORK VISITS AND FUND RAISING

In Phase 2 which began in 2002 we reorganized to become more effective facilitators of change. BioRegions formed a new nonprofit 501 c 3 tax exempt organization, BioRegions International (BRI), able to parallel the MSU BioRegions Program, which remains as an informal network of faculty and students. BRI thinking led us to designate Core Focus Areas based on what local communities thought would be most useful. The current Core Focus Areas include Education, Environment, Health, the Arts and Traditional Knowledge, Whole Community, and Business. During this second phase MSU BioRegions continued to organize yearly work trips and BRI developed fundraising and promotional activities including a "Mongolia for Montanans" trip for interested supporters and importation of a shipping container load of gers to sell across the United States.

The following noteworthy projects address the Core Focus Areas that BioRegions developed during Phase 2.

Education

Ger Schools

Darhad Valley Coordinator Mishig has "kids-at-heart." He realizes these young people of the Darhad Valley will soon comprise the fabric of society as productive and engaged citizens. Mishig also understands the lifelong value of classmates and friends from school. Whenever there is a need, Mishig seems to always have a friend or classmate from his school days ready to help. In this case Mishig had noticed that after the "forced attendance" for school children under socialism, there now were students falling through the educational cracks because parents with increasing numbers of herd animals wanted children at home during the winter school season when many herder families are dispersed and spread out into their winter grazing pastures. Mishig took us to visit Baatar-bagsh (teacher), an elementary school teacher who had just retired and moved back to the countryside north of Renchinlhumbe to herd with his wife and extended family. Baatar and others we talked with around the valley said that numerous children were missing out on school because of family obligations to care for their herds. So BioRegions hired Baatar to teach basic reading, writing, and arithmetic over the winter where he and other families stayed with their animals. In the spring I attended the end-of-school-year test administered by a teacher from the Renchinlhumbe school. The students arrived at Baatar's ger by horseback, took the test, received their certificates and left, hopefully with enough basic math and reading skills to get along. We also sponsored another

Renchinlhumbe school teacher to teach a summer "ger school" for families living in remote areas. This project was greeted with enthusiasm at a meeting where we sat with the family members in a circle as they discussed what each could contribute, like a ger for a school house and food and transportation for the teacher. BioRegions provided initial salary. Now this practice is more widespread and involves kids getting ready for kindergarten as well. After BioRegions, and also Boojum, gave initial encouragement, additional ger schools appeared and over time local governments have stepped up to provide these services.

We came to these successes by listening to the needs of local people, by suggesting a new paradigm, and by providing some initial support, but insisting on substantial participation and buy-in by the local recipients. We are learning while local people have tremendous capabilities to provide the community services they think most important, there is great power in having a trusted outside entity provide the initial thrust, to light the first spark, for the fire to take off.

We understood the urgency of their food and agricultural system but first addressed the community's perceived barriers to the "good life," the desired quality of life. Addressing food, agriculture, animal production, nutrition would come later.

Environment

Mongolian Sands

The Darhad Valley is far from the immense sand dunes of the Gobi Desert. But the valley was once filled with a 300 m deep lake dammed by glaciers. I take pleasure in pointing out the ancient glacial lake shorelines on valley footslopes and noticing huge ripple marks made when the glacial dam burst and the lake drained north to Russia, an event similar to the flood of Glacial Lake Missoula which created the Channeled Scablands in eastern Washington. As the water drained, the Darhad Valley was left with abundant sand and silt deposits susceptible to wind. On a windy day parts of the valley fill with rolling clouds of silt and sand from sand dunes often associated with lake and river shorelines. Now these are grazing areas for the sheep and yaks that is food security for the people of the Darhad Valley. Expansion of the sand dunes decreases grazing areas and so jeopardizes the community's food security. This was one of the first environmental issues Mishig shared with BioRegions, "How can we stop the advancing sand dunes?" Mishig drove us north of Renchinlhumbe to New Spring, a huge spring upwelling amidst spruce bogs, and surrounded by a field of sand dunes and their eroding upwind source areas. As we considered Mishig's query, what to do? I shared that overgrazing might be one of the causative factors, for the sand dunes themselves are fed by a pockmarked and eroding landscape upwind. Maybe protecting the land from overgrazing and spreading brush on the upwind source areas to stabilize them would help.

Mishig called for a local community meeting. We met one evening at the New Spring sand dunes, about 40 members of the community arrived, all on horseback, coming to see what's new. After hearing me explain the details of the fence idea, the local leader suggested the community try the idea. They agreed to provide the labor to construct a fence around one active sand dune as well as to pile brush in the upwind source area. BioRegions provided funds for fuel to transport wood materials to the dune site. We appointed a local family to guard the fence against theft or damage, and left with excitement.

The community constructed a fine fence around the sand dune and we were anxious to see the results a year later. What a surprise 1 year after construction of the fence later to find the fence largely dismantled. We had neglected to consider that the herders who constructed the fence including the local family designated to guard the fence would be spending the winter elsewhere, in their winter grazing areas, leaving the fence open to vandalism. By focusing just on demonstrating a tool, the fence, but neglecting to predict what could happen over the winter, we failed, at least short term. **RED FLAG WARNING #3 to learn from: While the idea, the agreement of the community is important, it is important to look beyond the immediate to how the situation might change in the near future**. The sand dune project continued for 6 years while the fence was intact. Increases in grass production were estimated at 30% by visual calculations and exact measurements were taken from clipping studies. Since this is a migratory culture, the people who built the fence were not there during the winter season. They were in their winter camps far away, but other people were living nearby. The fence was disassembled by people wanting to use the wood. Although the fence was in place for approximately 6 years, it was in poor repair after the first 3 years.

Although the project may be considered a failure at this writing, a Japanese scientist has experimented with other sand stabilization fences and a local environmental officer, Renchindaava, has fenced and planted willow saplings on dunes elsewhere in the valley. So again, by taking an initial try at a problem, BioRegions has stimulated interest and creativity from others. And in 2016, Mishig again is suggesting that BioRegions tackle the sand dune situation. After learning from our failures to consider the "whole situation," we may be in a better position to make useful progress. In the meantime, local awareness and interest have been aroused.

Health

Summer Field Clinics

Similar to the pattern of ger schools developing in summer, the summer medical field clinics were started thanks to the nudging of BioRegions students and faculty. Again, these processes started with repeated visits and careful listening provided by the circle format. Always a leader, and always ready to serve his patients and community, Dr. Purevsuren had a history of innovation in pushing the envelope of effectiveness for rural medicine in the Darhad. In Phase 1, BioRegions started the practice of visiting the hospital and school in Renchinlhumbe. At that time Dr. Purevsuren directed the hospital, and was always willing to take time for a visit and a discussion of health and hospital improvement needs. Transportation is often a challenge in the Darhad because of poor or nonexistent roads, river crossings, soft wetlands, and steep terrain. For the most productive grazing, which allows for proper recovery time from initial grazing, herds must be constantly on the move, so people are dispersed throughout the valley. Dr. Purevsuren wanted an effective way to reach this dispersed population during the summer, so he asked for a grant to purchase a locally made large tent. BioRegions agreed, and thus the mobile field hospital was initiated and continues today in the Darhad and other places in Mongolia. A relatively small investment, suggested by a trusted local source, provided impetus for a demonstration activity which worked and was then replicated.

The Renchinlhumbe Hospital doctors also expressed frustration that they had no way to provide prescription drugs for patients who would have to wait for several days to obtain drugs from the provincial capital. BioRegions provided a $200 loan to purchase an initial supply of the most frequently needed medicines such as antibiotics like penicillin and amoxicillin, which the doctors dispersed at replacement cost and paid back a year later. Now this has grown into a private pharmacy and people's needs are being met.

This focus on medical issues of the Darhad Valley opened a natural connection with students and faculty in our medical school first-2-year program (WWAMI) here in Bozeman from Phase 3 beginning with 2007.

In 2007 BioRegions invited Susan Gibson, MSU faculty instructor in the WWAMI[2] Medical Education Program, to join our field team. In that year we had a premedical student suggest that we offer health screenings. This began an ongoing relationship with WWAMI and with RiverStone Health in

---

[2]WWAMI (www.montana.edu/wwami/) is a regional medical education program for Washington, Wyoming, Alaska, Montana, and Idaho in which medical students attend the first 1 or 2 years within their home state and finish their MD degree at the University of Washington Medical School in Seattle, Washington.

Billings, Montana. WWAMI supplies second year medical students, RiverStone supplies a third year resident doctor, and the Mongolian National University of Medical Science provides a medical student. This team often now includes a US physician in private practice along with students majoring in nursing. The team also provides simple recorded screening, consultations, and most importantly, annual training sessions for Darhad Valley health professionals.

Boiling Duration and Water Quality for Tea

There also were opportunities for individual students to put the food and health issues together with culture to obtain important results. Here is one example.

Loren was insistent about my taking her on as a graduate student. She had just completed the land reclamation field course with Mongolian Teki Tsagaan as a classmate, and now was passionate about participating in our environmental work in the Darhad. With a strong interest in human health and the environment, Loren wondered if people might be getting sick from coliform bacteria in the water used to make milk tea every day. She sampled the water of spring, stream, and lake sources and then boiled it for different lengths of time. To test for coliform bacteria, Loren needed to incubate the water sample petri dishes at warm temperatures overnight. Nighttime temperatures in May are often below freezing in the Darhad, so she had to bring the samples into her sleeping bag to keep them warm enough. Loren's work showed that a longer boiling time would ensure against coliform contamination. Now, how to share this knowledge? Loren asked how we could make posters or present the data using PowerPoint. In a great leap of technology, she went to the local market, purchased some off-white muslin cloth, and used felt-tip markers to sketch the condition of petri dishes under different boiling times. Her "poster," hand drawn, could be easily folded up, transported on horseback, and then be displayed on any ger wall.

This is a good example of how students observed local needs and used simple technology to get answers directly and immediately to the people who need the answers to improve their quality of life. The health component of the BioRegions Program began with stories like this in Phase 2 and flourished in Phase 3.

The Arts and Traditional Knowledge

With the publishing of the book, Visions of the Blue Valley (Poulsen, 2008), the children of the Renchinlhumbe School of the Darhad Valley, a winter boarding school for children of nomadic families established that they are writers and expressive artists as well as horse racers, sheep herders, students, and nomads. They showed in this their striking artistry, in color and design and in stories. Tension, uncertainty, and danger during the migration multiple times each year come across clearly from the written images painted by the children, 6—16 years old. This is one example of the synthesis that BioRegions has inspired.

Festival of the Darhad Blue Valley

In the Darhad Valley our students were jiving with countryside teenagers over The Beatles. I had just been high up in a grassy valley when a Mongolian teenager rode by singing the most appropriate Mongolian songs for such a place...rich in melody and feeling. As in cultures around the world, song is a great communicator and expresser of humanity across Mongolian culture. In return for having filled a request for funds for a sound system and traditional clothes, the school music teacher from Renchinlhumbe and his music group would each year come over to our camp and give a mini-concert of traditional songs accompanied by various traditional stringed instruments like the "horse head fiddle." The music was moving in the way it connected us with this vast and beautiful valley, the grasslands, and especially the traditional herding life. These experiences, especially juxtaposed for that moment in time with the rock music, led us to suggest an annual festival for the traditional arts including music, dance, oratory, and artisanship. Our hosts agreed and the first Festival of the Darhad Blue Valley occurred in 2006 at the large log cultural center building in Renchinlhumbe.

People came in their finery, singers and musicians presented an exuberant outdoor performance in the spring sunshine, artisans displayed their work, and BioRegions reciprocated with blood pressure checkups by BioRegions nurse and informational health posters while reading glasses donated by Lions International and re-cleaned winter clothes and snow goggles from the lost and found of Bridger Bowl Ski Area (near Bozeman, Montana) were distributed. The festival has since rotated around the four Darhad soums (counties) annually. BioRegions supports the festival with funds for prizes and transportation so that outlying soums can afford to send participants. For several years we felt a growing "handout mentality" as officials got used to coming up at the last minute to request additional funds. This stopped when BioRegions made it clear that the festival should be organized and largely paid for through local initiative and "ownership." Initially local organizers wanted to include the traditional activities of horse racing, wrestling, and archery. But BioRegions firmly insisted on it being a festival of the arts, whereas racing, wrestling, and archery have always been, and will continue to be, highlights of the annual Naadam Festival which is similar to the US Fourth of July celebration.

Several times the festival has tried to "go big and go outside," with mixed results. In 2015 the festival "went big" with higher amounts of government and private enterprise funding which brought in professional performers from Ulaanbaatar when the provincial government decided to expand it into a tourist attraction at a spectacular riverside setting. While all had a good time with more of a regional festival, the provincial government did not continue funding. The bigger version actually seemed "other focused." The locals decided to continue the smaller more localized festival. I have always been concerned that the festival uses an electronic sound system which often malfunctions and results in more noise than music. In 2016 we again had a smaller local festival and afterward, having again suffered through a malfunctioning sound system, we suggested having the festival with just "natural" music and song. Small and natural is often more reaffirming of the local people and their talents than the larger version of the arts festival, even with just a larger sound system. The small, nonelectric version of the festival is a better way to build place-based self-confidence. There was high agreement, and we hope this will occur next year. Several takeaway lessons remain.

LESSONS LEARNED

- An "outside" idea can become owned locally if supported by soft persistence and inclusion of local responsibility and accountability.
- For an outside idea to be adopted, the "locals" must be given free-rein to place-base the idea.
- The "give me" mentality can be subdued through openly stating the rules of engagement up front after trust has been established.
- Food and animal agriculture in the Darhad depends on a healthy environment and cultural knowledge which are celebrated, memorized, and shared in traditional song, dance, and artisan skill. Like language, these elements are key to people's lives. Recognition through practice is an effective way to actualize them.

These takeaway lessons have application far beyond Mongolia. They also apply to the issues of student learning, research activities, policy making, and international development addressed in this book.

Whole Community and Business

Major events and/or large inputs of funding, particularly if it is nonspecific, often attract people and organizations who do not share the same desired quality of life or ethical framework as the community with which one is working. We have learned from Easterly (2006), an economist formerly with the World Bank that "The big plan is often not the best plan." Indeed, the best plan is often no plan at all. The disadvantage of growth too fast was seen in the arts festival example. In this story we learn of another red flag to watch for in developing a community-based holistic process that includes the food and agricultural sciences.

Twenty-Three Families of Bayanhangai Valley

Officials from an international regional environmental quality and community development grant suggested that BioRegions works with herding families of the Bayanhangai Valley User Group. Recently they had banded together to construct a fence across the entrance to their winter camp grazing lands to keep others from using up the winter forage. Their de facto leader, Nyamrenchin, had been to several training sessions sponsored by this outside grant, and seemed enthusiastic to work with BioRegions to improve grazing and forestry practices and establish artisan projects to provide for community needs while bringing in outside income. We thought this would be a great project and devoted funds from a National Geographic grant to the project. At one of the first group meetings, Nyamrenchin was absent because he was away training his race horses. This may have been a first warning sign. **RED FLAG WARNING #4 to learn from: Participation and focus were lacking but we proceeded on in a high-financial stakes project.** During the initial phases of our project, Nyamrenchin pleaded for BioRegions to provide an electrical generator and power tools so he and his colleagues, skilled craftsmen, could make furniture during winter months for local sales. Nyamrenchin pushed us, saying that he needed to produce tangible results from our project to get the rest of the families on board. We honored his request at considerable financial and labor cost, and the Bayanhangai group reciprocated by constructing a cabin to serve as a workshop and showplace for finished products. We had great fanfare and celebration over these early accomplishments. But the group and the cooperative projects disintegrated over subsequent years, partly because key members monopolized the generator and power tools.

We look back on this as a partially "failed" project, because there was not sufficient buy-in and also because the project started with a substantial investment in capital equipment which generated debt. Instead of generating self-worth, it may have produced the opposite. How does this experience relate to this book about the gaps in today's societies between technological wealth and traditional ecological wealth? It illustrates one of the prime mistakes of top-down decision making often done under a quick time frame without full consultation and participation by an entire group. We had tried hard to listen and become involved. We did have great involvement and a certain degree of reciprocity, but not enough. BioRegions' vision of the Bayanhangai User Group working together to guard and improve the environment and diversify their livelihoods dissolved into a "White Elephant" investment. While the project was of some benefit, the full potential was not attained.

PHASE 3: DEEPENED LISTENING EXCHANGES

During the summer of 2016, MSU graduate student Badamgarav (Badmaa) Dovchin and Darhad Valley Coordinator Mishig spent time talking with Darhad residents about their traditional skills. As they reflected on the past 20 years, they came to realize that the traditional richness of skills and knowledge in the Darhad Valley is dying out and not being effectively transferred to succeeding generations. The culture seems to be changing. These issues bring us back to the evolution of BioRegions as a listening, learning, and thinking organization, a prime focus point for this book.

Reflections

In the third phase of BioRegions, which we are in now, there has been a deepening of relationships as both sides learn more clearly what the other really wants. This phase began in 2009 and grew as our mutual trust and respect, our roots and relationships extended and deepened within the Darhad culture. Over these years BioRegions and our local hosts and collaborators have been able to listen more deeply to each other across many repeated visits, shared experiences, and brainstorming to define effective projects. In 2014 Turbat, BioRegions Coordinator for Bulgan Province (east of the Darhad Valley) said, "You just need to show us examples from your country or other places, and then we can take the ones we like and implement them in 'Mongolian style'."

In 2014 Jargalsaikhan, chief environmental officer for Huvsgul Province, Mongolia (Darhad Valley is in Huvsgul) told us, "our system is so fragmented into disciplinary departments that on the ground, one officer often is ignorant of what another officer in a parallel organization is responsible for. *We need to get away from these disciplinary silos and work together at all levels of government and science to solve problems.*"

In 2015 MSU graduate student Badmaa Dovchin and I attended a conference "Building Resilience of Mongolian Rangelands: A Transdisciplinary Research Conference" in Ulaanbaatar. The conference was a great sharing of problems and approaches, but was in an urban hotel with hardly any, but some symbolic, participation or buy-in by the herding community who live daily on the rangelands. We came away realizing the huge gap between science and application and wanting to establish an "extension service" to help bridge the gap. In 2015 we also visited with Dr. Gombojav at the Mongolian

University of Life Sciences (formerly the Agricultural University) who advised us that a great and important challenge is to work to disassemble the existing disciplinary and administrative silos at the community level so that the disciplines can interact with each other and with the community to effect lasting positive change. In 2015 Tumursukh, Director of the new Parks and Special Protected Areas told us "BioRegions is on track to be an effective organization with the right approach to blend environmental quality with culture, but you need to be present for longer periods during the year." During this same time both Tumursukh and several local government administrators told BioRegions, "we mostly want your assistance in broadening the outlook and mind-set of our community residents. We want this a lot more than infusions of money or material items."

Mongolian and Native American Students Working Together

These Phase 3 deeper listening exchanges led BioRegions to increase focus on Indigenous Research Methodologies which by nature are very participatory, based on personal relationships, and take repeated iterations and interactions leading to longer term adaptations. Much of this work is being led by graduate student Badmaa Dovchin. A former Boojum Expeditions guide Badmaa and Ulaanbaatar Coordinator Sunjee became interested in Holistic Management. Badmaa, working with BioRegions since 2008, wanted her MSU PhD research to be applicable to the Darhad, and to make real differences for the environment and the people. Hence she uses the Holistic Management process to listen to what people are concerned about, for instance grazing examples to stimulate proactive decision making and subsequent projects. Deeply aware of her Mongolian heritage yet fascinated by contemporary science, Badmaa is well positioned to help bridge the gap between realities of traditional countryside herding life, modern science and technology, and globalization changes.

The Darhad Valley economy is based on animal agriculture. A symptom of incomplete decision making with domestic grazing animals is over utilization of pastures with subsequent environmental degradation and loss of productivity. Badmaa's skills and her approach are at the nexus of the culture of her people and a sustainable future for her people. This is a good example of how the holistic process includes culture with equal weight as part of the whole along with Western science and technology while recognizing the changes that globalization is bringing.

Badmaa's research doesn't point fingers at "bad grazing practices" or "too many animals," rather it asks, *"What are the most important quality of life values we have?"* and *"How can we maintain those values?"*

The emphasis, Badmaa observes, is, "How can we rural Mongolians adapt? How can we adapt beyond the simple cause and effect analysis, which still remains an important component, but not the ending piece."

An important step deeper into Phase 3 comes from the 2016 BioRegions trip which included, along with a Health and a Community Team, a group of Native American university students led by Dr. Kristin Ruppel, a professor in the Department of Native American Studies, MSU, with the task of learning how two cultures, their own Native American cultures and the Darhad culture, use language, cultural experiences, and traditional knowledge as applied to Western Science and other Science, Technology, Engineering, Math (STEM)-based fields. In a month of field visits in all four Darhad soums (counties) led by Mishig and Badmaa, I saw the melting of language barriers when we got beyond the "greeting and sharing a cup of tea" phase. Our group would often have a scheduled meeting and agenda which got our students together with Darhad people, but after time the meeting would get off track. For example, Badmaa convened a soil and pasture quality improvement workshop for herders in the large log Soviet-era cultural center building in Renchinlhumbe. She and her two student assistants skillfully shared technical information on soil quality and grazing practices, and planned to administer a postworkshop survey. However an older herder and local leader mentioned that he collects and documents information on sacred sites for his community. Our Native American students heard this, and the formal meeting dissolved into a long and frank discussion about sacred sites on both sides of the Pacific Ocean. Our Native American students and Mongolian elders sprawled out on the wooden stage of the cultural center, sharing deep histories and value-sets of their two cultures, and plotting how these values could be captured on maps with Global Positioning Systems (GPS) and Geographic Information Systems (GIS) technologies. They also shared empathies, with Native Americans admiring the Mongolian culture where herders still live daily in their gers while the Native Americans just live in teepees during ceremonial days, and where Mongolian herders realized how deeply Native American culture and lifeways have been altered by the dominant culture, and that they too could be vulnerable to similar fragmentation and exploitation in this era of globalization. People from both cultures shared common values of deep connection with Nature and landscape (traditional values) as well as fascination and the desire to apply current science methods (GIS, GPS, nutritional analysis) to shared issues dealing with daily living in their landscapes.

Darhad elders came to us saying "I am dedicated to documenting and sharing our traditional knowledge which seems to be dying out, and at the same time I want to be involved in improving our systems of landscape (especially grazing and forestry) management, and we need to inventory our resources and place this information on a map using GIS technologies." Our students were thrilled at this prospect of using a science-based tool to understand more about the importance of place as a context for cultural values and practices. When these cultural practices are then blended with science tools, an interactive and participatory process to improve environment and society is created. Focusing on recognizing nonmonetary wealth, the wealth of culture and place creates a way of decision making with a vigorous and effective feedback loop that includes an economic component as a resource or a form of production, but not as the sole outcome. This feeds directly into Badmaa's research objectives which include documentation of how participation in learning about and documenting the natural wealth of a landscape may expand the mental viewpoints and decision-making capabilities of participants. This process blends traditional knowledge with Western scientific tools.

BioRegions is now well into Phase 3. Through a cooperative relationship with a private company, Mongolia River Outfitters, BioRegions helps with environmental awareness, and health improvement projects within communities which guard the gateways to important riverine habitat and fisheries. Badmaa works closely with several Herder User Groups on using the Holistic Management process to learn about grazing ecology and improve herd quality, both steps they can make to reduce overgrazing and soil erosion. Badmaa and Mishig plan to use the Blue Valley Festival and a school contest to encourage students and their families to document and showcase traditional knowledge and skills. Native American–Darhad people will continue to discover and share common bonds and work together toward solutions to the detrimental effects of globalization. Holistic Management remains as a fundamental process to bring traditional knowledge and contemporary science and technology to the decision-making process as demonstrated by a host of small projects which bring people and ideas together through the participatory listening and learning process.

Including Health in Food and Agriculture

Health and wellness have been a continuing theme through each phase of BioRegions. We were reminded of this just as the 2016 BioRegions group was leaving the Darhad Valley we received a request to stop in Ulaan Uul. It was late on a warm sunny Sunday afternoon and our students wanted to get to camp after a long set of delays and uncertain river crossings. Mishig prevailed and we searched for the home of Dr. Purevsuren, now head of the intersoum (regional) hospital. There he came, putting his shirt on, an infectious grin spread across his broad face. He ushered us into his family home, his wife offered tea and cheese, and he proudly showed us a large portrait of him receiving a medal of honor from the President of Mongolia just 2 days before in the nation's capital to commemorate his success in securing funds for hospital improvement.

Leadership and decision making for the BioRegions group is not linear. There is a complex web of human relationships and power structures at work. BioRegions must interface positively with the rural communities we work with. We tend to rely on Mishig as the Darhad Valley Coordinator. He knows the people and is a respected community connector. From the US side, I, as overall director, and Susan Gibson (Health Team coordinator) play key roles, because they bring in outside professionals and students and also influence monetary decisions. Badamgarav Dovchin (Badmaa) also plays key roles. She and Mishig work intimately together, she is now well known in the Darhad after some eight-field seasons there, including several fall, winter, and spring season trips, and she also connects intimately with the US side, having lived in Montana for numerous school-year durations.

So in this case, Mishig heard about Dr. Purevsuren's award, passed the information on to Badmaa, Cliff, and Susan, and we jointly made the decision to stop by in spite of it being on the late end of a long day. I, as Director, is ultimately responsible for these decisions, but they are made carefully with utmost consultation and participation. Hence as an organization BioRegions intends to "go slow" in building relationships in order to be able to "go fast" when quick decisions and actions are needed. We were also relying on our experience with Mishig in being able to get us to a desirable camp site before dark closed in. Students were not happy with the delay, but in the end got a chance to visit a Mongolian home and celebrate and relax with a family. This they will remember long after forgetting about getting to camp late.

The close connection of any food-related, agricultural production project with local health and wellness services signals that those involved from the "outside" recognize the "wholeness" of the community and are respectful of the local people. It was an important early decision of Cliff and the BioRegions group that health should be an integral part of this project which initially in Phase 1 was focused on sustainable grazing practices. It is reminiscent of the strong message that the village of Sanambele, Mali, recognized when Malian national agricultural scientists and US agriculturalists came to hear their barriers to the good life, even if it was about health and not directly related to agriculture (Chapter 5: Listening With Subsistence Farmers in Mali). Using the holistic process (Chapter 3: Decolonization and the Holistic Process), no matter how informally we use it, reminds us to keep health a partner with food and agriculture.

It is with the use of medicinal plants that health clearly intersects food and agriculture. Mark Johnstad, now a BRI board member and international development consultant, said that human health care is often the most effective entrée into a community. For BioRegions the "Health Team" has become a critical mainstay. Starting with the pharmacy and the field hospital tent, our activities have grown to more proactively address causes rather than treating effects. From nearly the first encounter with Darhad Valley, health professionals, they, and other citizens have shared their knowledge of traditional medicine. The hospital doctors provided us with a display of plant-based medicines and locals like Mishig know where to find plant remedies for many ills, but they also rely strongly on modern medicines. In 2015 and 2016 the health program expanded to provide training on a province-wide basis for two provinces. Besides providing encouragement and moral support to low paid local health professionals and encouraging citizens to adopt healthy living practices, this program builds immense good will.

The Holistic Management process hovering in the background is constantly asking us to focus on the most sustainable solutions while respecting cultural values.

As with martial art, Aikido, positive synergies require actions by both parties. In numerous cases our Mongolian coordinators and collaborators said, "You can't just talk and plan, you need to put something on the table to begin, you need to show some tangible action to get our real attention and participation." I first met Basbish when he brought samples of medicinal plants to our student group. Some years later, on our way home, we stopped to visit Basbish and his family at their summer camp. Just a few hours south of his camp we were turned back by a flooded river. The bridge was washed out and it was too deep to cross. We had no choice but to return to Basbish's camp and wait several days for the waters to subside. This setback turned into a great opportunity to live with Basbish and his family, prepare and share meals, and together and go deeper in our collaboration. Finally he said, "I would like to write and publish a book on the medicinal plants of our valley. Would you help me publish it? And I need photographs too." We agreed, supplied him with a camera to photograph the plants, and then worked to fund and publish the book (Armstrong et al., 2016). Now community members can read about the medicinal plants of their place in their own language.

The Yellowstone Connection

Another story, still developing as this is written, started around 2010 when local people demonstrated against free-for-all "Ninja" gold mining in which individuals were hand digging mining trenches, impacting soil and water quality, degrading the environment, and creating cultural tensions. In response the government established several new Strictly Protected Areas and a new National Park with regulations designed to prohibit mining and enhance habitat along with supporting the small ethnic group of reindeer herders. BioRegions wanted to promote useful exchange between people and organizations such as the Greater Yellowstone Coalition, the US National Park Service, and these new parks, given that Yellowstone National Park would have useful lessons for these newly established preserves. Although BioRegions does not have a formal cooperation agreement with the US National Park Service, in 2014 with US National Park Service personnel and nonprofit organization ecologists volunteering on their own time and expense to come, meet the Mongolian park rangers and community members, and start to share examples and lessons learned from Yellowstone. This visit was met with great enthusiasm by the Mongolian side.

Around the world there are many examples of local distrust of government agencies. In the United States, Grand Teton National Park came into being when far thinking locals as well as well-financed out-of-area land owners realized that this landscape might serve a higher value if held in public trust rather than remaining in private hands. The subsequent establishment of a National Monument, later converted to a National Park, created great angst among many locals. In 2016 there is a similar situation in Montana with creation of the American Prairie Reserve to convert a large landscape from private ranching and farming to free-ranging bison habitat. In the Darhad case, citizens lobbied the national government to control out-of-control artisanal gold mining, which was destroying environmental quality and impacting local society. The government then established two new Special Protected Areas and a National Park, with mining prohibited and many other traditional activities, such as hunting and wood gathering, closely controlled. Director Tumursukh and his ranger staff were charged with enforcing these new environmental rules, which often interfered with the activities and ways of life of people in the buffer zone communities. Strict enforcement and lack of direction for alternative lifestyles and income streams have led to suspicion, resentment, and mistrust.

In 2015 BioRegions returned to find that the Park people said the local citizens would not attend their information meetings and the local citizens said the Park people would not attend their meetings. The local government officials appealed to BioRegions for assistance. Luckily one of the BioRegions visitors was Scott Christensen of the Greater Yellowstone Coalition. Scott had prepared a slide show depicting the developmental history of Yellowstone National Park and surrounding communities. We took Scott and his slide show out to Bag 1. This was about 20 minutes' drive north of Ulaan Uul, the park headquarters town. There we spread the word for herders to gather at a cabin out on the green grasslands of the Darhad Valley. Within an hour, the cabin was full of doubtful people, all with hostile body language and suspicion toward the Park and its employees. Over the course of Scott's 20-minute slide show, we saw the hostile body language melt into acquiescence that, "well, maybe the parks would provide benefit for us, and they won't restrict our activities as much as we thought, and maybe we can actually get along."

So, by first listening, then sharing examples from other similar, yet different BioRegions, we were able to make great initial progress. As we left, the Park tourism director was huddled over a map with these people planning out routes for tourist horse trips. This story is far from finished, but it is similar to stories of the rancor between park and locals during the early days of Yellowstone National Park when the US Army had to be called into bring order to lawlessness. A similar story is now unfolding in Montana with establishment of the American Prairie Reserve which is purchasing ranchlands to convert to wild lands among great suspicion and fear on the part of local ranchers, small town community members, and county politicians. More listening and dialogue seems to always help bridge these gaps.

OUTSIDERS AND THE IMPORTANCE OF LISTENING

Those involved in agricultural systems, but who are not on-the-ground farmers, particularly smallholder farmers or subsistence farmers often try to control human decision making related to the resources that create consumer food products. The colonial model, top-down decision making, usually led by outsiders, has often resulted in loss of human dignity and self-sufficiency, along with ecological simplification and unintended consequences such as presence of invasive species or less healthy food choices than the traditional diet.

"Outsiders" may be extension agents from the provincial or national system in country or disciplinary experts who are often scientists and technicians from another country. As with many experts the knowledge flow tends to be one way from top to bottom. In the holistic process approach, knowledge is open and shared, and the perspectives of people directly working with the land and the community are just as important as those in the more traditional top-down positions, including outside experts. In addition to this disciplinary disconnect between top-down, highly trained experts and those working directly with the land, there is often a cultural difference expressed by a reticence of local people to trust outsiders. It is necessary to reverse the top down, "let me tell and direct you" approach to one of saying "we come here to learn about your culture and your environment and your issues." In the listening and learning process, we will likely discover issues to work on together. And most likely the solutions to such issues lie in combinations of local knowledge (Traditional Ecological Knowledge) and multiple outside disciplines. Another way to pose this is that the holistic approach to scientific adaptive management requires application of reductionist and holistic solutions within a broader context of environment and culture. "Soft" listening, over repeated contact opportunities, is prerequisite to trust and acceptance needed to start the holistic process.

CONCLUSION

This chapter is a prime example of the intricate ballet of give and take that comes from careful, respectful, appreciative listening and action that can be taken between "outsiders" and "insiders." With hierarchy removed and the vertical flow of information not used, as we saw happening in these many Mongolian examples, we have observed: health issues are solved; STEM education and the arts flourish in rural pastoral communities, even among the nomadic herders of the Darhad Valley; traditional ecological knowledge is preserved; and food production in rivers, gardens, and pastures thrives.

References

Armstrong, S., Montagne, C., Kherlenchimeg, N., 2016. Traditional herbal medicine and community health in Mongolia's DarhadValley: creating a holistic balance in the face of globalization. <http://bioregions.org/media/darhad_health.pdf>.

Dunkel, F.V., Coulibaly, K., Montagne, C., Luong, K.P., Giusti, A., Coulibaly, H., et al., 2013. Sustainable integrated malaria management by villagers in collaboration with a transformed classroom using the holistic process: Sanambele, Mali and Montana State University, USA. Am. Entomol. 59, 15—24.

Easterly, W., 2006. The White Man's Burden: Why the West's Efforts to Aid the Rest Have Done so Much Ill and so Little Good. The Penguin Press, New York, NY, p. 436.

Hodges, G., 2003. Mongolian Crossing. National Geographic Magazine. <http://ngm.nationalgeographic.com/ngm/0310/feature5/>.

Meadows, D.H. (Ed.), 2008. Thinking in Systems. Chelsea Green Publishing Co, White River Junction, VT.

Nishida, H., Jamsran, T., 2009. Darhadyn Wetland in Mongolia: Synthesis Investigation on Ecosystems. Mongolia Ecology Information Center, Japan (Alex Company, Japan. ISBN 978-99929-1-745-8)

Poulsen, W. 2008. Visions of the Blue Valley by the children of Renchinlhumbe School. 2008. . In: English and Mongolian. 2008. . People's Press, Woody Creek, CO.

Prince, S.J., 2010. The Written River. English and Mongolian. Dartmouth University Press, p. 53.

Further Reading

Cajete, G., 1994. Look to the Mountain: An Ecology of Indigenous Education. Kivaki, Rio Rancho, NM.

Cajete, G., 1999. Igniting the Sparkle: An Indigenous Science Education Model. Kivaki Press, Skyand, NC.

Cajete, G., 2000. Native Science: Natural Laws of Interdependence. Clear Light Publishers, Santa Fe, NM.

Kolb, D.A., 1984. Experiential Learning: Experience as the Source of Learning and Development, vol. 1. Prentice-Hall, Englewood Cliffs, NJ.

Wilson, K., Morren Jr., G.E.B., 1990. Systems Approaches for Improvements in Agriculture and Resource Management. Macmillan Publishing Company, New York, NY.

CHAPTER 8

Listening Over Power Lines: Students and Policy Leaders

Hiram Larew[1], Florence V. Dunkel[2], Walter Woolbaugh[3], and Clifford Montagne[4]

[1]U.S. Department of Agriculture (retired), National Institute of Food and Agriculture, Center for International Programs, Washington, DC, United States, [2]Department of Plant Sciences and Plant Pathology, Montana State University, Bozeman, MT, United States, [3]Manhattan Middle School, Montana Pubic Schools, Manhattan, MT, United States, [4]Department of Land Resources and Environmental Sciences (retired), Montana State University, Bozeman, MT, United States

CONTENTS

Introduction 187

Case Study 1: Bringing USDA Into the University Classroom: Two-Way Learning Through Mutual Listening 188
Why Bother? 188
Getting Started 189
Planning and Holding Classes 190
Impacts 193

Case Study 2: Doing Intercontinental/ Intercultural Science With Middle School Students 194
Finding Partners 195
Science Trunks 196
The Research Area ... 198
The Research Process 199
The Research Design and Data Collection .. 200
Data Analysis and Conclusions 201
Information Dissemination and the Global Research Symposium 203

INTRODUCTION

This chapter explores how a rarely used channel in agricultural classrooms—one linking students to policy leaders—can significantly benefit both groups through a two-way flow of knowledge and ideas. Establishing and then building such "power line" connections between grass roots and grass tops is proving powerful in promoting mutual learning—learning that can occur when students are given the chance to interact with leaders who typically are older, not on campus, and are variously experienced with and responsible for guiding programs that are relevant to classroom discussions. And while it may sound pat, it is surely true: The complexity of opportunities and challenges of current and future food security requires listening across such disparate stakeholder communities. Such bridges are first-step requisites to problem solving.

Undoubtedly, others have used such links to foster classroom learning. Guest lectures and interdisciplinary team teaching have undoubtedly been popular for nearly as long as classes have been offered. They are viewed as helpful in connecting theory to real-world practice (Henneberry and Beshear, 1995) and to breaking down barriers between stove-piped topic areas. Guest lectures may be combined with other teaching tools such as internships to promote preparation for work (Pennington, 2004). Team teaching has been used to enhance participatory learning in the natural resources and agricultural sciences (Karsten and O'Connor, 2002; William et al., 1999).

187

Incorporating Cultures' Role in the Food and Agricultural Sciences. DOI: http://dx.doi.org/10.1016/B978-0-12-803955-7.00008-0

How to Accomplish This Activity in Your Area . 204
Tips and Suggestions 205
Reflections.................. 205

References 206

Further Reading 207

Power Line Resources.......... 207

Included in this chapter are two examples of power line listening. We offer these case studies mindful of how varieties of examples may enrich our listening processes, and how our experiences may further help those who are curious to wade in.

In the first study, we showcase the benefits of a fairly distinctive model in which semester-long interaction with an off-campus guest policy leader participated virtually throughout the semester, and as a result became vested in the course and learned alongside the students. This first case study focuses on how a link between a Montana State University (MSU) transdisciplinary classroom and a US government official boosted classroom experience and government food security policy insights. While this relationship is ongoing, and thus, evolving, the lessons learned thus far may prove useful to instructors in other disciplines and on other campuses as they design such elements into courses.

The second set of examples illustrates how American secondary and middle school students and stakeholders from beyond the classroom can link with students and stakeholders from literally around the world. Together with local research scientists the group can effectively engage locally and internationally about important food security approaches and interventions.

Underlying these examples is the important premise that students have much to offer those "in power" and, on the flip side, that students may also learn and grow by rubbing shoulders with current leaders who are in decision-making positions. Narrowing and even closing what often seems like a communication chasm between students (future leaders) and current leaders isn't necessarily easy or comfortable to do, but we have found that it typically results in all kinds of ah-ha awakenings and innovations inside classrooms and government offices.

CASE STUDY 1: BRINGING USDA INTO THE UNIVERSITY CLASSROOM: TWO-WAY LEARNING THROUGH MUTUAL LISTENING

Why Bother?

To the American public, and especially to college-aged students, the federal government is another planet inhabited by aliens who speak, think, and act in ways that are foreign and often threatening or even malicious. And to those in the federal government, students may seem otherworldly, facile, self-centered, disengaged, and IT-implanted. So, why would students ever want to engage with federal government leaders? And, what possible benefit could a federal employee derive from listening to undergraduate or graduate students? Why would water ever want to mix with oil?

After years of tip-toeing around such hesitancies, the instructor (Dunkel) for a course titled AGSC 465R: Health, Poverty, Agriculture: Concepts and Action Research, and a US Department of Agriculture (USDA) employee (Larew) decided to buck traditional biases, and to test the hypothesis that by participating in classroom sessions, a federal employee could contribute to student learning, *and* bring what he learned from students back into government programs. The course seemed especially prone to such experimentation because of its core message about the value of listening across cultures. And so, starting in 2012, students taking AGSC 465R began to visit with Larew in real time during every class by phone.

But, before describing the several outcomes that resulted from these experimental classroom linkages, the early steps and formative principles that led to such a partnership should be described for any prompts that would be useful. The partnership didn't happen quickly, haphazardly, or without planning. At the same time, it wasn't fully predictable. And, not surprisingly, each term and each class during each term was different and the model for listening changed accordingly.

Getting Started

A key early element in what became the link between students and Larew was a longstanding, rich professional interaction between Dunkel and Larew. As a result of Dunkel's research in Africa, and Larew's work at both the US Agency for International Development (USAID) and USDA, their paths had crossed often and for years before the classroom experiment ever began. And they shared a number of professional interests such as pest management in Africa, teaching methods at MSU, and the importance of indigenous insights and knowledge. Here's the key point: Discussions only started to shift toward a possible role for Larew in AGSC 465R after a platform of trust, respect, and mutual professional interests had been built. Comfort levels between instructor and Larew were high before Larew ever joined a class taught by Dunkel.

As the idea of Larew's role in the classroom surfaced, two factors quickly came into play. First, because he was employed by a grants-making part of USDA, *ethical and conflict of interest issues loomed*; how could he be involved in an MSU classroom without raising concerns that he was providing advantage or insights to a single campus? Any actual or perceived favoritism would be unacceptable. To address this concern, Larew worked closely with his Agency's ethics officer, agreed to receive no compensation at all from MSU, and agreed not to discuss any pending grants-related information with the class or with others at MSU. He pledged to only provide insights to students on information, trends, insights that were fully in the public domain. And, he indicated his willingness to join classrooms at other campuses if he was asked to do so. (He subsequently served as guest lecturer at other universities and colleges.)

Upon his retirement in early 2015, he worked with the ethics officer to ensure that his involvement in the classroom was sanctioned. This typically meant that Larew was to state clearly whenever appropriate that he was a USDA retiree and that his views did not represent those of USDA. Once retired, he also became a courtesy faculty appointment with MSU's Department of Plant Sciences and Plant Pathology—the department in which Dunkel belonged. This unpaid appointment would have been problematic while Larew was employed by USDA but was not so once he retired. And, soon after retiring, he traveled for the first time to the MSU campus as a volunteer (travel costs were paid by MSU). While visiting campus, he offered workshops, chatted with students, faculty, and staff, and visited field sites that had been showcased in AGSC 465R.

A second early outset issue arose: Because Larew was located in the Washington, DC, area, not on campus in Bozeman, Montana, the issue of *remote connectivity* to the classroom surfaced. How could he most effectively "attend" the class? After considering the variety of e-connectivity tools available, it was decided that phone call-in would be the most reliable, easiest, and most affordable real-time option. The downside was that by not using technologies that included visual connections, students and Larew were unable to see one another; facial expressions and body language had to be inferred. MSU provided a call-in line for Larew and others who were off-campus to use during class hours. Typically, audio fidelity was such that students could clearly hear Larew and vice versa. Time zone differences caused minimal inconvenience; the class started in the late afternoon in Bozeman, which meant with a 2-hour time difference, that Larew was able to join the class during his early evening from home. The key point: Technologies now allow visitors a virtual presence in classroom; while not as effective as an in-person visit, if carefully designed, an e-presence can enrich classroom discussions and the visitor's thinking.

Planning and Holding Classes

Beyond these early administrative and logistical issues of phoning in, the question of Larew's role in the classroom was the topic of several planning discussions. Dunkel and Larew, for example, agreed that each student would be asked to submit a bio-sketch at the beginning of each term which would be shared with Larew to help "introduce" him to each student. They also agreed that, as the instructor, Dunkel had the lead, experience, and mandate to develop the syllabus, work directly with the students, guide and manage the classroom discussions, grade performance, and the like. Larew's role was to augment instruction and guidance given by Dunkel in this lecture-less, open discussion, and action-research-focused course. Since the format for the discussion was texts-and-critics, Larew also read all reading assigned to the students, parts of eight books (Savory and Butterfield, 1999; Norberg-Hodge, 1991; Bennett, 2004; Ayittey, 2005; Easterly, 2006; Chambers et al., 1989; Yunus, 2007; Weist; Chief Dull Knife College, 2009; Smith, 1999) and several peer-refereed journal articles (Straus, 1977; Halvorson et al., 2011; Dunkel et al., 2011, 2013). Further, they agreed at the outset that he would join the class 15 minutes or so after class began. This delayed entrance allowed students to settle and for Dunkel to go over any housekeeping or logistical issues with the students. Once he called in, Larew remained on the line with the class for about an hour of the 3-hour class. He listened to Dunkel's discussion with the students which covered assigned readings or in-class films, highlights from field trips, progress on individual action research projects, and the like.

The phone-in arrangement and Larew's designed involvement in classes usually worked well; as he listened to students' reactions and input at the outset of each class, he would prepare responsive thoughts so that when, about 45 minutes into class, he was asked by Dunkel to comment, he could speak (usually for about 5 minutes) about what he'd heard the students saying, and raise questions about the reading assignment (What was the most unexpected observation that you made? Was your work with another culture more or less challenging than you'd imagine it would be?, etc.). Now and then, he'd offer encouragement if it seemed one or more students were stymied, confused, or needed perspective.

Because of his off-campus perspective, Larew was sometimes able to offer a slightly unexpected point of view, one that would give the students pause. For example, after a film about intercultural conflicts over water rights (Dunkel had shared the video with him before the class), he noted that most of those interviewed in the video were male. He asked the students if they thought that female stakeholders would use the same or different approaches in handling such environmental conflicts as their male counterparts. He emphasized that understanding and interpreting such gaps in stakeholder input is an important skill and can impact government programming. In other class sessions, he noted aloud the importance of student reflective journaling, and how it helps to self-assess attitudinal shifts. In this case, he suggested to the students that recognizing one's own inclinations and changes in them is an extremely important life skill. And, when asked by students about USDA's position on issues such as the value of indigenous knowledge, biotechnology, land rights, or food security, he described the wide array of programs—reflective of a democratic government's scope—that respond to the opportunities and concerns in such issue areas. He typically steered away from defending or criticizing such USDA positions, but instead explained how policy decisions are made within a large Executive Branch-managed and Congressionally overseen Department.

In addition to students' questions, Larew also—and perhaps, most importantly—tuned in as students described to Dunkel and to each other their passion for the meaning of cultural differences, for the land, for indigenous foods. Those strong signals from the class affected his own thinking about if/how government agricultural education programs account for and respond to them. He realized, in essence, how powerfully engaged students were. He began to respect, as he never had, their level of conviction. On his job, he also started to challenge peer professionals in the widespread notion that students were uninterested in agriculture or the issues that shape it. In essence, he began to discern the value of student input and insights.

To prepare for each class, Dunkel (or, during her sabbatical, guest instructor Jason Baldes) and Larew would visit between classes. They walked through the upcoming week's discussion and reviewed Larew's role. Such plans were often adjusted once class started and the discussion's tone and dynamics became clear. Both Dunkel and Larew remained flexible, and would hold back on preplanned interventions and/or would grab hold of unexpected opportunities as each class discussion unfolded. The overall classroom environment and tone were also important in accommodating call-in. For example, by providing a light meal at the end of each class that usually featured indigenous foods, Dunkel created a comfortable classroom experience which cued up the students for other innovative aspects of the class. Also, over the course as Larew became a more expected and familiar "presence," although his entrance into the class was the "Wizard-of-Ozian" mode of a voice coming out of a black box in the center of the room. In discussions, students typically became less shy of sharing their thoughts with him or of asking him questions. And as he learned more about each student from hearing them (there were typically 8–12 students in each class), he became more comfortable directing a question or comment to a particular student. He also began to anticipate each student's perspective and was able to track an individual's progress from week to week.

Between classes, Larew was occasionally contacted by students via email with questions, and he responded as quickly as possible. On another occasion, a student from the class attended an international meeting with Larew and Dunkel. These meetings of the Association of International Agriculture and Rural Development (AIARD) leaders were a good example of speaking to power. The student was able to join in conversation with practitioners in agriculture and rural development as they (we) worked to influence the thinking and actions of those in the legislative and administrative branches of the US Government and elsewhere through drafting recommendations (AIARD 2017) and visiting Capitol Hill. These "outside of classroom" interactions were useful to Larew in fine-tuning the terms and examples he used in class. And, during his postretirement visit to campus, he further developed a feel for the students when he informally met several who had been in class over recent years.

The end of each semester's course culminated in a "Share the Wealth" symposium conducted by the students to highlight their action research projects. Before Larew joined the event via the black box in the center of the room, students hosted faculty, students, family, friends at a reception in the atrium of the conference room. For a half hour, each student gave extemporaneous descriptions of their research depending on who was visiting their poster. Following an opening honor song and drumming from the Eastern Shoshone instructor, Hiram greeted guests. Dunkel provided a few minutes of history and thank yous and then all class members gave a well-practiced 10-minute oral presentation in front of an audience made up of class members, other students, and faculty and staff from MSU and Tribal Communities. Larew provided questions during and closing comments as well. Dunkel shared the posters by email attachment with Larew before the Symposium, and he joined the session by phone. By exercising their presentation skills, students grew in confidence. In fact, beyond the information that was shared by student presenters—which was typically considerable—their poise was in contrast to the hesitancies and shyness they commonly exhibited at the beginning of the term. Knowing that they were speaking to a policy maker in Washington, DC, as well as to tribal members in person and through the video recording, students stepped up to the plate with their very best analysis of the situation.

Impacts

At the outset and ending of each course, students were asked to self-assess their awareness of the course's topics—health, poverty, and agriculture. They also were asked to comment on their understanding of the government's role in shaping those topics. Results from these pre- and post-course assessments provided a hint of the course's impacts. For example, as a result of the course, student understanding of food security issues became more nuanced and thoughtful, and most class members became appreciative of cultural-based differences that impact those issues. Threaded throughout the course was a focus on listening to and respecting differences. This emphasis seemed to impact student openness to differences. And, while they might continue to disagree with government actions, students acknowledged that the course had provided a clearer picture of how government policies are developed, the weaknesses and strengths of those policies, and ways to affect them through the democratic process. Such impacts will very likely last well beyond the students' enrollment at MSU.

Drawing upon a longstanding professional partnership and low-overhead connectivity tools, this case study illustrates the how-to's of adding government perspectives in real time to classroom discussions. It also highlights the value of doing so. While the topics covered in this case class were specifically focused on transcultural dimensions of health, poverty, and agriculture, the case study has wider relevance. Courses in environmental planning, pharmaceutical regulations, or promoting artistic innovations would benefit from such linkages. By encouraging students and federal government officials to listen to each other, chances are increased that students will benefit from learning about priorities, civic decision making, and program design at the national level. And as significantly, national policy leaders will hear directly from the next generation of citizens, policy shapers, and program implementers about what matters.

CASE STUDY 2: DOING INTERCONTINENTAL/ INTERCULTURAL SCIENCE WITH MIDDLE SCHOOL STUDENTS

Collaborative projects involving real-world problems and applications must be based on local needs; then these experiences become valuable for students. When teachers in classrooms can integrate this with their course curriculum intertwined with, and contributing to, addressing the school system's prescribed standards of learning (SOL), the project becomes that much more valuable, and, simply, makes possible a learning experience as described in Case Study 2. A number of steps need to be addressed to complete such a project. Finding partners, addressing common science goals, engaging partners in the local community, developing research areas, sharing data, and bringing together and completely sharing final outcomes are all part of this challenge. To help understand how all these components can come together, we present Case Study 2.

With support from USDA CSREES Secondary Education Grant in 2005–08, five schools participated in a cultural-based collaborative science project. Partner schools were from West Africa, Asia, and North America, including a tribal sovereign nation: two schools from Mali; a school in a nomadic community in Mongolia; two schools from rural Montana; and one school on the Northern Cheyenne Tribal Reservation, also in Montana. Such a selection of schools meant that not only would students see how different cultures respond to different science issues, but students could start to view and experience the global nature of the world. A first step is to engage the participants. Until trust is established and relationships begin, collaboration is a challenge.

Finding Partners

As with the USDA NIFA Center Director linkage in the university classroom that we explored in Case Study 1, the middle school partnership began with a similarly longstanding and rich professional interaction between Dunkel and Woolbaugh. As a graduate student, Woolbaugh learned that his professor, Dunkel, shared his hands-on, inquiry-based, hypothesis-testing approach to teaching science. Woolbaugh already was a master science teacher in a rural Montana public school. Following graduate studies, he added to his career an adjunct faculty position in the Department of Education at MSU. Recognizing the important role of research scientists-in-the classroom when youth were making life decisions about career paths, Woolbaugh engaged Dunkel as a partner with his seventh and eighth grade students.

To begin the collaborative local research he took his students down the road from their school in western Montana to listen to the local seed potato farmers discuss their issues and walk with them in their fields. Dunkel introduced her new seventh and eighth grade research partners to her university research team working with plant-based management of the sugar beet root maggot (Dunkel et al., 2010) and the sugar beet farmers she worked with in eastern Montana. She literally brought the sugar beet farmers into the middle school classroom along with a local scientist from the USDA ARS, Sidney, Montana, laboratory via speaker phone for a serious conversation. Instantly, agriculture and food production and the case for sugar in one's diet became a hot topic and possible career opportunity to these teenagers (Woolbaugh and Dunkel, 2003).

Together, Woolbaugh and Dunkel listened to the discussions and watched the enrichment that crossing power lines had had in the lives of both the young people and the farmers. Next, Dunkel suggested that this richness could be taken to a completely higher plain by connecting his students with their age-mates in schools where she worked in Mali, and where her collaborators worked in Mongolia (Montagne) and on a Native American Reservation (Madsen) in Montana. Again, these were longstanding relationships. Montagne was a soils professor who espoused a holistic process in understanding the complexity of a nomadic herding community in the Darhad Valley of Mongolia. Madsen was a science instructor (and initially the Agricultural Science Department) at Chief Dull Knife College, a tribal college of the Northern Cheyenne Nation and a 1994 Land Grant Institution.

Woolbaugh engaged longstanding colleagues, also middle school teachers and their students from the neighboring town, another science teacher, and a social studies teacher.

Science Trunks

Since the project was dealing with middle school age students, it was decided that concrete artifacts might be the best way to begin relationship building. A trunk composed of items that reflected the school, community, culture, and experiences of each community was put together. The items in the trunk were selected by the students so this experience was very good in helping to formulate student reflective skills as to what other students from other countries might wish to see about their area.

The *Mali* trunk was actually a handwoven basket made from local reeds. It contained examples from African life cultures, games, historical artifacts, and school life. Textiles made from local cotton plants, woven, and decorated with earth and clay to make a cloth called bogolan. The Lame Deer, Montana, trunk contained samples of *Northern Cheyenne* culture and regalia, as well as a drum made with local cowhide, sinews, a stone, and a chokecherry stick. Videos of Native American dances, along with explanation sheets of these dances, sample clothing, various Native American historical implements, stories involving Native American culture, and sample beadwork were some of the items. The Belgrade, Montana, and Manhattan, Montana, trunk contained various items reflecting school spirit, pictures of local super markets, some miniature farm implements, and also a series of pictures and examples of local plants along with some descriptions of the various diseases and insects that infest them.

The *Mongolia* trunk arrived over the summer (due to shipping opportunities) and that was reviewed the next year. This trunk was a real trunk constructed from local trees in the Darhad Valley, carved and painted inside and out by the community artisans. The trunk contained fur samples of several animals, information as to Mongolian culture, games, plus a history of the school for nomadic children in this part of rural Mongolia. Most popular were the festival clothes of a man and a woman in the trunk.

The trunks produced several valuable outcomes. They helped secondary school students reflect upon the area they are from plus students were confronted with the challenge of creating a trunk that would communicate what some of the important points were from their local area. Class time was spent in all schools going over the contents of the trunks and the trunks remained as a resource for use with future classes.

We soon discovered that people made the trunks come alive. When the Malian trunk was opened in Northern Cheyenne classrooms by Cheyenne who had actually been to the village in Mali, it was exciting to hear the stories from a Cheyenne point of view, but excitement was at a whole different level when a Malian actually came to grade school classrooms on the reservation. When Yacouba Kone, a Peace Corps trainer of the Bambara ethnicity, opened the Mali trunk there was a hush in the room while he one by one pulled out the charcoal, the tea, the sugar, and all the homemade tea-making equipment, including the fan made in the village of local grasses. Yacouba demonstrated how he as the father of the family makes tea in his grand family, turning the tea, aerating it until it is just right, and shares it with everyone.

Joan Montagne, an artist, who works hand-in-hand in Mongolia with Cliff, her husband, and soil scientist shared the Mongolian trunk with Belgrade, Montana, students. Her stories of watching the artisans making the items in the trunk, the bridles, and saddlebags, and select the medicinal herbs helped students become aware that necessities of daily life, like food, clothing, and shelter, can be provided by people using traditional knowledge to craft these items from sources easily available within the landscape of the community. This becomes an important realization in a globalizing era in which so many daily use items come from outside of one's control. Furthermore, items of cultural practice, like a sheep anklebone game, do not have to depend on higher technology. This "grounding" process provides a base of interest and observation upon which to build further science investigation. The Belgrade students shared that white settlers, their ancestors in their valley, would have used technologies similar to those being used by the Mongolian community when they began to grow potatoes in the 1870s, and that science, as a form of "adaptive management" over the years (Graham and Kruger, 2002; Johnson, 1999), had simply expanded capabilities to grow more potatoes with less human labor.

Another spinoff of this food and agriculture collaborative research in secondary schools was used by language teachers in the village in Mali, in a private city school in Mali, and in high school classrooms in Montana. With all the students and teachers going back and forth, a letter exchange was developed between with high school French students and the Malian French students. Letters were hand carried between schools to avoid the high cost of postage. French is required beginning in fourth grade for all Malian students, rural and urban.

An unexpected consequence was the involvement of a grade school in Virginia who found us from the student website www.montana.edu/mali. Learning about Malian history, geography, and resources is part of the standards of learning (SOL) for third and fourth graders in Virginia. This group of students in Orange, Virginia, took the opportunity to learn in the classroom from Dunkel when she visited the area. Students spent the year preparing booklets and art materials to helping the Malian village students learn English and about Virginia and US culture. English is a required subject for all students in Mali beginning in seventh grade, even the subsistence farming village school.

The long-term importance of these "trunk experiences" is that instantly students felt connected to students of their age who had prepared a box or basket or actual carved wooden trunk just for them, but half way around the world on another continent. At that moment the "trunk" was opened, students could see agriculture, food, range animals, art, language, and community issues in one place. They could see how science is part of the whole. They could see how people were part of the whole. Art, music, and other cultural expressions are part of this whole, both in their community and on the other side of the world. Gregory Cajete validates this process in noting the importance of integrating world views which in turn integrate "spiritual, natural, and human domains of existence and human interaction" (Cajete, 1999). Now the students had learned about each other as people and their specific place they occupied on the Earth.

The Research Area

The next step was for each group to develop a research focus in their own community. The project lasted for 30 years so multiple groups of students experienced the process and outcomes. Each year students in each classroom worked on a local science research project that was developed by the students in collaboration with local scientists, and local stakeholders. Problem areas emerged and were defined. Students learned that science often happens because stakeholders face challenges and scientists can develop research questions that help to address these problems. Rarely will one research question answer the entire problem, but through a series of questions, the combined results began to help to answer the question. This process helped simulate actual research conditions in which scientists test related hypotheses for several years. Each school developed their research question in different ways.

Manhattan, Montana, schools studied soil conditions related to seed potato production, and this topic was selected by having students visit a local farm that is walking distance from the school. As the farmer talked with the students he mentioned that he had four different fields where he grew his seed potatoes, and that one of the fields seemed to produce more potatoes than the others. The students asked him why that might be so, and he replied that he didn't know. This later led students to want to investigate why this one field produced more seed potatoes. Students learned that good science research develops out of problems faced by the stakeholders.

Belgrade, Montana, schools studied water quality. The school already had some significant data on one of the main streams that flow through their community. Since the community was one of the fastest growing areas in Montana, the students felt that looking at water quality would be a valuable project.

Malian schools looked into impacts of cowpeas on improving children's nutrition as well as malaria preventative practices in the village. Forty percent of rural Malian children are estimated to have kwashiorkor or cognitive and physical stunting. Malaria was a significant health hazard. The rural school that we collaborated with was in Sanambele, Mali. Of about 250 children under 5 years of age in this village of 750 people, about eight children died each year from cerebral malaria. Village students wanted to address this.

Mongolian schools were interested in the process of dissemination of information to its population. With the remote nature of the Mongolian school, the students felt that not everyone had access to valuable information about their place-based issues. Their project addressed that need by focusing on community environmental issues in which people were wantonly disposing of trash, allowing overgrazing to create rampant soil erosion, overharvesting local forests, and poaching fish. Their town water supply was being contaminated by feces from the large number of yaks, cattle, goats, sheep, and horses grazing the community commons.

The entire process with all the schools also had strong cross curricular components with three social studies teachers involved, a French language/culture teacher, a global social studies teacher, and a teacher of Native American studies as well as two biology teachers. Five teachers and secondary school administrators visited Mali or Mongolia, two lived in the village of Sanambele, Mali, and each carried one or more of the trunks back and forth. Students learned that strong science involves strong language, math and writing skills in addition to having historical and cultural implications. Such approaches help to validate the entire educational curriculum.

The Research Process

It is important that students are led through the same process scientists would use in addressing a research question. Once the research question was developed, students did modified literature reviews as to what had been done in the past in regard to their research interest. This was approached in a "jigsaw" fashion whereby some students became "experts" in some areas of the research, and then they shared with their fellow students what they found out. After this it was time to formulate a testable hypothesis. Throughout this process these seventh and eighth grade students recorded data and observations in actual science laboratory notebooks, and these pages are peer reviewed and signed off on by fellow students. Students are taught that science writing is a detailed process so that accurate descriptions are essential. The next process with many of the groups was to develop a standard operating procedure (SOP) as to how the research question might be answered and how the hypothesis might be tested. During this time, scientists at the university were consulted and shown the students work thus far. A lot of contact with scientists, farmers, engineers, etc. was done through phone conferences in some cases, a speaker phone was brought into the classroom so students could address preselected questions with the expert. All of these techniques helped students realize they were part of an actual science research project. All of these activities were worked into the regular class curriculum and time frame. The research activities were embedded into the class content. Once results were obtained (in some cases a year later), these results were shared with stakeholders that included fellow students, the farmers, scientists, and community members. Students learned that sharing information, specifically, communicating effectively is very important in science.

The Research Design and Data Collection

Once a research question was developed in each school, the next step was to decide on a research design, and to begin a literature review. This was approached in a "jigsaw" fashion whereby some students became "experts" in some areas of the research while others became experts in other areas, and then they shared with their fellow students what they discovered. Using official, bound, automatic carbon notebooks, each student first recorded a hypothesis that they thought might be the outcome of the research question. This step is important because it begins to give students ownership of the research. Next, students were divided into laboratory partners to begin work. A challenge throughout a project in which there is one hypothesis and a class of 25−30 seventh and eighth grade students is how to manage the students. There was one research design, one SOP, one analysis, and conclusion, yet large numbers of students are working on the project. Using the same type of approach that is often used in research laboratories solved this challenge. Laboratory partners work together on each phase of the project and bring their ideas back to the large group for sharing; then the large group decides on the best approach. This process began with the research design, and once a research design was selected, the individual laboratory partners went to work writing out SOPs. These should read like a set of directions. Anyone who tries to duplicate the experiment should be able to read the SOPs and repeat the experiment. Students were surprised how challenging writing user-friendly SOPs can be. A couple of words left out means important steps of the experiment could be missed. As each team of students wrote a different part of the experiment, they often traded SOPs with another team and the other team tried to perform that phase of the experiment just by reading the directions. Using this technique, detailed and well-thought-out SOPs were written.

Once the experimental design was completed, each student team carried out the experiment. Because there were multiple teams of students in the class, multiple repetitions were made for each phase of the experiment. Students were given the task of not only keeping data in their laboratory notebooks, but also keeping data in electronic storage, both in a personal file and in a class master file. Just as scientists often have witnesses sign each page of their laboratory journal after a day's work, students modeled this in the class setting. It was explained that this was done in case there were ever any legal needs to prove who performed the experiment on which day, and what was done. It was impressed on students that one reason this might happen is because some scientific discoveries merit a patent and approved patents can lead to income for the scientist or their institution or both. Such techniques help to even further engage the students in the research process.

In Mongolia, students did not have to really hypothesize solutions as a first step, they simply had to point out some obvious solutions such as providing some garbage collection bins, asking the local officials to regulate grazing near the water supply, build a protective fence around the spring, request reestablishment of traditional forestry practices leading to a forestry plan, and call attention to the fish poaching while exploring ecotourism potential through catch and release fishing. This provided opportunities to compare and contrast Native Science processes with Western Science processes and appreciate the values of each.

Thus, the first step in recognizing culture in the food and agricultural sciences was taken by these youth, identifying their own culture, their own land, their own materials used for food, fiber, and energy, and becoming aware of place-based issues and practices. Because of the trunk and teacher exchange the groundwork had been laid for now entering into the worldview of recognizing differences in food, fiber, approach to place-based issues. The final step into appreciating these differences as neither better nor worse, just different, was now possible. As these young people mature, this experience could lead to the ideal of their being able to move seamlessly between cultures, adapting to cultural differences (Bennett, 2004) in use of resources, in production of food and fiber and ways to keep healthy.

Data Analysis and Conclusions

At least once each week students participated in a research meeting. During these meetings teams reported results from their experiments. After each individual reported, other group members gave ideas and suggestions for the group members to consider as they continued their work. Often there were contradictions to some of the information. For example, one particular soil type might be growing potatoes faster in one student's experiment, and in another student's experiment the same soil type potatoes were not growing as fast. These moments provided excellent opportunities for the class to engage in conversations as to why this may have been happening. Such conversations sometimes led to new directions in the class work.

An emphasis was put on students presenting their data and asking students to provide ideas on how the data might be analyzed and what the data might mean. Often this process included student teams evaluating data, writing up their analysis, and sharing this with the class. The class could then decide which ideas should be included in the analysis, and which ones should be eliminated. As students worked through this process, they began to actively voice "we don't have enough repetitions to prove that" or "we are making a big leap in saying that; we probably don't have enough proof." This kind of critical thinking and insight are exactly what teachers hope for in a research project. Throughout the process, university scientists were called upon for questions, decisions, interpretations, and several times they actually came to the classroom to meet with the students as *colleagues*, not as a "sage-on-the-stage." Many times this exchange happened via email with students directly emailing and asking questions of the scientist.

As analyses came together, conclusions were made. The Belgrade, Montana, middle school using macro-invertebrate information plus chemical water analysis found that water quality was very similar entering the town's water drainage system as it was when it left. It appeared to students that water quality at that time wasn't impacted by the current community size. The rural Manhattan, Montana, middle school found soil type yielding high amounts of potatoes had high amounts of nitrites, had a sandy texture, and also contained sufficient amounts of calcium carbonate. Middle school students working in an urban school in Mali discovered that adding powder made from leaves growing on neem trees, *Azadirachta indica*, just outside their classroom to stored cowpeas preserved the protein longer. Students working in a subsistence farming village school in Mali took this idea home to their parents and found that their parents thought the neem powder might make the cowpeas taste bitter when cooked. These parents suggested that they used ashes left from the cooking of meals each day. A Malian cowpea scientist suggested cowpea storage learned from his mother included using shea butter (an expensive cosmetic ingredient in the United States and Europe) to coat the cowpeas for protection from insects during storage and then add flavor to the cowpeas during cooking. As with all good research, students were left with many more questions than they had answers. Students made a next-step list of other questions that should be investigated, and at this point, they were ready for the final step in the research process; dissemination of the information.

Meanwhile, in Mongolia students in the middle school for a nomadic herding community used simple data for their place-based research. They identified the erosion areas in the Darhad Valley around Renchinlhumbe by name, by their distance from town, and by size, and then listed specific responsibilities for (1) local government officials, (2) herding families, and (3) themselves. The students took on the tasks of providing information for the others on erosion and helping the community to build fences for their herd animals.

Information Dissemination and the Global Research Symposium

Teachers emphasized to students that one of scientists' tasks in research is to publish the results so others can understand, challenge, and make good use of the results. Scientists help to create knowledge and they do this by addressing issues and problems, creating research questions, experimenting, observing, and finding answers. The next step is to share that information and have it critiqued. Students in this research project accomplished this several ways.

One way was by creating a poster display of their work. This poster display helped to show their research in a visual, lasting way. Other students and parents who came into the classroom read about the research and its outcomes. The poster display became a lasting artifact of the work and was hung in the hall of the school to show other students what was happening in science class. A poster display at school is a motivational recruiting tool for students to enroll in science classes.

A second way students disseminated information was by a global science symposium. This was set up at the local university in a videoconference center. All of the stakeholders were present to participate in this live video experience, and this included students from urban and rural schools in Mali (West Africa), students in urban and rural US schools, Malian and US university scientists, and scientists from the USDA funding programs in Washington, DC. Students in Manhattan, Montana, and Belgrade, Montana, boarded school busses at 7 am to travel to MSU in Bozeman. Students from the village in Mali were picked up in a van with one of the plant scientists from the capital, Bamako, and driven to USAID Bamako headquarters where they were taken through security to the conference room in the USAID "tower." Most of these students had never been to Bamako. It was a 2-hour journey by car. Students from the urban school in Bamako traveled with their teachers by city bus to the same USAID conference room. Meanwhile, Dunkel had flown from Bozeman to the USDA NIFA (then CSREES) headquarters in Washington, DC, and was seated at a conference table with a variety of scientists, all program officers at USDA NIFA. Among the program officers was the Director of the Center for International Programs, Hiram Larew. Students from the schools presented their information to each other. Renchinlhumbe, Mongolia, students had to send a time-delayed presentation since they were 12 hours ahead of Montana time. A question and answer period followed. First the USDA program officers posed questions. Then students questioned each other. It felt like a typical research-based, global videoconference about food, agriculture, and health implications. But then, one realized the presenters were youth, middle school students. The learning among youth and Elders around all the connected tables was astounding.

A final way was to prepare a summary of findings and disseminate this back to the people who were concerned about the initial problem. In one case, this was a potato farmer so a one-page summary was delivered to him so he could see what the students discovered. In another case, information discovered about cowpeas was delivered to a group of Malian and US scientists who were also working on this problem, and in a third case, parents and community members received information as to the outcome of the water quality analysis during a school board presentation.

How to Accomplish This Activity in Your Area

Case Study 2 illustrates a process that was successful for five secondary schools, a grade school, one university, and various other stakeholders. There are many possible approaches to this same outcome. The basic idea is to form collaborative partners, engage them with some type of activity like the trunks, blogs, discussion groups, etc., and then go through steps of the research process followed by a sharing session with all the stakeholders. If financial support can be obtained, that is much better as expenses would be covered. A key component is to enlist, inform, and see support for everyone who might be involved—teachers, Elders, other community members, business members, university personnel, and scientists. By doing this you will better promote your project and students will tend to have a stronger learning experience. It is also possible to do small parts of a global exchange. Students are impressed seeing information from neighboring schools, or schools from minority areas, in addition to having school partners from other countries. A desire to accomplish some of this is all that is needed.

Tips and Suggestions

Tips and Suggestions for Doing Culture-Smart Research Projects in Your Classroom:

- Have students select a research question from a local issue or problem. This helps teach students that good research questions are developed to solve real, current problems.
- Have frequent collaborative meetings so that students can share what they learn as they proceed with the project.
- Have students maintain laboratory books with precise SOPs. This process helps to teach them the importance of descriptive, accurate writing.
- Have students do multiple repetitions of the experiment. Good science happens because experiments are repeatable and results are consistent.
- If possible, form cooperative ventures with other schools, even if it is on the other side of the world. The motivational factor and the accountability factor become more important when this is accomplished.
- Have students share the results of their work in multiple ways. It might be a blog, a website, videoconference, poster display, publishing in the school newspaper, a DVD, or school website. Results of good science need to be published in peer-refereed journals.

Reflections

Learning isn't limited by classroom walls. This chapter illustrates how learning can be significantly advanced—and real-world problems taken up—if the effort is made to open classroom doors to the world of communities, leaders, and stakeholders.

University students can advise national and international policy makers. Middle school students can become research partners with university professors. University students, yes, even Liberal Studies, Modern Languages, and other Humanities students, can gently broaden the framework of their professors in the food and agricultural sciences.

Surely, such a process isn't always predictable; it's often messy and can be confusing. But, there is no more powerful way of eliciting cognitive dissonance and through it, wide-eyed learning by students and classroom guests, than through real-time, action-focused classroom engagement with the world.

References

AIARD, 2017. SMART Investments in International Agriculture and Rural Development: Recommendations to the New Administration and Congress. Association for International Agriculture and Rural Development. http://www.aiard.org/uploads/1/6/9/4/16941550/smart_investments_final.pdf.

Ayittey, G.B.N., 2005. Africa Unchained: The Blueprint for Africa's Future. Palgrave, Macmillan, NY, 483 pp.

Bennett, M., 2004. Becoming interculturally competent. In: Wuzel, J. (Ed.), Toward Multiculturalism: A Reader in Multicultural Education, second ed. Intercultural Resource Corp., Newton, MA, pp. 62–77.

Cajete, G.C., 1999. Igniting the Sparkle, An Indigenous Science Education Model. Kivaki Press, Skyand, NC, p. 49.

Chambers, R., Pacey, A., Thrupp, L.A., 1989. Farmer First: Farmer Innovation and Agricultural Research. Bootstrap Press, NY, 218 pp.

Chief Dull Knife College, 2009. We, the Northern Cheyenne. Red Bird Press, Bozeman, MT.

Dunkel, F.V., Jaronski, S.T., Sedlak, C.W., Meiler, S.U., Veo, K.D., 2010. Effects of steam-distilled shoot extract of Mexican marigold, *Tagetes minuta* (Asterales: Asteraceae), and entomopathogenic fungi on larval *Tetanops myopaeformis* (Roder). Environ. Entomol. 39, 979–988.

Dunkel, F., Shams, A., George, C., 2011. Expansive collaboration: A model for transformed classrooms, community-based research, and service-learning. North Am. Coll. Teach. Agric. J. 55, 65–74.

Dunkel, F.V., Coulibaly, K., Montagne, C., Luong, K.P., Giusti, A., Coulibaly, H., Coulibaly, B., et al., 2013. Sustainable integrated malaria management by villagers in collaboration with a transformed classroom using the holistic process: Sanambele, Mali and Montana State University, USA. Am. Entomol. 59, 15–24.

Easterly, W., 2006. The White Man's Burden: Why the West's Efforts to Aid the Rest Have Done So Much Ill and So Little Good. Penguin Press, New York, NY, 436 pp.

Graham, A.C., Kruger, L.E., 2002. Research in adaptive management: working relations and the research process, USDA Forest Service, Research Paper PNW-RP-538.

Halvorson, S.J., Williams, A.L., Ba, S.H., Dunkel, F.V., 2011. Water quality and water borne disease along the Niger River, Mali: a study of local knowledge and response. Health. Place. 17 (2), 449–457.

Henneberry, S.R., Beshear, M., 1995. Bridging the gap between theory and reality: a comparison of various teaching methods. NACTA J. 39 (4), 15–17.

Johnson, B.L., 1999. The role of adaptive management as an operational approach for resource management agencies. Conserv. Ecol. 3 (2), 8, <http://www.consecol.org/vol3/iss2/art8/>..

Karsten, H.D., O'Connor, R.E., 2002. Lessons learned from teaching an interdisciplinary undergraduate course in sustainable agriculture science and policy. J. Nat. Resour. Life Sci. Educ. 31, 111–116.

Norberg-Hodge, H., 1991. Ancient Futures: Learning From Ladakh. Sierra Club Books, San Francisco, CA, 204 pp.

Pennington, P., 2004. Professional development in agriculture: opening doors through creative leadership. NACTA J. 48 (4), 27–30.

Savory, A., Butterfield, R., 1999. Holistic management: a new framework for decision making. Island Press, Washington, DC.

Smith, L.T., 1999. Decolonizing Methodologies: Research and Indigenous Peoples. Zed Books, London.

Straus, A., 1977. Northern Cheyenne ethno-psychology. Ethos. 5, 326–357.

Weist, T. A History of the cheyenne people council for indian education, billings, MT, pp. 227, 2003.

William, R.D., Davis, S.L., Cramer, L.A., Stephens, K., Greswell, R., Stephenson, G., et al., 1999. Team approach to teaching participatory group process involving natural resources and agriculture. J. Anim. Sci. 77, 163–168.

Woolbaugh, W., Dunkel, F.V., 2003. Community, university, and school research: Marigolds for Montana. Sci. Teacher 70, 46–48.

Yunus, M. (with Karl Weber)., 2007. Creating a World Without Poverty: Social Business and the Future of Capitalism. Public Affairs. New York, NY, 261 pp.

Further Reading

Ba, A.H. 1972. Aspects of African Civilization (person, culture, religion). Chapter 1, Chapter 2, and 5. Questions posed by students of A.H. Ba. Translated from the French.

POWER LINE RESOURCES

Several organizations, groups, and individuals influence the agricultural sector through policies, funding, and thought leadership. The following is a partial list of organizations and individuals that may be tapped for policy leadership insights useful to an agricultural, food security-focused classroom. Importantly, trusting relationships with such groups and individuals should be built well before classroom partnerships are undertaken. These relations are most fruitful if they are members of the teaching and learning discussion team and not helicoptering in as a sage-on-the-stage.

Farmers—Those in America and elsewhere who produce food or fiber (plant or animal) using conventional and/or nonconventional methods on large or smaller farms.

Government officials in the United States—County, state, national, tribal leaders. These may include colleagues from various departments and agencies, e.g., agriculture, trade, international relations and assistance, emergency food relief, natural resources, health/nutrition.

International organizations—United Nations Organizations (e.g., Food and Agriculture Organization) and The World Bank, and the Consultative Group on International Agricultural Research (CGIAR), Association for International Agriculture and Rural development (AIARD),and government representatives from other donor (e.g., Japan, Britain) and/or recipient (e.g., Indonesia, Kenya) countries.

Private sector—Beyond farmers, those processing, packaging, storing, transporting, marketing, and/or trading agricultural product. Also, those in the agrochemical (fertilizer/pesticide) and biocontrol industries.

Charitable organizations—Foundations, religiously affiliated groups, relief organizations, and those involved in promoting health and food security in the United States and/or overseas.

Legislators—US senators, Congress women and men.

CHAPTER 9

Listening With Students*

Greta Robison[1], Jason Baldes[2], and Florence V. Dunkel[3]

[1]Department of Earth Sciences, Montana State University, Bozeman, MT, United States, [2]Wind River Native Advocacy Center, Washakie, WY, United States, [3]Department of Plant Sciences and Plant Pathology, Montana State University, Bozeman, MT, United States

CONTENT

Case Study #1: Learn Communication Patterns 211

Case Study #2: Recognize Unique Backgrounds and Personal, Community Missions 214

Case Study #3: Connect Reality With Action 218

Case Study #4: Recognize Wealth of Insights Foreign Students Bring to Classrooms and Policy-Making Organizations 222

What Students Want 233
Create Personal Connections 233
Close the Gap Between Teaching Methodologies and Real-Time, Complex Issues 234
Provide Cross-Cultural Immersion 236

Amy Wright is an enrolled member of the Tlingit. I met Amy on the first day of class in my undergraduate capstone course, AGSC 465R Health, Poverty, Agriculture: Concepts and Action Research. She was a junior majoring in Liberal Studies with an Environmental Studies option. After several weeks in class and weekly individual research mentor meetings, it was clear to me that she was a scholarly student interested in sustainable management of traditional berry patches in Hoonah, Alaska, home to the Tlingit People. "Why didn't you major in Organismal Biology or in Horticultural Sciences here at MSU?" I asked. Amy paused for a long silence and finally shared "I didn't see that the College of Agriculture would be giving me the things I needed for life. They seemed to be all about plants and science and not people. I thought a major in Horticultural Sciences would put me in a greenhouse for the rest of my life." During the semester as a student in AGSC 465R, Amy delved into one of her favorite subjects, salmon berries. She conducted a thorough review of the nutritional and botanical literature. In a series of interviews with her Elders, she learned about the key role these berries played in her cultural history and in the potlatch, and other traditional practices of her people. Amy shared this all with her class and the next semester's class when she became the teaching assistant (TA). Her classmates and I learned for the first time about the Tlingit People, salmon berries, and other foods, like the "fruits of the sea," limpets, algae, and others. By the end of her four undergraduate years, though, Amy's academic background had not prepared her for what she

---

*With reflective log comments from Amy Wright, The Tlinget Nation, Hoonah, Alaska; Sierra Alexander, The Northern Cheyenne Nation, Lame Deer Montana; Gizelle Peynado, Kingston, Jamaica; Aedine Ndi Peyou,Yaounde, Cameroon; Akihiru Kuriki, Nagoya, Japan; Dawn Delaney Aimsback-Falcon, The Amskapii Peikuni Nation, Browning, Montana, United States; Jamie Sowell, Cave Junction, Oregon, United States.

Incorporating Cultures' Role in the Food and Agricultural Sciences. DOI: http://dx.doi.org/10.1016/B978-0-12-803955-7.00009-2

Encourage Passion...237

Response From
Florence Dunkel........238

Conclusion.........240

References 240

Further Reading 241

Summary
Illustrations.......241

really wanted to do: return to her home in Hoonah and work on improving salmon berries in both managed and wild patches. Berries are about health and culture for the Tlingit people. When the horticulture program did not meet her needs, Amy tried to obtain a place in the nursing program at MSU after a 5th year making up the missing chemistry and math, her postbaccalaureate studies did not lead to a place in the MSU nursing program.

Listening with this student as an undergraduate proved to be too late. Listening with her during her postbaccalaureate studies was probably not holistic enough to help Amy navigate academia to find the exact mix of fundamental skills and specific knowledge so that she could return to her ancestral home and have a productive career based on her Western culture Environmental Studies and basic science preparation. By listening with, we mean being appreciatively interested in Amy as a whole person. To the Tlingit, you are first a member of this Native American nation and only second an individual with unique goals and dreams. To listen with means to let the student teach the professor and other students about their culture. If horticulture had been more people-oriented and sensitive to the various uses and deep meaning of berries in non-Western cultures and appreciative of the various cultures represented in the classroom, Amy would likely have stuck with her original passion, horticulture. Amy was expecting that her Western culture introduction to the food and agricultural sciences would have recognized these connections.

To provide the framework and the atmosphere for culturally meaningful guidance and to receive critically important information, a course or program should, at its very core, accept and appreciate values of other cultures, particularly of non-Western cultures.

There is much wealth to be shared in courses and programs that celebrate the cultural component in food and agriculture. First, had the holistic process been engaged earlier in Amy's undergraduate years, this mistake may not have happened. Second, Amy's story illustrates the opportunity for two-way learning. Amy shared important knowledge about place-based food, environmental sustainability, and the interconnectedness of berries and culture that was new to all of the students that semester as well as to the instructor. For success with students like Amy, this exchange of knowledge in an appreciative venue is needed at each level of undergraduate development *and* within the institution's administration as well. Faculty need to be equipped with a well-developed ethnorelative world view to offer these courses within a sphere of acceptance and appreciation. As we learned in Chapter 8, Listening Over Power Lines, even policy makers in Washington, DC, can become a listening ear and an active participant in these classrooms where important truths are shared and ideas are posed.

In this chapter, we listen to other former students of AGSC 465R as they describe their experiences and desires for learning during their undergraduate and graduate years. How the course was designed to create a learning environment to teach skills and concepts, and how it exactly works teaching skills and concepts is discussed in Chapter 3, Decolonization and the Holistic Process. How the course works to apply these skills and concepts in the communities-of-focus is covered in Chapter 5, Listening With Subsistence Farmers in Mali, and Chapter 6, Listening With Native Americans. Chapter 9, Listening With Students, is the students' response to these methods and a summary with specifics of how the students see it working in their own case.

We explore ways to listen with students of the Millennial Generation and with the Next Generation. In this listening process, students share how they want to engage with their coursework to have immediate impact on their lives and careers. Two hundred and eighteen, students participated in this process of listening and sharing. Each student created and documented their own unique learning experience based on the course content, the specific concepts and skills required for course completion. We selected a diversity of guides as representative for this chapter. They are: Sierra Alexander, Jason Baldes, Greta Robison, Gizelle Peynado, Aedine Ndi Peyou, Akihiro Kuriki, Dawn Falcon, and Jamie Sowell. They also describe how they like to engage with faculty. These former students will present us with ideas to bridge the gap they perceive between what younger generations would like to have in order to encourage the passion factor in our current teaching methodologies. How can the increasing diversity on campuses—cultural, racial, religious, economic, gender orientation—positively affect education, particularly in the food and agricultural sciences? What can students, faculty, and policy makers do to benefit from differences, and even learn from the inevitable moments of misunderstanding and awkwardness?

CASE STUDY #1: LEARN COMMUNICATION PATTERNS

Sierra Alexander is an enrolled member of the Northern Cheyenne Nation. From the moment the first class of AGSC 465R began in the new semester, Sierra was not only a student in the course, but also a site mentor for all of us. She had just finished a semester of Holistic Thought and Management at MSU with Dr. Clifford Montagne and had been guided to take AGSC 465R by Kurrie Small, a fellow student who is an enrolled member of the Apsaalooke and a site mentor for AGSC 465R. In class 3 weeks later as we were preparing to make a 4-day immersion visit to Sierra and Kurrie's reservations, we were in discussion, seated as we always do, in a circle. I looked at Sierra and asked her to explain the "pause time" that we will experience when speaking with the Northern Cheyenne. I further asked her to help us understand if the Apsaalooke have the same speaking pattern.

An expression of bewilderment instantly came across Sierra's face. It was an awkward moment for me as the professor trying to model culturally sensitive actions. I tried quickly to explain to the students seated in the circle with me a little history of my decade of working with the Northern Cheyenne. I am Sicilian and our pause time is zero; we often begin speaking before the other person has finished speaking. This was considered a sign of closeness and friendship. To leave a long pause time was a sign of disinterest or even anger I learned as I was growing up. In discussions with the Northern Cheyenne, I often felt uncomfortable waiting for responses when we were in a discussion. In stark contrast, I noticed I did not experience this discomfort when speaking with Michele Curly, the Dean of Academic Affairs at Chief Dull Knife College, the tribal college of the Northern Cheyenne. One day I shared this feeling with Michele. She laughed. "I am French and married into a Northern Cheyenne family. As a child we were always finishing each other's sentences. This is also tough for me. I have to work hard to wait for the Northern Cheyenne pause time." The next time I brought AGSC 465R students to Chief Dull Knife College to meet with Michele before the main meetings with the Northern Cheyenne faculty, students, and College President, I asked her to explain this. This time Michele explained to all of us the depth of the pause time difference for her. She said, "I have even thought of bringing an egg timer with me to administrative college and tribal meetings, to help me keep silent until the proper time had elapsed!"

By now, the awkwardness in my own classroom had subsided and Sierra's expression of bewilderment had turned to a quiet, confident smile. "I didn't realize" she said "that my pause when speaking was normal for me as a Northern Cheyenne. I had always thought I was just shy, especially in my classes at MSU. Now I have learned another good thing about myself."

The key to a successful interaction with students is just knowing communication patterns of your student's culture or that of your colleagues at the negotiating table.

This knowledge can open the path to a deeper level of communication. In Northern Cheyenne culture, speaking too quickly after the previous speaker may indicate the next speaker, talking so quickly after the first, is a rash person who does not think things through before he or she speaks, or is showing disrespect for the importance of the other person or of what they had to say (Elliott, 1999). In Anglo, Euro-American cultures pause time is very brief; often people speak on the end of the first speaker's last sentence (Kochman, 1990). Interrupting another speaker is unbearable rudeness in Native American cultures and in most indigenous cultures, and may lead to severe social consequences if the person interrupted is an Elder. When interacting with members of other cultures in which appropriate pause times are shorter, Native Americans may have to be rude (by their own standards) in order to participate in the conversation at all (Basso, 1988, p. 12) or as Sierra chose to do appear as shy or not understanding. This is a stressful experience for the person who feels forced to violate their own standards and self-concept in order to be heard (Elliott, 1999; Elliott et al., 2016).

Consensus is easier to reach in negotiations when interactions are "decolonized," particularly when mutual respect is exemplified in speaking patterns.

In Bambara culture, such as in Sanambele, Mali (Chapter 5: Listening With Subsistence Farmers in Mali), responding *while* the first person is speaking is considered polite and is a sign that you are listening intently. For formal presentations, a spokesperson is actually chosen to provide "prescribed active listening sounds." Understanding this specific communication pattern was important for all of us, faculty and students gathered one night in the pitch darkness of this small farming village. The Sanambele Women's Association with whom we had collaborated for a decade was about to formally accept the Community Service-Learning award from the President of MSU for their work with MSU students. The award included a commemorative crystal statue and a cash award and had been brought to the village by us. The entire community was invited for the announcement and speeches. They gathered after dinner, most bringing their own wooden stools to sit on the earthen courtyard surrounded by the one room mud huts at our host's home. It was 7 p.m. and in complete darkness in this nonelectrified village, a few women had flashlights. This was quite an honor to be paid to the village. The President of the Women's Association was to represent the women and give the acceptance speech. There were several additional speeches all extolling the good relationship of the village with MSU. Overlaying the rhetoric, coming out of the complete darkness every few minutes during the speeches was a rhythmic "unhun." If our hosts had not explained this custom, these interruptions in the speeches could have been perceived as scary or rude and the significance of the moment could have been lost on us.

The respondent, giving the prescribed interruptions during the speeches, is a position of honor in this culture. That night the official respondent chosen was not of the noble caste, but a member of the potter caste, and she was to represent all the nonnoble castes in the village. This was a significant egalitarian action. Knowing who was giving this response every few minutes and what this meant in village dynamics was as impressive as the speeches themselves. Experiencing this complex communication pattern may well have prepared me years later for understanding the importance of the pause time and the "finishing the end of my sentences" discussion with Sierra and with my French origin mentor on the Northern Cheyenne Reservation.

How far is this case study from a course in Nutritional Biochemistry, Family Consumer Science, or Plant Physiology? Not too far when one considers the faculty member leading a class sets the communication pattern, and if that pattern squelches contributions from some students, many effective teaching moments are lost. How relevant is this pause time discussion for a USDA food scientist in a policy meeting with Native American tribal representatives or when visiting a remote African farming village?

Space, arrangement, distance between people speaking to each other, and voice pitch also play a role in creating a culturally inclusive tone in any setting.

Pause time and turn-taking are not the only communication characteristics that have basic relevance in the classroom, the boardroom, or negotiating table. In Chapter 7, Listening Within a Bioregion, we discuss the importance of the circle in decolonizing interactions and in reaching consensus. The normal range of voice pitch, for example, with Spanish speakers is narrower than it is for native English speakers; often pitch and volume that are part of "normal" conversation in English are only present in Spanish in the "angry" range of conversation. Consequently the Spanish-speaker may experience the European American as arrogant or intimidating. The English speaker may experience the Hispanic as shy, lacking self-confidence, or think the Spanish-speaker is mumbling when they are only speaking in the range that is "normal" for them (Olquin, 1995). These comparisons of cultural value systems are not meant to stereotype individuals or cultures; rather, they are meant to provide generalizations, observations about a group of people, from which we can discuss cultural difference and likely areas of miscommunication.

Once a comfortable setting for communication is established, the deeper level of personal goals and of individual student's connections to family and community can be shared.

CASE STUDY #2: RECOGNIZE UNIQUE BACKGROUNDS AND PERSONAL, COMMUNITY MISSIONS

Jason Baldes is an enrolled member of the Eastern Shoshone. Early on in meeting others for the first time, Jason usually volunteers this story, his story. This story is an example of what a professor, an advisor, or a policy maker needs to listen for in their first meeting. From where does Jason come and what experiences of life have shaped him and literally brought him to this university, this major, this course, this position? Listen carefully.

When I was 18, a trip to East Africa changed my life. Many experiences I had there, are fresh in my memory and will remain a part of my visceral thought process for the rest of my life. The migration of wildebeest is currently the largest migration in the world with 1.5 million individuals. When I saw the sheer number of animals on the landscape it was quite overwhelming. However, what was more overwhelming, was to consider that that number of animals was small in comparison to the number of buffalo, or bison, that inhabited the Great Plains of my homeland. After six months traveling in Kenya, Tanzania and Uganda, I gained a new-found appreciation for my home, heritage, culture and community.

Experiences in East Africa drove me to become involved with a range of issues upon my return home to the Wind River Indian Reservation. The Big Horn River case is the longest running court battle in the state of Wyoming and it concerns the control of water on the Reservation (Flanagan and Laituri 2004; Edlund 2001; Gunn Carr, Hawes-Davis 2000). Uranium mill tailings placed on the Reservation decades ago have contaminated ground water and increased cancer rates for nearby Reservation residents. Jurisdictional rights, sovereignty and self-determination of the Eastern Shoshone and Northern Arapaho tribes is continually challenged and undermined by state and local government. In a recent survey, three of the greatest issues faced by community members are housing, substance abuse, and access to adequate health care. This broad range of issues is far from being settled and now more than ever, young people will be faced with the challenges in the near future. Two thirds of the enrolled tribal Reservation population is under the age of 30.

When I returned from Africa, I worked with others to launch two non-profit organizations to address many of the concerns mentioned above. After several years however, it became apparent to me that an education was critical in order to have the necessary abilities to make an impact in my home community. Montana State University was a good fit for me as a Native person, I thought, because of the acceptance and welcoming of Native Americans to the university and the Bozeman community. When I decided to attend the university, Elders' advice to me was to take advantage of opportunities, however, to be careful to hold on to my identity as a Shoshone person, and not to become what the university wanted me to be. With this advice, I always chose to work on projects that were of benefit to my home community, so that I would always maintain my ties with the Wind River Reservation. Although many topics need attention, two issues that I have stayed focused on are buffalo and water. The discipline in Land Resources and Environmental Sciences made a good fit for me in my college career.

Thankfully Jason was fortunate to find professors who accepted and appreciated his world view while immersed in Euro-American academia. Jason was also well-grounded in his culture. By the time I met him as a student in AGSC 465R, he was a well-known traditional singer and drummer on campus and in the community. It was perfect timing for him to take this course.

In the fall of 2011, I had the opportunity to take a course called Health, Poverty and Agriculture from Dr. Florence Dunkel. The class was unlike any other I had taken at the university due to the holistic teaching and learning style. The course was so important for me, that after taking the course I remained involved for all of the subsequent semesters of my college career. I became a TA (teaching assistant), co-instructor, site mentor and finally instructor for the course (for the semester Dr. Dunkel was awarded a sabbatical leave to begin this book). The holistic nature of the course encourages students to be teachers as much as learners, and places emphasis on real world application of student projects. In addition, and of great importance personally, is the requirement of students to visit and interact with the third world countries of the United States, Indian Reservations. Even though there are seven Reservations in Montana, few people have ever visited one, and most have a very limited understanding of the unique relationship that Native American tribes have with the federal government. Sovereignty, trust responsibility, self-determination, and treaty rights are the foundation of agreements made with Native Americans in the founding of this country. This basic knowledge, or lack thereof, is perpetuated with the lack of opportunity for understanding. This course makes that opportunity available for students.

The students are asked to step out of their comfort zone and self-reflect on personal belief systems and perspectives. This provides opportunities for multi-discipline problem solving. Readings, videos, class discussions with the teaching team, and reservation visits are designed to bring students out of their comfort zone and create cognitive dissonance for them early in the course so that through self-reflection and mentoring the dissonance is transformative before the end of the course. Few courses combine multiple colleges from within the university to work together, on the ground, in real world contexts. It is as likely for a student from the mathematics department to work with someone from health, as biology student with one from art. Few courses offer that opportunity. Students transition through various levels of intercultural competency by selecting communities-of-focus to interact with throughout the semester (s). The communities chosen vary as greatly as the diversity of the students who take the course. Few college classes can admit that real cognitive change occurs within the students perceptions over the short course of a semester; Health, Poverty and Agriculture (AGSC 465R) provides the experience.

Listening with students can unleash amazing wealth for building curricula in the food and agricultural sciences.

During the first several weeks of AGSC 465R, each student must select a community-of-focus. Jason, of course, chose his own Reservation, the Wind River Reservation as his community-of-focus. The topic was food—the bison. The research was perceptions of his people regarding the proposed reintroduction of free-ranging bison to the Wind River Reservation. Over the many semesters of his teaching assistantship and instructorship in the course, Jason shared much of his wealth of knowledge with us about bison and a new section of the course was added in resource sharing, allocation of water rights, and environmental conflict. As Jason describes, this listening can also foster a deeper level of how to build relationships with those who are not like us.

Jason completed both his BS and MS degrees in Land Resources and Environmental Sciences in the College of Agriculture at MSU. He describes his present situation as follows:

> Upon returning home to the Wind River Reservation, I have been hired as the Executive Director of the Wind River Native Advocacy Center. The organization was founded to give Native Americans in Wyoming a voice for empowerment in self-determination of health, education and economic development. The Eastern Shoshone and Northern Arapaho people have many obstacles ahead of us, and an education provides the framework to address problems within our communities. Not until we as a society can work with and across disciplines, amongst cultural differences and perspectives, and within multiple jurisdictions, can we solve real world problems, whether local, national, or international. Tribes from Reservations in Montana and Wyoming (and others) have unique issues and problems but similar circumstances and obstacles. Many problems cannot, and should not, be solved by working on our own. Working with non-tribal members, organizations and others from off the Reservation develops relationships that can last a lifetime. Cross cultural awareness, respect and appreciation enhances the communities in which we live. Building relationships with those that are not like ourselves, diminishes misunderstanding and builds trust. Without courses like Health, Poverty and Agriculture, many of those relationships are never conceived.

Jason earned two degrees in the Western Science process, and contributed to faculty and students the opportunity for a deepened understanding of the role of culture in food and natural resource management. He contributed his time to build relationships on campus especially in AGSC 465R where he was hired as instructor for a semester during the professor's (Dunkel's) sabbatical. Jason made positive changes in cross-cultural awareness, respect, and appreciation similar to that depicted in the film, *McFarland U.S.A.* (Disney, 2015). The respect, trust, and deepened understanding resulting among his AGSC 465R students was documented in the standardized, confidential, formal evaluation system of Montana State University (Knapp Evaluation Forms May, 2015). Jason is now engaged in a similar process as he brings into conversation the two disparate Native American nations that share the Wind River Reservation with non-Native Americans managing resources for the State of Wyoming.

CASE STUDY #3: CONNECT REALITY WITH ACTION

Greta Robison is Celtic, Swiss, and Menorcan (a small Mediterranean island belonging to Spain between Spain and Corsica) in origin. She was born and raised in Bozeman, Montana. Her education through high school was from public school in Bozeman with a semester exception when she was able to travel to Mexico, Guatemala, Honduras, and El Salvador during high school. She later attended Global College of Long Island University.

> I went to Global College of Long Island University (LIU Global) for the first two years of my undergraduate education. Graduates from this small liberal arts program earn a degree in Global Studies after they live abroad for all four years of their studies. During the two years I spent in the LIU Global program, I lived in Costa Rica, Thailand, Taiwan, and Australia. With each of these location there was a required core set of courses that were required. For example in Australia required courses in LIU Global were: History of Aboriginal People; History of the Australian People; Permaculture and Theoretical Economics of Food Security; Coastal Issues; and an independent research project. In each location, food and agriculture and their ties to Indigenous people and the immigrant populations were relevant and studied in the classroom. These studies were performed while students were immersed in the host culture(s). Each semester involved different locations, different sets of cultural norms, and different food and agricultural issues.
>
>   I left LIU Global after two years because I could not afford the high price of tuition, and I missed STEM (science, technology, engineering, math) courses and technical skill classes. I returned to my hometown and began to study geography and geographic information systems (GIS) at Montana State University (MSU). While these courses gave me the technical STEM skills I had not received from LIU Global, I noticed a stark difference in teaching style and class structure. My previous classes all included immersion, language, fieldwork, and one-on-one mentoring for self-directed

research projects. At MSU I felt I was learning specific skills so I could make major contributions to projects and research programs which excited me. However, I was not asked to apply my skills in way that I found meaningful. I experienced a lack of real world situations such as working with the communities who were actually going use the results of my research. My program at MSU did not teach cultural competency, but rather technical skills as if the two were mutually exclusive. My classes did not relate how my learning could be applied to change the world.

The one place I found the rigidness of disciplines challenged was in the Native American Studies program, where I received a minor. These courses, which were often taught by Native people, were the only ones that challenged me to relate my work to the non-academic world. There was more openness to interdisciplinary work and a constant emphasis on cultural difference, history, and language. Interconnectivity in Native Science (Cajete 1999) was a common model in these courses and it encouraged innovation and learning. Having worked with indigenous groups around the world I saw clearly that Native Studies, and especially Native Science, was relevant in STEM curricula and necessary to understand if we wanted to be active participants in the changing world. Being a Caucasian woman from Montana, I also found it imperative that I learn these 'histories and herstories' my education had failed to teach me. I felt uncomfortable, I made mistakes, and I challenged my view, but through this I began to touch the surface of cultures I had grown up beside but never been taught about. When I was in high school in Montana I learned little to nothing about the Indigenous people of Montana (granted this has somewhat changed now that Indian-Education-for-All is a Montana State law).

While my Native Studies classes kept me engaged and humble it was up to me to make connections between history, language, culture, and the STEM sciences. AGSC 465R Health, Agriculture and Poverty, was the only class I found that saw the connections, taught it, and asked us as students to use our skills to create change. The class was the balance I'd been seeking, and failing to find, in my academic career. LIU Global required constant cross-cultural awareness and stimulated learning through experience and participation, but did not provide STEM-based courses. MSU did just the opposite. AGSC 465R required that we apply specific skills, such as Geographic Information Systems (GIS), plant propagation, understanding of the human immune system, or architectural design skills across cultural and physical boundaries. In 2014 this class, AGSC 465R, was the only class at MSU that brought students to a reservation.

It is fitting though, considering the colonial, destructive history of academics in Indian country, that the best venue for this work between two cultures would be through science—Native Scientists and Western Scientists—an important exchange. It was only possible, though, because we as 'Western scientists' were willing to learn history, language, and culture, breaking down the western disciplinary model and acknowledging the validity of Native Science even if its methods were at times different. We recognized that this made us uncomfortable.

In AGSC 465R, I was asked by the Apsaalooke to make suitability maps for Kurrie and Tracie Small, Apsaalooke sisters who were working through the holistic process (Chapter 3) to improve community health and connect youth and elders while encouraging good nutrition (Chapter 6). The maps I made were used to help select a native berry site where chokecherries, June Berries, buffalo berries, and native plums could be grown. As a part of this project, and my internship the following summer, I visited the reservation many times. I ground-proofed the maps with the Small sisters on the sites I was mapping. I met with tribal elders, and I met with the families of Lodge Grass, Montana.

Adoption of holistic teaching methods in classrooms, particularly in the food and agricultural sciences, would be advantageous not only for student learning, but also for our changing world.

This was a revolutionary experience for me. It was the first and only time at MSU that I had the opportunity to work cross-culturally, and one of two times I saw my skills used outside of academia. Especially considering MSU is a land grant institution, this strikes me as odd. AGSC 465R was a true combination of practical STEM skills and effective teaching. While the class is not perfect and the teaching methods are still being discovered and developed, it is one of the most rewarding classes available at MSU and certainly the one that has the biggest impact on the students and host communities. Using the holistic process would mean beginning with the expressed desired quality of life of a community. Then, it would mean to work toward solving any roadblock to this quality of life by using the combination of skill sets of the students, including those being taught in the specific courses. In effect, it means the community decides on the action taken, neither the instructors, nor the students.

The most difficult balance to strike is the one AGSC 465R takes on headfirst: working across cultural boundaries to improve quality of life using specific skills from Western and Native Sciences. While it is difficult for the students, teachers, TAs, assistant professors, and administrators, it is what the younger generation craves from their education. I have been in countless conversations with students frustrated with the higher education system, and their frustration is not because they dislike their major(s) or feel they aren't learning anything. The frustration is that no matter how much you are interested in what you study, as a student you are rarely taught, much less given the opportunity, to apply your knowledge to the world. We know we are learning, but we feel we are learning the notes of a tune but not the melody.

Greta is making a powerful point here. Some would say students should actively search out opportunities to apply their separate packages of skills and knowledge in the real world. Yet, we have observed that some behavior patterns such as the colonizing actions and comments and the ethnocentric world-views are difficult to break or relearn both for faculty and students. Decolonizing methodologies such as the holistic process are best learned with a guide-by-the-side. This guide needs to be an instructor who has an ethnorelative world-view and who knows how to identify colonizing behaviors.

Younger generations want a challenge in higher education, but it is not harder tests or more readings. They want to be challenged to take action and use what they learn beyond an academic realm.

If universities fail to open more avenues for this sort of multidisciplinary work I fear academia will remain a bubble where those who think in a non-western, academic way will feel out of place and unable to make an impact. This would be a tragedy because it is in this exchange of knowledge between the academic culture of U.S. universities and the thousands of other cultures in the world (some more close by than we may think), that we will be able to discover, innovate, and attack major global challenges we must confront within our lifetimes.

The younger generations are ready with their desire for a challenge in the real world. They still want guidance in integrating all their skills. These new generations are asking for a fundamental change in the way courses are taught in the food and agricultural sciences. This fundamental change requires professors and administrators to acquire new skills. How do policy makers nudge this process of acquiring new skills through in-service training or continuing education?

How can administrators and professors create changes that ensure they are listening to *all* students, not just those who are initially comfortable speaking their needs due to their cultural and value-driven norms being similar to that of the professor? Can more courses in the 50 US Colleges of Agriculture be designed to engage with local and foreign communities in collaborative research using the holistic process? Most professors involved with the USAID Innovation Platforms or the former Collaborative Research Support Programs (CRSPs) have developed long-term relationships with indigenous communities related to food and agriculture. The Mali Extern Program, which became AGSC 465R, emerged from a CRSP in the year 2000. A model for designing a course from one's own collaboration network is detailed in Dunkel et al. (2011).

What resources and incentives do administrators need to provide to hasten this transformation? How can other USDA program officers and perhaps World Bank, Peace Corps, US Agency for International Development (USAID), the United Nations Food and Agricultural Organization (FAO), the Association for Public and Land Grant Universities (APLU), Board for International Food and Agricultural Development (BIFAD), and Association for International Agriculture and Rural Development (AIARD) develop listening channels with students such as AGSC 465R has had with USDA National Institute of Food and Agriculture? Students have a lot they would like to ask of career government officers and leaders in these organizations. Students also have a lot to share with these program officers and administrators. To be able to ask and offer ideas within the classroom in real time and to become a known entity to a policy maker such as Dr. Hiram Larew (Chapter 8: Listening Over Power Lines: Students and Policy Leaders) is, for some students, the highlight of their years in higher education. On the other hand, it is an ear-to-the ground opportunity that seems useful for policy makers.

CASE STUDY #4: RECOGNIZE WEALTH OF INSIGHTS FOREIGN STUDENTS BRING TO CLASSROOMS AND POLICY-MAKING ORGANIZATIONS

We easily recognize as "foreign" those students who come from across an ocean or a clearly defined border in North America. We faculty, administrators, and policy makers are often more forgetful about recognizing as foreign students those from sovereign nations contained entirely within the border of the United States. What is very difficult to recognize are indigenous pockets of Caucasians within the United States. In Case Study #4, we explore the special needs, rich observations, and suggestions these foreign students have and share with us. As academicians, administrators, policy makers, we should be grateful for opportunities to interact with these students. They are a rich source to guide us in recognizing culture in the food and agricultural sciences, and indeed for simply recognizing and appreciating cultural diversity.

Foreign students in colleges and universities have special needs as well as special contributions to be made to US higher education.

They are also keen listening posts to improve our government programs in international food security and related health issues. The Chicago Council on

Global Affairs recognizes this in their annual call for US and international Next Generation scholars to attend their Washington, DC, April conference on specific global issues. Another example is the Association for International Agriculture and Rural Development (AIARD) annually selects US students to participate in the global issues conference in Washington, DC, each June. Ten next generation agriculture leaders currently completing degrees at land-grant and other US universities are brought to Washington, DC, for the Future Leaders Forum. At the Forum national leaders in rural development are given a chance to hear the wisdom and point of view of the students, and students are given the chance to participate at AIARD meetings with members on current rural development issues (AIARD, 2017).

In Case Study #4, we explore multiple student situations. Some of these stories are from folks easily recognizable as foreign students and others are clearly not "foreign students." Both the across-the-ocean foreign students and the not easily recognized foreign students, I have observed, find it is during this time abroad (in the United States) or outside their community, that it is safe to ask questions about their own nation and their specific communities within these other nations.

Gizelle Peynado was a Jamaican preveterinary student at MSU. In some of our first discussions, Gizelle noted as she described her family and community that many Jamaicans lived in the United States. Even her parents and extended family where choosing to leave Jamaica. Jamaica has the second highest incident of brain drain in the world (Haughton, 2013). Her friends in other US universities were making similar observations. It was from her perspective of being across an ocean, in a "safe" classroom where she could question *why* her country to where she very much wanted to return and raise her children, was creating a brain drain. Over 20% of Jamaicans worldwide live in the United States and almost half (40%) of the total number of Jamaicans live outside Jamaica.

First, Gizelle sought to understand and then, together with her classmates also seeking education outside Jamaica, explore what they could do to help their country change into a place these educated Jamaicans would want to return. Through her research, she found in in-depth interviews with 17 in the Jamaican diaspora, would really rather return to Jamaica to work and raise their families. Gizelle concluded, with her Jamaican classmates, the answer lay in large part with the unbalanced economic base causing economic stagnation. Her research seemed to point to the food insecurity created by the burden of repaying World Bank and International Monetary Fund loans and reliance on agricultural imports rather than subsisting on their own agricultural production and improving their own agricultural sector as a key factor in the decision to leave Jamaica.

Aedine Ndi Peyou, a cell biology and neuroscience major from Cameroon, began her cultural introspection in the first week of class as she read excerpts from Bennett (2004) and Norberg-Hodge (1992). She summarized the start of her journey as follows:

> The concept of ethno-relativism is quite eye-opening. There are so many ways of thinking about culture that I don't even realize. As I read, I tried figuring out where I was on Bennett's scale (2004) of intercultural development—somewhere between adaptation and integration. I am the former, but I also experience the feeling of not really feeling culturally belonging anywhere. I am really learning how much Western culture influences so many things to the point of diminishing—minimizing other cultures. Ancient Futures (Norberg-Hodge 1992) has shown me that.

Development does not equal westernization.

By the second week of class, Aedine understood her needs as a diaspora participant, a Cameroon citizen living in the United States while receiving her career training. Aedine summarized the start of her discovery process:

> After reading Ayittey's Africa Unchained (2005), I realized how the issue of Africa comes from Africa itself. I am part of the "cheetah" group and part of the diaspora. As a cheetah, how can I help initiate change that is positive in Africa—in Cameroon! My own president has been in power for over 20 years. He changed the constitution so that he could be in power for longer. It is now a dictatorship. The people in leadership positions in the government are all wealthy. It is true, as Ayittey (2005) stated so well, that this is how Cameroon operates. All the money is retained by people working in the government. So I am learning and I hope to come up with ways that I can positively work in Cameroon with simple, non-elite solutions. I don't know what that looks like yet.

Aedine recognized herself as one of a group of young leaders whose job it would be to guide her agricultural-based country out of a dictatorship without resorting to relying on a colonial power. This is a "tall order."

Health issues, food security, and governance come together at the nexus of traditional ecological knowledge.

The challenge Aedine faced during her undergraduate education was to find a safe forum and guidance from her professors to consider all of these parts of her world at the same time, not in disciplinary pieces. She needed at least one of her professors to provide guidance while she explored the nexus, a professor who could advise from a transdisciplinary position, not from a disciplinary piece.

Within the first month of the semester, all students together visit the Northern Cheyenne and Apsaalooke reservations and live on the Northern Cheyenne reservation for 4 days. To prepare for this visit each semester, Jason Baldes presents the Wind River water conflict and shares the film made on his reservation by his family and fellow tribal members. After this class session with Jason and watching the film, Aedine continued her cultural journey with this observation, "I learned the value of not making assumptions in policy-making."

A big problem arises when we write policies based on definitions that do not apply to all cultural groups.

> <u>All</u> groups, even minority groups, need to be included in the decision—making process. The holistic process is great for doing just that— determining the desired quality of life and forms of production of each group involved. I hope that I can be an advocate for the holistic process, especially in my country where there is little sense of decolonization (*suppression of dominance by the former colonial powers or by Cameroonians adopting ways of the French or English*) in the cities.

On Day 2 of her first reservation visit, we asked Aedine to specifically help us understand what colonizing language is and how to decolonize our speaking and writing when we are working with Native Americans there on the reservation and in class. Cameroon was most recently colonized by both France and England. Aedine grew up in the French section of Cameroon. She made this observation about her own project on perceptions of health care in Cameroon:

> When approaching my own culture about the project, I realize that I have to even relearn how to properly, respectfully ask questions, and, motivate people to answer. I want to know how to help or allow people to achieve their desired quality of life in the midst of (the on-going atmosphere of) colonization.

Aedine's final thoughts were:

> I am glad that I'll be able to carry what I learned from this class to every single community I will help. By using the holistic process I can find out what the desired quality of life of the people I serve really is. I can start from the roots when it comes to helping my own community. I will keep all the concepts I learned from this class—Bennett (2004), Ayittey (2005), the

Ladakh village (Norberg-Hodge 1992), the reservations—all these readings and experiences add on to my knowledge base to understanding cross-cultural leadership.

Aedine became a TA in AGSC 465R, graduated with a BS in Cell Biology and Neuroscience, and now is in Physician Assistant school. She completed her clinic rotations in Guatemala and Tanzania and she and her husband intend to eventually return to her home in Yaounde, Cameroon where her family lives. Aedine's research project focused on designing and conducting in-depth interviews with 100 citizens in the city of Yaounde and the traditional village of Aedine's grandmother to determine perceptions of health care and specific health care practices in rural and urban areas in Cameroon. Desired quality of life was explored with the 100 participants as was their perceptions of health, health care delivery, and the role of traditional medicinal plants and Western culture remedies in their own process of keeping healthy. During the semester, Aedine contributed to local understanding of the health care system there in her own community. Her most surprising results were in the abandonment of traditional medicinal plants in the rural community and the high level of appreciation for these plants in urban areas. Aedine may not be on the road to a specific career in Plant or Animal Bio-Science or Nutritional Biochemistry, but with her holistic approach to health care, certainly diet and food choices and other health-choices such as medicinal plants will be a central part of her practice. Aedine engaged with her course-work in this original research course on health, poverty, and agriculture in exactly the way she needed to in order to have an immediate impact on her community, her own life, and her career.

Akihiru Kuriki: This story is about an agriculture cropping systems major in the University in Nagoya, Japan. Aki came to MSU as an exchange student in 2013. He was a junior and wanted to learn some agricultural production ideas that he could make use of in Japan after graduation. He had one more year back at his university in Nagoya before he was "on his own." He wanted to live in Nagoya, the same city where he grew up, a metropolitan area with intensive farming on the perimeter. He did not know where to begin to figure out what would be a profession in this location that would have to do with food and agriculture, and not be a factory or desk job. In AGSC 465R, he chose his home town as his community-of-focus and proceeded to conduct a series of in-depth interviews using the holistic process with his family and friends in Nagoya, Japan. The age distribution of his sample was bimodal, young people his age and elders his parents' and grandparents' ages. His research was conducted in Japanese. What he discovered is that young adults his age were not interested in producing their own food, but did like to eat nutritionally trendy foods in restaurants or prepare them at home. Elders, he interviewed were interested in producing some of the traditional foods, particularly herbs and vegetables, in their own gardens. The reflective log and the individual weekly mentor meetings allowed Aki's specific career development needs at that moment in the United States to be satisfied. At the end of his exchange year he emerged with a clear idea of how to carve out a career niche he could be passionate about in his home town, farm-gardening traditional herbs and Japanese vegetables. Aki was ready to complete his final undergraduate year back in Nagoya focused on what skills and market analyses he would need to complete to become a successful farmer in his home-town metropolis.

Sometimes it is not so much academic help or technical skills development that a student needs from higher education, as it is a buffering process to create a learning environment in which all views, gender orientations, and ethnic origins are appreciated and celebrated.

Dawn Delaney Aimsback-Falcon describes her home and introduction to higher education as follows:

> Nestled in the eastern slopes of the Rocky Mountains adjacent to Glacier National Park and the Bob Marshall Wilderness, lies a small community within Pondera County, Montana. With its rustic scenery and picturesque mountains in the background, the community is a hidden gem known to locals as 'God Country.' This is my home in Heart Butte, Montana. Twenty-six miles to the northeast of Heart Butte is the main town of Browning, Montana and the location of the local community college. It is this college that set my path to engage in classes that center around general studies and my culture. The experience at my local community college allowed me to see an alternative view when it came to science. I embraced the local knowledge that was shared with my peers. To engage with Elders on knowledge that is passed from generation to generation through oral traditions of storytelling shaped my approach to education. This gave me the push and the strength to receive my Associates Degree and to continue on with my education at Montana State University (MSU).

Dawn is an enrolled member of the Amskapii Peikuni, a part of the group of Native Americans that non-Natives call Blackfeet. Dawn is also part African-American. From early years, she encountered ostracizing while growing up both on and off the reservation among the Amskapii Peikuni because of her "other ethnic origin." Still she was not entirely prepared for the experiences at MSU. In classes for her minor in Native American Studies, Dawn learned "how people come to believe false information regarding a minority race" and "what non-Natives perceive about Native American First People; lifestyle, culture, contemporary Native people, and Native American Policy."

Dawn reflected on her first week of classes at MSU.

> I assumed that my undergraduate classes would respect other people's values and culture. I would soon figure out that people are quick to respond with uncalled—for comments that center on others cultural practices...In one of my undergraduate classes I was astonished by the comments of the students when the discussion turned to Montana First Native People...to the demise of the culture and way of life of the Plains tribes...I felt uncomfortable...First, I was upset that I did not respond to the negative comments, and second, I had never partaken or heard of this type of academic conversation at the community college, The Blackfeet Tribal College. Third, it was a very weird experience for me, I felt embarrassed for the students and the professor...This incident with the offensive discussion, was my first perception of the university. That was a freshman level class. As the semesters went on and I took higher leel courses in my majors of Native American Studies and Environmental Studies, the conversations were more mature, and I got along better.

Some Native Americans would have given up after this first experience at a Western culture university. Dawn was strong and had also experienced some negative response to her own African origins from her own people, the Amskapii Peikuni. This strength-through-accepting her own differences coupled with her well-grounded traditional up-bringing and her interest in the preservation of sacred sites of her people played a role in her being chosen to be part of a cultural history and language project the MSU Department of Education had underway with communities in the Altai region of eastern Russia. Being with Indigenous people other than her own tribe and half way around the world was a confirming experience for how Dawn wanted to engage with her coursework to have immediate impact on her own community among the Amskapii Pikuni. Bolstered by the connections she felt with the Altai people, Dawn began her senior year at MSU with a spirit of activism. She had a new-found confidence from the summer immersion with the Altai people and she was ready to put the core readings of AGSC 465R[1] to work for her own tribe as well as to fulfill her own personal goals. In AGSC 465R, she noticed the

> Indigenous peoples point of view...was presented through a mindset that engaged students to see through a non-Western view...a nonlinear approach. The professor introduced an alternative way to do research, using a Traditional Ecological Knowledge point of view. This insight sparked a light inside of me! This frame of thought is unique within higher education. The respect that is reflected through the learning process allows misconceptions to be dealt with through educated conversations that pertain to the given culture, lifestyle, gender, or religion. Once this was established,

we began to see the barriers broken on subjects that embrace an
alternative view, using the holistic process.

Dawn wanted to engage with her coursework in an accepting, non-threatening environment so that she could use the coursework to have an immediate impact on what she cared most about: preservation of sacred sites of the Amskii Pikuni; and reintroduction of free-ranging bison on her reservation, the Blackfeet, nestled in the Eastern slopes of the Rocky Mountains. This kind of course, this specific learning environment, most of all, as Dawn observed,

gave me the motivation to speak more and be more attentive to
issues...to not hold back on your beliefs. Speak up and make a point and
hopefully it will get across to the person with whom you are
communicating. Do not let people define who or what you are about. Let them
define you by how strong you are to be different and unique. Remember why
you left your family and home in the first place,—to make your reservation
better as a whole. This mindset shapes the whole class, learning to break
barriers that define wealth and happiness. This course is very important to
MSU because it allows students to engage in conversation and research within
indigenous nations. It allows us to use common sense skills. The class comes
alive, literally! Students see and engage in topics that are of great concern to
the local people within contemporary times. Universities can learn from this
class. AGSC 465R Health, Poverty, Agriculture: Concepts and Action Research
will strengthen the minority quota and set a new tone for academia—preparing
their students to engage with local minorities on issues that are key to the
holistic view.

After her graduation from MSU-Bozeman, Dawn became USDA Extension Director at Blackfeet Community College in Browning, Montana.

To engage and partake with Indigenous people allows for misconceptions to be put to rest.

Jamie Sowell is a Caucasian who grew up in rural Oregon and had never lived abroad or even traveled widely in the United States until her senior year as a university student. Why were her needs as a Sustainable Foods and Bio-Energy Systems major (cropping systems option) similar to those of a foreign student such as Aki or Aedine or a Native American such as Dawn, Jason, Amy, and Sierra? During winter break between semesters and in the first week of class, Jamie digested the first assignment in AGSC 465R. She read two books, *Holistic Management: A New Framework for Decision Making* (Savory and Butterfield, 1999) and *Ancient Futures: Learning From Ladakh*, which readers find immerses them in an ancient culture living in harmony with the land at 13,000 ft in the Himalayan Mountains (Norberg-Hodge, 1992). She also read a book chapter on development of intercultural competency (Bennett, 2004), part of the core reading.[1] By the time Jamie came to her first mentor meeting with me during the second week of class, she had already decided how these concepts were related to her own desired quality of life, which appeared as seamlessly tied to that of her family as that of an Indigenous people, where one is a member of the community first and secondly, much less important, an individual. In this course and in the holistic process, desired quality of life is the starting place for our conversations and ultimately for the research process with the student's community-of-focus. Jamie was ready to engage with her coursework to have immediate impact in her community and on her life and career.

Jamie's analysis after the first week of class was:

> Holistic management requires a lot of communication and cooperation to be properly implemented. It seems to work best with smaller groups and organizations. I think it would be interesting to see if it could work on a community scale. I would like to try to approach the management of my family's property using a holistic goal. I think that most of my family would find it interesting and participate, but I fear that a few would refuse. However, I want to talk to my grandmother about it.

With respect to her own development of intercultural competency, Jamie revealed that she was in the acceptance or minimization stage of Bennet's ethnocentric/ethnorelative development continuum. This was a surprising observation since Jamie's personal experience in foreign countries and with other cultures was nonexistent or extremely limited. Jamie, however, recognized that key relationships in her extended family had been helpful in her development as a culturally integrated community member and Sustainable Foods major. Jamie wrote: "This article made me think of some of the conversations I have had with Tiazza, my foster aunt from Morocco." Jamie was culturally astute. In her Animal Science (ANSC 222) Livestock in Sustainable Systems course, a small class exercise made a strong cultural impact for her. Of this she writes,

> My ANSC 222 professor shared with us photos of families from around the world with their groceries for the week. Below each photo were the figures on how much and what percentage the family had spent on meat, fish, and dairy as well as on prepared/fast food. The healthiest-looking families tended to spend less in general and had more fresh or dried food. Also, they spent 0—1% on prepared/fast food. I found this very interesting. It made me think of this summer when I spent about $150 a month on groceries from the Community Food Co-op and how most of what I bought was fresh or dried as well as semi-local and organic.

In Week 1, we were not only seeing Jamie making connections with her own family and food consumption styles, but she was beginning a deep analysis of her community and how this affected her own desired quality of life. These were the broad connections she made in Week 1 of the course, "Stereotypes within communities, families, and society…have power." Jamie shared with us a prevailing stereotype in Cave Junction, Oregon, her hometown. "For example," she wrote, "Outsiders familiar with the area consider everyone in Cave Junction a 'stoner,' especially in the Takilma area. Why is this? Why do stereotypes get established and how are they perpetuated?" She then made the connection between her community and connection to the land for production of food.

> Ancient Futures: Learning from Ladakh (Norberg-Hodge, 1992) drove home that when a community loses their connection to the land and becomes materialistic that the quality of life deteriorates. Native species need to be protected and Native food crops should not be replaced with GMOs and hybrids. For a healthy community the members should help each other without compensation. I wonder if it is possible for a former mining and logging community (such as Cave Junction, OR) to achieve even a fraction of that which the Ladakhi people have, in terms of community connectedness and self-reliance. Can the principles of the Ladakhi survive along with modern amenities like computers and CDs? I don't think I can give up music (CDs/Youtube) for complete self-reliance.

This passage is even more important when one understands that Jamie was an exceptional scholar, allowed to take the course designed for juniors, seniors, and graduate students even though she was only a sophomore, and very shy, even in this class of just 18 students. Without the individual weekly mentoring sessions and sharing of the reflective log this set of rapid connections she made at the beginning of the semester, and the action research she conducted during the semester may not have happened. Jamie chose Cave Junction and its 1875 residents as her community-of-focus for the course. She was the eighth generation of her Euro-American family to live there and held one of the most common last names in the area.

By Week 5, Jamie was seeing the drama of her home town superimposed on the concepts of the course. Ideas were coming together as she used the holistic process (Savory and Butterfield, 1999) to begin in-depth, electronic interviews in her home town. She wrote,

> And more chaos ensues. I am almost afraid to go onto social media sites anymore. I learn about what is going on in the Valley when I do. This is how I learned that a former classmate got shot, two men died in a hit and run, a homicide on the street a friend lives on, a teen jumping in front of a truck.

I feel so helpless. It will be years before I can do anything that might help. For now all I can do is gather information and watch from thousands of miles away. Some other things that happened, but Dad beat the Internet to informing me: the post office burning down (arson), the bank I use getting robbed...our irrigation pipe getting stolen from underneath blackberries...I can see, especially now, why we have such a bad reputation [in Cave Junction]. The thing is this is the worst it has ever been. Dad told me about some drug/biker wars in the 60s but that is the closest to what things are like now. What caused things to go downhill like this? I know that there are little rays of hope within the community. I just don't know how to show the rest of the world this. What both readings for this week (Dunkel et al. 2013; Halvorson et al. 2011) had in common was the bottom-up approach. For the people of Sanambele, Mali (Dunkel et al. 2013) this was eliminating cerebral malaria that their kids were dying from. For the four Malian villages in the inland delta of the Niger River, it was childhood deaths from diarrhea (Halvorson et al. 2011).

At the end of the semester, we saw Jamie using a bottom-up approach bring together these various threads; food, community drama, family goals, and desired quality of life. After this study in AGSC 465R, Jamie completed her coursework in Sustainable Foods and Bio-Energy Systems and spent a year in Portugal studying language and additional major and minor requirements. When she returned for her graduation ceremonies at MSU, 3 years after she began this holistic process in her own community of Cave Junction, she made an important announcement surrounded by her extended family, her professors, and department head. Jamie announced that she will return to her home, and implement the action ideas suggested by her holistic research there as part of AGSC 465R in addition to beginning the long-term development of an olive farm. The Sustainable Cropping Systems course at MSU that employed similar individual approaches to complex, but real, issues, had given her curricular space and time to explore the long-term environmental needs, nutritional benefits, and economic sustainability of olive cropping. Jamie has a distinct plan now to dispel local stereotyping with, among other actions, olive farming on her family farm. Jamie's undergraduate education met her needs, her family goals, and may as well stimulate an impressive community revival in Cave Junction, Oregon.

Simply retuning our ears, opening other channels of listening, and requiring individual mentor meetings with reflective journaling will help students like Amy to have successes like Sierra, Jason, Greta, Gizelle, Aedine, Akihiru, Dawn, and Jamie.

The take-home messages from Case Study #4 require extraordinary, in-service preparation by professors, administrators, and policy makers. Rewards are great in individual student success and in locally solving complex issues. Preparation is different, perhaps for some it will seem difficult, but not impossible.

Students want coursework that is directly useful in solving issues personally important to their own career, family, community, now, immediately, at this moment.

WHAT STUDENTS WANT

Create Personal Connections

Each of these stories is distinctly different and yet, tell of personal and academic growth through Dr. Dunkel's and others teaching methods. Without the disciplines of each student's tale being told in reflective log entries and in mentoring conversations and this being understood by the professor, much of this learning would be lost. It is only through one-on-one mentor meetings that personal connections can be made between student and teacher to a point where they are able to teach and learn from each other. This type of connection is essential if research is going to be impactful and sustainable. From this connection, combined with the use of the holistic process (Chapter 3: Decolonization and the Holistic Process), and reflective journaling, a trusting advising environment emerges. Such an environment is ideal for learning. The following summary comments are a synthesis of suggestions of the 218 MSU students who have conducted this community-based research using the Expansive Collaborative Model and the holistic process.

Two coinstructors further enrich the course with personal connections: Dr. Hiram Larew, recently retired as Director of the Center for International Programs at the US Department of Agriculture National Institute of Food and Agriculture in Washington, DC (Chapter 8: Listening Over Power Lines: Students and Policy Leaders), and Jason Baldes, Executive Director of the Wind River Native Advocacy Center, Ethete, Wyoming. Hiram's presence in the classroom brought new meaning to the course. While we were focusing on the complexity and uniqueness of each person and culture and we were working with through our independent research, Hiram was a force that kept our work in context. Jason gave us the example of a real, live "Cheetah" with whom we could watch and to whom we could pose questions. Ayittey described the Cheetah (2005) as a way forward for his continent, Africa, and how, in conversations with us in Bozeman, he agreed was the way forward for Native American Nations as well.

Greta summarized this as follows:

> While I worked with a few members of the Apsaalooke Tribe on a specific
> task, I was having discussions with a major decision maker in global food
> politics about the issues I was facing on a global scale. Many times I made a
> mistake in how I interacted in Apsaalooke culture, and I felt disheartened or
> as if I was not meant to be doing this research. And then, speaking with
> Hiram in class, I came to realize that this discomfort and making mistakes
> happens at levels much higher than my own experience—that USAID, the
> USDA, these large, almost intangible branches of government, make these
> cultural mistakes. And that drove me to continue my research, and to see
> the importance of it. If I wanted to make a global impact, I was going to have
> to be okay with being uncomfortable and wrong now and then, so that if I
> was in a position to make decisions that could impact entire countries, I
> would have the perspective to imagine the ways my plans could be taken
> differently. Hiram helped me see that it was through cross-cultural
> work that anything, in regards to food and food safety, gets done on a
> global scale.
>
> The second take away from Hiram's presence in the class was that it, the
> USDA, and in some ways the Federal Government, seemed like something
> real. During my time taking the course and being a Teaching Assistant for
> the course I heard countless students comment about how surprised or
> inspired they were that he was there. For many, myself included, the
> knowledge Hiram shared paired with his acute interest in our thoughts
> and research was a motivator. To converse with him made me think for the
> first time that I, or my voice, may have a place in the USDA or another
> Federal Agency.

Close the Gap Between Teaching Methodologies and Real-Time, Complex Issues

Millennials have observed a gap or disconnect between teaching methods used in food and agricultural sciences and the ways that students, the Millennial Generation, the younger generations, and adult learners, want to learn. Food and agricultural sciences, not unlike other STEM disciplines, have a designated "wiggle room" for "alternative teaching" and cross-disciplinary work. For example, presentations on the Apsaalooke Let's Pick Berries project were not discouraged. Faculty and students were not told it was inappropriate. It was acceptable for someone else to organize and implement. Plant Sciences can be taught through addressing a complex and real problem. All the basics of this discipline and all related disciplines can be incorporated into the teaching process. There is no program, no professor I have encountered that is against cross-disciplinary work, encouraging cross-cultural exchange, or engaging students in their education on a personal level. Academics and future employers smile upon mentoring, international exchanges, and collaborative projects. The real issue is that, while these ideas are attractive, many professors teaching relatively small, seminar-style courses see them as "extras" and administrators are reticent to assign a TA to these capstone. They are happy to have students take charge of their education, but when it comes to transforming curriculum to be more cross disciplinary, working with other cultures themselves in their classes, or finding time to personally mentor each student, these seem like outlandish and unrealistic changes. Ideally, these ideas work best in smaller classes. Some suggest it is time to put the large class size 100 to over 600 on the examining table, particularly if one wants to recognize culture in the discipline. However, in Chapter 7, Listening Within a Bioregion, we explored the power of the circle as a way for middle-sized classes or other groups of 40−60 people encourage discussion by promoting openness, transparency, and trust. For larger groups, especially in the over 100 size gatherings, multiple circle discussions with reporter sharing captures some of this personal connection and individualized learning.

No matter the class size, we want our classes to engage and challenge us, we want to work across disciplinary and cultural lines, and we want to see results from our work.

This is where the gap between how students are being taught, and how they want to be taught lies:

> One report I found when researching this chapter explained how students of the Millennial Generation are less engaged and less concerned about grades and classes, and that short-term immersion was a solution to this issue (Anderson, 2006). As a recent graduate who has engaged with hundreds of my peers at multiple universities, I take offense to the idea that Millennials are less engaged or are just in school for a degree. I would call us intelligent, capable, and highly engagable individuals. The issue is, our education both in high school and university programs is not relevant to the world. Our education does not even teach us how to take out a loan, fill out our taxes, or rent an apartment, much less how to engage in the world either with indigenous communities or any communities or with local or national government policy makers. We are faced with issues such as food deserts, global warming, gender inequality, poverty, racial injustice, and political systems ridden by corporate influence, just within our own country. Many of these concerns extend beyond to a global scale. Of course each generation has their challenges, but ours are looming now

and we do feel the pressure. Finding ways to handle these issues will require innovation, collaboration, and transformations of thought.

If we seem unengaged in our education, it is not because we are uninterested in our world, but because we *are* interested, and our education seems irrelevant to solving these "grand challenges" and wicked problems. Of course technical skill and basic knowledge are necessary—we cannot work to end malaria if do not understand the disease and its transfer. But we want more than that in our education: we want to be able to take our skills and collaborate; we want to have in impact outside of the university; and we want to see the fruits of our labor in more than a degree or a grade point average.

Provide Cross-Cultural Immersion

Considering students are customers of universities, educators and administrators should work to meet their needs, not just to increase enrollment, publication numbers, or profit. How ridiculous will we all feel when our universities are making high profits, publishing like none other, enrolling thousands of students, and pumping out graduates, but we are unable to feed our country and world. It is time to listen to what students want in the classroom, and not just allow it, but also make it easy for students to do. Cross-cultural and transdisciplinary work should be a norm, not an exception. Immersion should not be a thing only students going on exchanges on their own accord do, but something integrated into the curriculum. Regardless of your major, but especially if you are studying food and agriculture, cross-cultural work and student engagement is a requirement for educating students who: (1) understand current issues in their disciplines and (2) have the tools to contribute to a better world in a way that is sustainable scientifically, culturally, and personally.

Of course forced cultural immersion could be problematic; not everyone wants to spend a semester or year abroad and requiring time abroad would cause harm to the host communities who would be receiving uninterested and possibly resentful and resistant participants in cross-cultural exchanges. In Chapter 4, Immersion, Case Study #2, however, we describe the program that has been underway for 40 years at the agricultural university of Morocco requiring for all majors, an immersion during each of the 4 years of undergraduate study. Classes like Dr. Dunkel's show that cross-cultural work comes in many forms, however. The United States is a diverse place and working with other cultures is a possibility in every state. There is a way for each person in a Land Grant institution to work on one cross-cultural project at some point in their 4 +

years of university education. Believing otherwise is a denial of the diversity of our country, and the sovereign nations that it surrounds. When Amy, Sierra, Jason, Gizelle, Aedine, Aki, Dawn, and Jamie shared their research progress in class each week or interjected their views in our concept discussions, it was a miniimmersion for all the rest of us, including the instructor. Many of these students went on to be TAs and Jason became a coinstructor.

With the rise of sustainable foods programs in the 1990s and 2000s, literature began to emerge on teaching methodologies for immersion programs. From the mid-2000s to present, pedagogy research has appeared in journals such as *Agriculture and Human Values*, *The Agricultural Education*, *North American College of Teachers of Agriculture (NACTA) Journal*, and *The Agricultural Magazine*. The general trend of this literature pushes educators to teach using cross-cultural and interdisciplinary methods (Dunkel et al., 2011) to meet students' need and interests, as well as to meet our future employment demand and global food needs in the next 50–100 years.

As discussed in Chapter 1, The Quiet Revolution: Where Did You Come From?, having enough well-educated agriculture students, workers, and teachers will be paramount in facing the "grand challenge" of feeding 9 billion people by 2050. A 2015 study published in the *NACTA Journal* acknowledged the future need for STEM workers in fields of food and agriculture, and asked why students select majors relevant to food and agricultural sciences (Hegerfeld-Baker et al., 2015). Their survey spanned 3 universities and of the almost 500 survey respondents, 79% of which were in their first semester of college and over 300 of which had chosen a STEM major related to food and agriculture (Hegerfeld-Baker et Al., 2015). Surveys revealed three statistically significant factors cause students to choose a food and agriculture related major: passion/enjoyment of field of study; financial gain and security; and high school courses (Hegerfeld-Baker et Al., 2015). Here we focus on the passion factor. Unlike high school courses and financial security, passion is something university administrators and professors have complete control over.

Encourage Passion

How, exactly, do educators ignite student's interest and passion for food and agriculture sciences? Research shows that student engagement, relevance of subject matter, and immersion are important. As Ronald Phillips et al. discuss in a 2008 publication on short-term international immersion for agriculture students, in a globalizing world it is increasingly important that agriculture students spend time abroad. With issues of food access facing developing countries disproportionately in the face of climate change, answers may be lying dormant in another culture half way around the world or in the nearest Indigenous community. "Exposure to the basic unmet needs of people around the world puts in perspective the value of research and teaching in agriculture" (Phillips et Al., 2008). This experience changes students' lives and truly engages them in their studies, igniting their passions, and hence keeping them in the food and agricultural science. The issue Phillips encounters, and that we have encountered in AGSC 465R Health, Poverty, and Agriculture: Concepts and Action Research, is that international immersion is often too large of a time commitment for students working toward a degree, taking years to be done well. If university tuition was inexpensive, such as at MSU, or free, immersion could be required, but because of the cost, the time-crunch of the university is a reality for many.

James C. Anderson II, the Director of Distance and Graduate Education and assistant professor at Virginia Tech in his study as a graduate student on minority engagement in agriculture, food, and natural resource studies, found "Providing meaningful experiences in agriculture, food and natural resources... that incorporate not only agriculturally related content, but strategies for academic, social, and career success is a method for increasing selection of agriculture, food, and natural resources as an area of study" (Anderson, 2006). He argues that a combination of affirmative action, opportunities for leadership, and relevance to future success will bring more diversity to the food and agriculture fields.

Considering that "minority" groups in the United States will soon be greater than 50% of the population (making their identification distinctly incorrect), it is important to discover how to encourage more diversity in the food and agriculture fields (Anderson, 2006). "Minority groups" often have different cultural perspectives, which as we explain in detail in Chapter 5, Listening With Subsistence Farmers in Mali, and Chapter 6, Listening With Native Americans, can have extremely positive impact on students, communities, and agriculture science itself. Collaborations with Tribal Colleges provide a wealth of knowledge to understand the process of Native Science and information on local foods and sustainable life styles (Cajete, 1999; Savory and Butterfield, 1999).

Response From Florence Dunkel

Students speak often in this chapter about the MSU course AGSC 465R Health, Poverty, Agriculture: Concepts and Action Research. This course actually began as an impassioned request from a student, Kathleen (Katy) Hansen, a sophomore Honors College student at MSU pursuing a double major in Agricultural Economics and Industrial Engineering. Kathleen was already a student leader who in the near future would become Vice-President of the MSU student senate and President of Engineers Without Borders. In a face-to-face meeting in my office, she kindly, but emphatically said that I must redesign the small course I offered since 2000 for Mali externs (students) who are given two semesters of research mentoring and cultural preparation for their holistic process-based work in villages in Mali. This redesigned course must be open to all students who meet the prerequisites, not just those students going to Mali, Katy insisted. The course must tackle the core readings that MSU faculty and Malian mid-career scientists and an engineer had been gathering on-campus after work to discuss.[1] Being a firm believer in listening to students, I agreed. After MSU approval, the course has been offered every Spring and Fall semester since 2008.

In summer 2008, Katy conducted the first solo (without me and other externs) externship in Mali. Her observations in Sanambele during a national malaria epidemic led to an intense collaboration between the 15 and 20 students in AGSC 465 each semester and the village of Sanambele, all focused on cerebral malaria (Chapter 5: Listening With Subsistence Farmers in Mali). Beginning in 2009, there were no more deaths from cerebral malaria (with two exceptions related to malnutrition [kwashiorkor] and travel to a neighbor village). By May 2016, 218 undergraduate and graduate students had completed AGSC 465R.

Listen to students. Take action. It is our conduit of service to the world. By listening to students, you will light a fire of experiential learning of the sort described by Irish poet and playwright William Butler Yeats, who won the Nobel prize for literature in 1923, "Education is not the filling of a pail but the lighting of a fire."

---

[1]The core readings for AGSC 465R were selected by a team of faculty and graduate students from: Plant Sciences and Plant Pathology; College of Business (both Finance and Management); Agricultural and Technical Education; Land Resources and Environmental Sciences; Environmental Engineering; Liberal Studies; and Modern Languages and Literatures. These were: Ayittey (2005), Calderisi (2006), Easterly (2006), Yunus (2007), Savory and Butterfield (1999), Bennett (2004), Chambers et al. (1989); and Norberg-Hodge (1992). On a weekly basis, they would gather to discuss the books over dinner at various faculty homes. From this process of sifting, winnowing, and heated discussions, the core readings for AGSC 465R were selected. In the ensuing years, these faculty wrote additional literature (Dunkel et al., 2013; Halvorson et al., 2011), which became part of the core readings and Calderesi (2006) was dropped. An environmental conflict set of readings was added by Jason Baldes. Nine authors or author groupings were selected and have remained fairly constant. The 10th author grouping is a set of culture specific readings. These are chosen by the community or site mentors for the specific site.

CONCLUSION

The key lesson of this chapter is that younger generation students have a desire to see their studies impacting people outside of academia and want to make the world a better place through the work they do, not just earn a degree. They should be encouraged to engage in conversations and projects that have real impact. Millennial generation students and the next generations can contribute to solution of local problems with guidance from professors and site mentors who are well-integrated into these communities and who share the younger generations eagerness for experiential learning.

References

AIARD, 2017. SMART Investments in International Agriculture and Rural Development: Recommendations to the New Administration and Congress. Association for International Agriculture and Rural Development. http://www.aiard.org/uploads/1/6/9/4/16941550/smart_investments_final.pdf.

Anderson, J., 2006. Insights for recruiting underrepresented individuals into careers in agriculture, food, and natural resources. Agric. Educ. Mag. 72 (5), 11−13.

Ayittey, G.B.N., 2005. Africa Unchained: The Blueprint for Africa's Future. Palgrave Macmillan, New York, 483 pp.

Basso, K.H., 1988. Athabaskan-English interethnic communication. Source unknown. Cited in Elliott, C.E. 1999. Cross-Cultural Communication Styles, pre-publication Masters thesis. Online at <http://www.awesomelibrary.org/multiculturaltoolkit-atterns.html>.

Bennett, M.J., 2004. From ethnocentrism to ethnorelativism. In: Wurzel, J.S. (Ed.), Toward Multiculturalism: A Reader in Multicultural Education. Intercultural Resource Coporation, Newton, MA, pp. 62−79, 406 pp.

Cajete, G., 1999. Native Science: Natural Laws of Interdependence. Clear Light Publishers, Santa Fe, NM, 339 pp.

Calderisi, R., 2006. The Trouble With Africa: Why Foreign Aid Isn't Working. Palgrave Macmillan, New York, NY, 249 pp.

Chambers, R., Pacey, A., Thrupp, L.A., 1989. Farmer First: Farmer Innovation and Agricultural Research. Intermediate Technology Publications. London, UK. Short Run Press, Exeter, Great Britain, 218 pp.

Disney, 2015. McFarland USA. Film 8039849, 129 min.

Dunkel, F., Coulibaly, K., Montagne, C., Luong, K., Giusti, A., Coulibaly, H., et al., 2013. Sustainable integrated malaria management by villagers in collaboration with a transformed classroom using the holistic process: Sanambele, Mali and Montana State University, USA. Am. Entomol. 59, 15−24.

Dunkel, F.V., Shams, A.N., George, C.M., 2011. Expansive collaboration: a model for transformed classrooms, community-based research, and service-learning. North Am. College Teachers Agric. J. 55, 65−74, December.

Easterly, W., 2006. The White Man's Burden: Why the West's Efforts to Aid the Rest Have Done So Much Ill and so Little Good. The Penguin Press, New York, NY, p. 436.

Edlund, E., 2001. Wind River: Study Guide. Bullfrog Films, Oley, 14pp.

Elliott, C.E., 1999. Cross-Cultural Communication Styles, pre-publication Masters thesis. Online at <http://www.awesomelibrary.org/multiculturaltoolkit-atterns.html>.

Elliott, C.E., Adams, R.J., Sockalingam, S., 2016. Multicultural Toolkit (Toolkit for Cross-Cultural Collaboration). U.S. Office of Minority Affairs, Seattle, WA.

Flanagan, C., Laituri, M., 2004. Environmental assessment of local cultural knowledge and water resource management: the Wind River Indian Reservation. Environ. Manage. 33 (2), 262−270.

Gunn Carr, D., Hawes-Davis, D., 2000. Wind River. High Plains Films. 35 minutes video.

Halvorson, S., Williams, A., Ba, S., Dunkel, F., 2011. Water quality and water borne disease along the Niger River, Mali: a study of local knowledge and response. Health & Place. 17 (2), 449−457.

Haughton, A., 2013. A look at Jamaica's brain drain. *The Gleaner*. May 29, 2013, Kingston, Jamaica, West Indies.

Hegerfeld-Baker, J., Anand, S., Droke, L., Chang, K., 2015. Factors influencing choosing food and agriculture related STEM majors. NACTA J. 59 (1), 34−40.

Kochman, T., 1990. Force fields in black and white communication. In: Carbaugh, D. (Ed.), Cultural Communication and Intercultural Contact. Lawrence Ehrlbaum Publishers, Hillsdale, NJ.

Norberg-Hodge, H., 1992. Ancient Futures: Learning From Ladakh. Sierra Club Books, San Francisco, CA, 204 pp.

Olquin, L. 1995. Keynote address to California State Education Conference, October, 1995, Sacramento, CA. In Elliot 1999.

Phillips, R., Magor, N., Shires, D., Leung, H., Mccouch, S., Macintosh, D., 2008. Student opportunity: short-term exposure to international agriculture. Rice 1, 11−15. Available from: http://dx.doi.org/10.1007/s12284-0089003.

Savory, A., Butterfield, J., 1999. Holistic Management: A New Framework for Decision Making. second ed. Island Press, Washington, DC.

Yunus, M., 2007. Creating a World Without Poverty: Social Business and the Future of Capitalism. Public Affairs, New York, NY, 261 pp.

Further Reading

Basso, K.H., 1970. To give up on words: silence in Western Apache culture. Southw. J. Anthropol. 26 (3), Autumn.

SUMMARY ILLUSTRATIONS

These photos summarize valued experiential learning opportunities and provide evidence that students can move rapidly from a passive student to an active learner and respond with action and creativity to a community's requests. Hundreds of additional photos are available at www.montana.edu/mali under Partners New Courses (Figs. 9.1−9.9).

FIGURE 9.1

AGSC 465R students gather at the Dunkel-Diggs farm to discuss the concepts of Ayittey (2005) and practice participatory diagramming (Chambers et al., 1989) in the field and with farmer interviews.

FIGURE 9.2

Northern Cheyenne History course students at Chief Dull Knife College (CDKC), Lame Deer, MT meet with students in their linked course at Montana State University (MSU), AGSC 465R, who they challenged during the semester to learn how to make dry meat and use it in a stew and to learn why this was important in Northern Cheyenne history. CDKC President Richard Littlebear joins with Elders, Kathy Beartusk, Mina Seminole, and George Nightwalker to test the dry meat stew created by the MSU students and their instructor, Dunkel.

FIGURE 9.3
Students in the linked course (Rachel Anderson, left; Gwen Talawyma, middle) and their Northern Cheyenne teaching assistant (Sierra Alexander, right) learn the dry meat cutting from Gwen, a traditional dry meat processor.

FIGURE 9.4
AGSC 465R students meet with Apsaalooke Kurrie Small, students' site mentor, a rancher, and former AGSC 465R student to listen to her debrief on the "Let's Pick Berries" project and to update her on their research at MSU (from left Tanner Mcavoy, Taylor Anderson, Danielle Bragette, Kyle Lavender, Kurrie Small, Florence Dunkel, Isaac Petersen).

FIGURE 9.5

Learning participatory diagramming developed their predecessors for the Sanambele village women to avert stunting in their children (Dunkel with two AGSC 465R students).

FIGURE 9.6

Village team, Karim Coulibaly (left) and Bourama Coulibaly (next), gather with AGSC 465R student Wendy Nickish (now a dentist in rural Montana) (far right), and her site mentor, Keriba Coulibaly (IER, USAID) (next left) to monitor local dry bed river in March 2009 for anopheline larvae whose adult stage can carry malaria.

FIGURE 9.7
From left, Keriba Coulibaly, Wendy Nickisch, and Bourama Coulibaly use equipment available in village to sample dry bed river pool for dry season populations of mosquitoes.

FIGURE 9.8
Dr. Florence Dunkel and Dr. Clifford Montagne engage with a transdisciplinary team of faculty (representing engineering, biology, and communications) on the University of St. Thomas campus, St. Paul, MT, November 2002 to learn to link the Holistic Management Model with service-learning methods and apply it in their classes at their respective universities and other US universities as they work together with villagers in Mali.

FIGURE 9.9
Prairie turnips, a sustainable, wild-collected traditional food, introduced to AGSC 465R students by Linwood Tall Bull, elder of the Northern Cheyenne Nation, Busby, MT.

Bridging the Gap Between Food and Agricultural Sciences and the Humanities and Social Sciences

CHAPTER 10

Two Cultures and a Second Look: Humanities and Food Sciences

Florence V. Dunkel[1], Khanjan Mehta[2,3], Alison Harmon[4], and Ada Giusti[5]

[1]Department of Plant Science and Plant Pathology, Montana State University, Bozeman, MT, United States, [2]The Pennsylvania State University, State College, PA, United States, [3](currently: Lehigh University Bethleham PA United States], [4]College of Education, Health and Human Development, Montana State University, Bozeman, MT, United States, [5]Department of Modern Languages and Literaure, Montana State University, Bozeman, MT, United States

CONTENTS

Case Study 1: Food, Storage, Marketing Research and Development of the National Food Quality Laboratory of Rwanda Combined Humanities With the Land Grant Mission...............253

Case Study 2: Sustainable Foods and Bioenergy Systems: MSU...257
The Design Team......258
Curriculum and Learning Outcomes ..260
Lessons Learned in the First 5 Years..............262
Student Response, Graduates, and the Future 262
How Can a University Encourage Integration of the Humanities and the Food and Agricultural Sciences?..................263

In Part II of this book we visited with the Elders in Indigenous communities on three continents in four nations (Chapter 5: Listening With Subsistence Farmers in Mali, Chapter 6: Listening With Native Americans, and Chapter 7: Listening Within a Bioregion). Each of these people, some would say, live in food insecure situations, or in a food desert, or in a situation where malnutrition among young children is commonplace. In these households we described, there was also abundant culture—artistic wealth, family-centered life, rich traditions, and, in some cases, strong Indigenous speakers of all ages. It was also observed that Native plants and local sources of protein are available and can be produced by villagers in their own environment without input of actual cash. It raises questions: Did we experience food insecurity? Were the people living in a food desert? Was the food grown in these areas lacking in essential nutrients, both quantitatively and qualitatively?

Also Part II presented how Washington, DC, career scientists are learning from and are inspired to learn more from direct linkages with these communities. When the guides are Millennial Generation and Next Generation students, this inspiration is even stronger. Middle school students on three continents are fine peer instructors of each other in their own local geography and health and environmental issues (Chapter 8: Listening Over Power Lines: Students and Policy Leaders). Not only can educators K through 20 make these linkages, but students within the classroom can also share their own background in Native foods and ways to produce or gather the nutrients and prepare a meal from the harvest (Chapter 9: Listening With Students).

249

Incorporating Cultures' Role in the Food and Agricultural Sciences. DOI: http://dx.doi.org/10.1016/B978-0-12-803955-7.00010-9

Case Study 3:
Graphic Arts, the
Humanities, and
Entomology........264

Case Study 4:
Humanitarian
Engineering and
Social
Entrepreneurship
Program: The
PSU 269
Introduction...............269
The Journey to
Affordable
Greenhouses 271

Closing
Reflections.........273

References 275

This final section of the book (Part III) is a call to action. The time has come for educators, professors, program officers managing funding opportunities, administrators, and nongovernment organizations to incorporate the role of culture, and therefore the knowledge imparted by the humanities, in their work in the food and agricultural sciences. As the world transitions to sustainable food production with health as the bottom line, there is more recognition that there are concepts and ideas saved by cultures that did not take the path of industrial-scale agriculture. Recognizing the cultural component (the people, the history, geography, literature, and arts) in cropping systems, plant breeding, plant pathology, entomology, animal and range sciences, land resources and environmental sciences, agricultural and technical education, food sciences, and nutritional science may help us "see" these ideas. To incorporate the cultural component into each of the courses in each of these disciplines, funding programs, research projects, or board room decision processes, scientists and policy makers will need help from colleagues in disciplines not usually included in scientific processes, such as literature, religion, philosophy, history, foreign languages, and the graphic and performing arts. These are the humanities.

The Stanford University Humanities Center (2016) defines humanities as "…the study of how people process and document the human experience. Since humans have been able, we have used philosophy, literature, religion, art, music, history, and language to understand and record our world. These modes of expression have become some of the subjects that traditionally fall under the humanities umbrella. Knowledge of these records of human experience gives us the opportunity to feel a sense of connection to those who have come before us, as well as to our contemporaries." All human cultures have these components, science and the humanities, in one form or another. Chapter 5, Listening With Subsistence Farmers in Mali, Chapter 6, Listening With Native Americans, and Chapter 7, Listening Within a Bioregion, were devoted to cultures whose, science, literature, and history are primarily oral. The keeper of this historical literature is a person and not a library building or Internet server. Traditionally in West Africa, the keeper of this historical information is called a griot.[1] Griots are

---

[1]The griot is one way that other-than-Western cultures integrate the humanities into their intellectual and social life and their natural world which includes the food and agricultural sciences. The griot is a serious profession, an advisor in traditional governments mainly focused on providing a broader, honest perspective on issues that cause problems in society as a whole and also in families. Some consider being a griot an art inherited from one's parents (James, 2016). In West Africa the griot, he or she, is a storyteller, an oral historian, trained to excel as an orator, lyricist, or musician. The griot keeps records of all the births, deaths, marriages through the generations of the village or family. Master of the oral traditions, the griot plays a key role in West African society. In most of West Africa, griots archive the family history and cultural history in songs. Griots originated in the 13th century in the Mande Empire of Mali. For centuries they have told and retold the history of the empire, keeping their stories and traditions alive. They tell their stories in music, using instruments such as the ngoni (lute), the kora, the balafon, or voice. The griot tradition was remarkably resilient in West Africa, seven centuries after its beginnings during the Malinke Empire which stretched from modern day Senegal to Timbuktu and Gao in Mali, including parts of Côte d'Ivoire.

dedicated to preserving the memory of society. In the academe, humanities serve this function now, but the usefulness of this role is viewed as marginal by the "hard sciences," the very location in academe where historical perspective, ethics, and cultural appreciation are urgently needed.

The gap between the humanities—the "literary intellectuals"—and "hard scientists" seriously disturbed C.P. Snow, himself a respected British novelist and a first-rate scientist in charge of Britain's scientific research recruitment in World War II. Snow, himself, navigated the gap with respect for both cultures, but warned his colleagues on opposite sides of the gap that this lack of appreciation between these two cultures was "To us as people, and to our society. It is at the same time practical and intellectual and creative loss." This statement along with other warnings delivered as a Rede Lecture at Cambridge University in 1959 created heated debate (Snow, 1963). As a young undergraduate, I had no context at age 20 to really understand. That small paperback remained with me, though, through all my life. Fifty-four years later, now, as a mature scientist and a professor myself, I am able to see how the gap of which Snow spoke and wrote can lead to time, energy, and research dollars lost, not to mention lives lost, without eliminating the situation.

In this chapter, we explore the role the various disciplines that encompass the humanities can play in understanding the crosscutting issues of food security, poverty, and the environmental footprint of agriculture. We specifically look at how science faculty have built bridges with their colleagues in the humanities, and how in so doing, they have been better equipped to address these complex issues. Already forming a bridge between these two cultures are the social sciences: sociology, anthropology, psychology, and economics. Collaborating with scholars who conduct research into the human experience is essential to understanding the wholeness of our world. As the Stanford University Humanities' Center aptly points out: "Through the work of humanities scholars, we learn about the values of different cultures, about what goes into making a work of art, about how history is made. These reservoirs of knowledge preserve great accomplishments of the past, help us understand the world we live in, and give us tools to imagine the future. Humanities, not only the sciences, help us understand our world (Fig. 10.1)."

Agriculture is identified as a natural science in Land Grant Universities and other agricultural universities. There are 75 US Land Grant Universities of which 23 are Historically Black Colleges and Universities (HBCUs) (www.aplu.org/members) and 33 tribal colleges (www.diversity.cals.iastate/land-grant-colleges-universities). There also are 60 non-Land Grant Agricultural and Renewable Resource Universities (www.aplu.org/members). The humanities and social studies are unlikely to be integrated with the mandated empirical knowledge being taught in the classrooms and laboratories. Most Land Grant faculty in the food and nutritional science disciplines are not

FIGURE 10.1

Graphic comparison of science and humanities. *www.hum.utah.edu*

adequately prepared and are seldom encouraged within their departments or college to reach out and collaborate with colleagues in the humanities and social science disciplines. In 1982, Vern Ruttan, University of Minnesota agricultural economics professor, aptly noticed that "...the structure of organization and incentives within the university tend to discourage rather than encourage such collaboration. When it does occur, however, it usually takes place within a somewhat looser coordinating structure..." (Ruttan, 1982, p. 315). Unfortunately this is still true 35 years later, though program officers at national agencies such as National Science Foundation (NSF), US Department of Agriculture (USDA), and US Agency for International Development (USAID) are taking leadership in encouraging an appreciative relationship between the humanities and the agricultural and food sciences. Ruttan reminded us that "Anthropology departments usually include a high proportion of their faculty whose professional interests are on rural communities. Students of agricultural history, agricultural geography, and agricultural politics are often found in departments of history, geography, and political science" (Ruttan, 1982, pp. 314, 315).

To understand the significance of developing a curriculum or a research program that provides training in the food and agricultural sciences as well as the humanities requires understanding of the Land Grant System and recognition of the limitations imposed by the academic pattern of fragmentation. The Morrill Acts of 1862 and 1890 created colleges of agriculture throughout the United States, but with their many technological successes, the Land Grant Colleges and Universities have not yet successfully merged agro (field) and culture (the totality of socially transmitted behavior patterns (Merriam-Webster, 2017)). The separation of academic disciplines into silos is in direct contrast to the ways of knowing of Indigenous people. The "reality" of Western scientists, including those in agricultural disciplines, is based upon analytical research, hypothesis generation, and statistical probability that puts minimal importance on information gained during the millennia during which Indigenous cultures have been observing and listening to the behavior of their fields, forests, sky, wind, rivers, plants, and animals, all the while telling stories to each new generation about their environment and how and why it is changing.

Yet, a few university faculty have bridged this cultural gap and designed programs integrating humanities with food science and other agricultural sciences. The following four case studies will highlight those who have built bridges to connect these two cultures in a Western culture academic setting. The reader will discover how to incorporate the humanities, the role of cultures, in food science. Case Study 1 is a short example of how the humanities were successfully integrated into a food, storage, marketing project of USAID Rwanda. Case Studies 2 and 3 are examples of curricular and community outreach opportunities to intertwine the two cultures, humanities and the food and agricultural sciences. Case Study 4 is an example of how the human dimension has been included in engineering and social entrepreneurship. Land Grant institutions, University of Minnesota, Montana State University (MSU), University of California Davis, and The Pennsylvania State University (PSU) are the sources of these studies which explore diverse food science projects with some relationship to the humanities.

CASE STUDY 1: FOOD, STORAGE, MARKETING RESEARCH AND DEVELOPMENT OF THE NATIONAL FOOD QUALITY LABORATORY OF RWANDA COMBINED HUMANITIES WITH THE LAND GRANT MISSION

Most of the 23 faculty from seven departments of the College of Agriculture and five departments in the Institute of Technology and the College of Liberal Arts of the University of Minnesota (U of MN)-Twin Cities had no idea where Rwanda was in Africa when I asked them to participate with me in writing the Food, Storage, Marketing proposal to USAID in 1983. Moreover, we had no idea what Rwanda's history or food situation was like, however they enthusiastically agreed to participate if funded.

We were the "dark horse" in this competitive race. Kansas State University was sure to win. They had a system for organizing workshops for food and storage managers in material-resource-poor countries to come to the United States to learn how food is produced in the United States and protected from postharvest loss. Surprised that we ended up on the short list, we still knew we had little chance. When the review team came for an on-site visit, however, it became clear that we were their choice. Why? Of course, we can never be sure, but looking back, we think it was in part due to our putting together of not-the-usual in-country team senior experts, but a group of six relatively young people: three graduate students, two former US Peace Corps volunteers, and a Belgian postdoctoral associate trained at a Belgian university requiring coursework and practical experience. All of the six young people chosen were fluent in French (FSI level 4), the colonial language, and seemed to understand the importance of also learning the Indigenous language, Kinyarwandan, that had survived in that area. This young team was committed to bring their families, cats and newborns included, to Rwanda and live there for 2−4 years during the lifetime of the grant. The 23 senior faculty served as advisors and came and went from Rwanda in short visits of 2−4 weeks.

It is true, we had an unusual USAID approach. Based on my positive experience in The People's Republic of China the previous 2 years, I was full of energy to try the model I had been taught on intercultural competency by the US National Academy of Sciences (NAS), Committee on Scholarly Communication with the People's Republic of China (CSCPRC), which basically was language, history, philosophy, politics, ethics with a particular emphasis on language and history, "no matter if you are a scientist." The NAS emphasized language and history over and over again. To me this was a clear message. As a result, we informed the USAID review team that we would offer our research team a strong predeparture training emphasizing language and history as well as collegiality and intercultural competency. We had assembled a team of four communication specialists including professors in intercultural relations, nonverbal communication, and psychology of intercultural communications and had outlined in the proposal culture-general workshops and individual predeparture and post-return sessions. During the on-site visit while seated around the table informally discussing our specific proposal, there was a distinct change in posture when we described our predeparture training. "Well," the review team said, clearing their throats in chorus, "USAID does not fund that kind of preparation. It sounds like a good idea, but our guidelines just do not allow for use of the funds in that way." The verbal and nonverbal message was perfectly clear. Several weeks later, a phone call to me from the review team leader confirmed the final decision about the entire proposal.

The dark horse won the race. The proposal was sent to the contract office. A few weeks after the phone call, in August 1983, USAID Rwanda sent to us in Minnesota the counterpart Chief of Party, the person who was to be the manager of the project in Rwanda, Phocus Kayinamura. Phocus was an engineer who lived in Rwanda and worked at the national government unit that managed grain and bean storage for the entire country. We began the planning together with Phocus in charge, interspersed with social gatherings of the team, our families, and our best attempt at Rwandan meals. From the beginning we incorporated the social part with the scientific goals. Our goals were to establish the national food and seed quality laboratories in Rwanda. In the process we were to develop a storage management plan for beans and sorghum and create long-term, large-scale strategic government storage.

By November 1983, we were ready to send a reconnoitering team from U of MN to Rwanda to continue the planning. We, as a faculty and student/post-doc team, had spent the intervening months not on planning, but on preparing. We had "French for Lunch" almost every day and once a week an intensive predeparture workshop with the finest minds on the campus in East African history and sociology as well as intercultural competency.

Dr. Josef Mestenhauser, a U of MN professor of international relations, opened each of the weekly workshops with the statement, "In cross cultural relations, we should be able to rethink everything we know. How will this relate to your stay in Rwanda? Here is an analogy. You will want to file things where you used to and you will have to do it in a different way. While you want to do that, you will file it wrong and lose it. When relating to a person, you may not have a file for it. This is important, too, when you are managing your time. You will be tired, sleepy, you will actually be emotionally tired. …Give your mind a break…write a letter home, make sure you have one good friend in Rwanda, a Rwandan. Get physical exercise. Take a walk (Mestenhauser et al., 1984)." During these 24 hours of intensive predeparture training, we also learned that Rwanda was reaching the carrying capacity of its land. Dry edible beans were the main source of protein and calories for the people in Rwanda and fuel supplies, the local trees, for cooking them were in short supply. This was the intense overt story.

East African historian at U of MN, Dr. Lansine Kaba, made sure we knew the "back story." "Remember," he emphasized in a more serious tone of voice, "The decolonization movement that surfaced in 1962 in Rwanda has never been resolved. The Tutsi, puppet leaders of the Belgians, were massacred in large numbers by the Hutu. The Hutu took over the government. This violent event has never been resolved. It is not spoken about in Rwanda, except to mention 'The Event' in a tone of voice not above a whisper. This uprising will happen again." And indeed it did, after this project had successfully accomplished all its goals and the National Food Quality Laboratory of Rwanda was launched. A trained group of 14 professionals, all Rwandans were leading the Laboratory and its outreach programs without help from the United States. April 4, 1994, the next "event" began with greater force and life loss. Some of the Laboratory scientists went home that eve to find their entire family massacred. Eventually, after hiding in attics of fellow scientists and fleeing with their children to refugee camps in Goma, all of the scientists returned to the Laboratory and the National Food Quality Laboratory of Rwanda survives now. Phocas with his Tutsi wife and their family, though, had to flee to Kenya and eventually Canada and likely will never be able to return to Rwanda.

Without this humanities professor on our team, we would have been unprepared for the second event, the 1994 genocide. Having had this foresight from Lansine, we took positive steps to protect the project and its results. We completed the project as rapidly as possible—in 4 years and published the eight resulting research monographs before the end of the project (Bylenga et al., 1987; Dunkel et al., 1986, 1988; Edmister et al., 1986; Hanegreefs et al., 1988; Lamb and Dunkel, 1987; Lamb and Hardman, 1986a,b). From the beginning, we focused on maintaining a strong esprit de corps in our Hutu and Tutsi team, and initiated a tradition of social gatherings/family dinners with the Rwandans at the homes of the US team. Likely it was development of this informal esprit de corps that literally saved the life of Ausumani Serugendo, the young director of the Laboratory at that time as he hid for 6 weeks in the attic of the home of the entomologist at the Laboratory.

African sociology professor at U of MN, Dr. Earl Scott, gave us simple advice, "African time is social time. If you forget this, nothing will happen. Social time comes first." Earl as well as the African students we had on the team reminded us that "In Africa, old age is a positive value. I see many young people on your team, including your Project Director. (I was 41 at the time.) Be careful that your ideas, results, and conclusions are sound and perceived with value."

Armed with all of these warnings, cautions, and survival suggestions, I chose my November 1983 reconnoitering team and departed with them. I took with me our team's French translator who had learned well all the technical language related to dry edible beans and postharvest protection. I also took a professor of international relations who specialized in nonverbal intercultural communication. To complement my entomology and microbiology background, I also brought an engineer specializing in food processing.

The most notable response I received during that first visit of any of us to Rwanda in 1983 was in the form of advice from Anastase Nitizelyayo, Rwandan Minister of Agriculture, "The peasant farmers," he advised me, "don't know anything about storage. Don't spend your time in research with them." We did not take this advice. By 1986 we published a monograph establishing that although the Food and Agricultural Organization of the United Nations reported postharvest weight loss of 20%−30%, we found in quantitative national surveys that the Rwandan subsistence farmers knew how to keep postharvest dry weight loss of dry edible beans and sorghum in the single digits, <5% (Dunkel et al., 1986, 1988).

To make use of the new information from that initial visit of the faculty who had just returned from Rwanda with the project director another 24 hours of predeparture training was held. To further prepare the team, we held an intensive preparation, this time in a wilderness retreat location cooperatively funded by us off-campus. This time we included our families.

The total cost of the 4-year project was estimated at $4,750,000, part contributed by the US government ($2,900,000) and part by the Rwandan government ($1,850,000). Together this team of professors from the humanities, the social sciences, and the agricultural sciences at U of MN worked side-by-side with their counterparts at the National Office for the Development and Marketing of Food and Animal Products (OPROVIA) specifically in the National Grainary of Rwanda (GRENARWA) and the national research institute of Rwanda (ISAR).[2] To what do we attribute the success? Close contact with the humanities, especially in language and history. Social scientists played a particularly important role, particularly in psychology and international relations.

CASE STUDY 2: SUSTAINABLE FOODS AND BIOENERGY SYSTEMS: MSU

The Sustainable Food and Bioenergy Systems (SFBS) BS degree program (http://sfbs.montana.edu/) is an innovative, interdisciplinary, and experiential program at MSU that promotes the sustainable production, distribution, and consumption of food and bioenergy. The overarching goal of the program is to develop graduates capable of addressing complex food and energy problems as the next generation of food and energy system leaders (Jordan et al., 2014).

---

[2]Within 4 years, the team published eight research monographs (Edmister et al., 1986; Lamb and Dunkel, 1987; Dunkel et al., 1986, 1988; Hanegreffs et al., 1988; Bylenga et al., 1987; Lamb and Hardman, 1986a,b). All in all it was considered a success by the stakeholders (Dunkel et al., 1988).

Although the goal and mission statement of the SFBS program does not address intercultural competency, culture general skills or culture-specific knowledge, these attributes are included in the elective part of the program. Students in the various SFBS options take a core curriculum of six courses designed to provide broad exposure to key principles of SFBS. The additional coursework in each option, however, is specifically designed to create more detailed and subject-specific knowledge in each area of specialization. Hence, students take 18 credits of complementary coursework in political science, economics, business, Native American studies, and engineering. Through these six elective courses students learn from disciplines that span the social sciences, humanities, and business. Each cohort of students has about 20–25 students for a total program enrollment each year of about 80–100 students.

Two popular electives chosen annually by about 30 of the juniors and seniors in SFBS provide culture-general training in the holistic process. For example, NRSM 421 Holistic Thought and Management and AGSC 465R Health, Poverty, Agriculture: Concepts and Action Research are two electives based on the holistic process. Additionally, research and outreach of faculty in this program are under way in several countries in Asia and Africa. Several of the faculty in this program are also collaborating with Native American nations in Montana. These international activities of the SFBS faculty have been providing intercultural research and outreach opportunities for the SFBS students that enrich the academic experience far beyond what is possible in the classroom.

The Design Team

The SFBS program began as a desire to integrate sustainability into the undergraduate agriculture degree programs. The approach was expanded when faculty in plant sciences and agroecology connected with faculty in health and human development. Collegiality seemed to be the key. The resulting program could then address issues related to both production and consumption, linking agriculture with consumer health. Representing the Medical Sciences and Health on the design team were nutrition scientists. From the beginning, it was up to the SFBS students themselves to include an international or intercultural aspect or theme to their training. Many students who took Health, Poverty, Agriculture made major contributions to food security, particularly in child nutrition in West Africa (Sanambele, Mali). Carly Grimm who we met in Chapter 3, Decolonization and the Holistic Process, as she communicated with the women of the village, may have been successful because of her work outside of this major with Dr. Ada Giusti, professor of French and Francophone Studies. We learned about Ada's immersion experiences in Mali in Chapter 4, Immersion. Ada's immersion experiences were primed by her professional background in the humanities, which she was able to transfer to her French students, such as Mary Hunt, whose major was SFBS.

It is important to note that when these students enroll in language courses, their learning reaches far beyond the acquisition of vocabulary and verb conjugation. At MSU (Dunkel and Giusti, 2012), as well as in many other US universities, foreign language classes are developed to comply with the *National Standards for Foreign Language Learning* in the 21st century. These standards, created in 1996 and most recently revised in 2016 when they were aptly renamed World-Readiness Standards for Learning Languages, focus on the following goals: communication, cultures, connections, comparisons, and communities (Fig. 10.2). Typically referred to as the five C's of foreign language education, they are described as follows: "Communication is at the heart of second language study, whether the communication takes place face-to-face, in writing, or across centuries through the reading of literature. Through the study of other languages, students gain a knowledge and understanding of the cultures that use that language and, in fact, cannot truly master the language until they have also mastered the cultural contexts in which the language occurs. Learning languages provides connections to additional bodies of knowledge that may be unavailable to the monolingual English speaker. Through comparisons and contrasts with the language being studied, students develop insight into the nature of language and the concept of culture and realize that there are multiple ways of viewing the world. Together, these elements enable the student of languages to participate in multilingual communities at home and around the world in a variety of

FIGURE 10.2
Diagram illustrating the five goals of the World-Readiness Standards for Learning Languages from foreign language learning.

contexts and in culturally appropriate ways" (World-Readiness Standards for Learning Languages: https://www.actfl.org/sites/default/files/publications/standards/World-ReadinessStandardsforLearningLanguages.pdf).

The SFBS curriculum development team included one faculty member from Plant Science and Plant Pathology, one faculty member from Land Resources and Environmental Science, and one faculty member from Health and Human Development in the specific area of Food and Nutrition. Additionally, the team included a graduate student who developed the introductory course as her graduate project, but who also was the current president of a related student organization. The student organization, Friends of Local Food, formed 1 year prior and was instrumental in launching a campus farm, Towne's Harvest Garden (http://townesharvest.montana.edu/), that would later become a cornerstone and critical hands-on experience in the degree program. An additional team member in Animal and Range Sciences joined the program once it had been approved and launched, creating additional program breadth and providing more opportunities for students. Although no members of the curriculum development team were from the humanities or used the holistic process in research or teaching, students had access to this knowledge base through electives.

In finalizing the degree program proposal, the team attended a shared-leadership workshop and continued to work with a leadership coach throughout the following year to establish the vision and mission of the program. Beginning the process by appreciating how to share leadership was a critical part of the early success of the program, as interdisciplinary projects are challenging, time consuming, and rarely rewarded in traditional ways by the institution. It was important that the team enjoyed positive collegial relationships and camaraderie that would outlast the work that needed to be done. It was also critical that members of the team respected one another's disciplines such that each person was willing to give up content in their own area in order to incorporate content from other areas.

Curriculum and Learning Outcomes

In designing the curriculum, the team had to determine the courses from each discipline that would be required of all students and that would serve as a solid foundation for learning more complex concepts at upper division levels. Additionally, the team had to imagine the new interdisciplinary courses that would need to be developed in order to create a cohesive degree from freshman to senior year.

In addition to required general education courses, the initial foundational courses included plant biology, economics, human nutrition, food fundamentals, sustainable energy, soils, field crop production, sustainable live-stock production, and community nutrition. New interdisciplinary coursework included an SFBS introduction course, a summer field course at the campus farm, a junior level internship, and a senior level capstone course. Curricular options, which would allow for specializing within the degree program, incorporated upper division electives, had defined career paths for students and could each be managed by one of the four collaborating departments.

The *sustainable food systems* option trains students in the natural and social sciences to evaluate and mitigate outcomes of complex interactions in the food system for human health and nutrition. This option focuses on the interconnections between production, policy, food security, and health. Graduates are empowered to address food and health challenges such as obesity, food insecurity and poverty, food safety, and vulnerability of Indigenous food systems (Wharton and Harmon, 2009).

Agroecology focuses on application of population principles and community ecology, environmental science, and cropland ecosystems. The *agroecology option* explores how crops and pest organisms interact with their environment, and the application of technology to efficiently and sustainably produce crops. The curriculum is based on the philosophy that to be able to successfully predict management outcomes and thus make informed recommendations, one must understand fundamental principles of evolution, ecology, soil science, agronomy, and pest management. One could argue that also understanding the history, religion(s), language(s), and ways of knowing of the individual communities (all subjects taught in the humanities) would further students' success in predicting management outcomes.

The *sustainable crop production* option is designed to train students in a broad range of principles including agronomy, soil fertility, plant genetics, plant physiology, greenhouse production, plant propagation, integrated pest management, and small business management. Both large- and small-scale food and bioenergy production systems are examined.

The *sustainable livestock production* option focuses on the biological understanding of animal agriculture and its continued presence in sustainable grazing systems as well as its potential role in sustainable farming systems. The option focuses on the science of animal production but expands students' learning to a larger system of understanding, including the role of domestic livestock in sustainable systems. In addition, students are exposed to the role of strategic grazing in landscape management as well as using livestock to manage potential waste streams from other industries.

Degree program learning outcomes entail that by the time students graduate with an SFBS degree, they be capable of systems thinking and critical thinking, have problem solving skills and practical skills, be effective communicators, have developed agency (the capacity to make choices and act in a society framework), and have a body of knowledge related to SFBS concepts. The SFBS capstone course syllabus has four main objectives with accompanying exercises including developing as professionals, learning how to work in a team of stakeholders, becoming competent in community outreach, and developing systems thinking skills.

Lessons Learned in the First 5 Years

The proposed program garnered support from a USDA Higher Education Challenge Grant, which was an important way to validate the demand for the program, to cover the costs of developing and piloting new coursework and to support the wages of a curriculum coordinator, and campus farm managers. Each of these components would have to be funded entirely by the institution at the end of 3 years. This program, focused on sustainability, would itself not have been sustainable had it not attracted a rapidly growing number of students who chose this major, thereby generating ample student credit hours, and positive press at the local and national levels. Advocating for support was time consuming and exhausting for the core faculty group. Ultimately, MSU funded a new faculty line to support the program, a permanent production manager for the campus farm, and ample sections of required coursework. This support ensured that the program would be sustained long-term. However, the relationships among collaborating departments and faculty continue to require careful and thoughtful coordination, transparent actions, and deliberate communication to continue to seek involvement from all parties in important decision making. Initiating new team members into the culture that was developed by the original team members is an ongoing need (Malone et al., 2013).

Student Response, Graduates, and the Future

The immediate response to the degree program confirmed a need and desire for the program from our most important group of university stakeholders—students. Student enrollment has stabilized at a manageable level, ensuring that students have optimal hands-on experiences on the campus farm, have the opportunity to take coursework taught by tenure-track professors, and have the opportunity to engage in meaningful research as undergraduates. Meanwhile, the SFBS degree program and the integrated Towne's Harvest Garden have both been nationally recognized (Jacobsen et al., 2012). Program faculty will continue to innovate and adapt the program to meet current needs based on students' and other stakeholders' feedback and program assessment. The program enjoys sustained institutional support and will continue to be integrated into additional degree programs. Meanwhile, graduates are making their mark in the food system as small-scale and sustainable growers, ranchers, food retailers, policy makers, directors of nonprofit organizations, and entrepreneurs. Many have continued on to related graduate programs at MSU and elsewhere. Some of these young professionals will become mentors for current students wanting hands-on experiences and internships.

The SFBS degree program is a good model for developing and maintaining a collaborative and interdisciplinary program that serves both student demand and *societal need*. This is a well-constructed program that focuses on satisfying local societal needs—indeed growing local, selling local, and sourcing products locally are fundamental components of this program. In fact, local is at the heart of MSU's SFBS. As a result, one wonders if this local approach to food literally focused in small town, rural Euro-American communities in Montana, has been guided by the faculty and learned by the student to be transformed to a culture-general knowledge so that they may use these skills to work in other communities such as Native American, Hispanic and at the global level. A recommendation would be to keep this question in mind as the program develops.

How can a University Encourage Integration of the Humanities and the Food and Agricultural Sciences?

How is the interaction of faculty, students, curricula, and research fostered between the humanities and the food and agricultural sciences? (1) Through the intertwining of teaching and research, including at the undergraduate level (a graduation requirement); (2) by incorporating cultures' role in agricultural sciences and other STEM areas; (3) through community engagement (including with Indigenous people); (4) by implementing experiential learning; and (5) through the development of transdisciplinary courses. The conundrum occurs when not all departments universally recognize these five attributes. If the department faculty do not recognize cultures' role in the food and agricultural sciences, cultures' role will not be recognized as part of excellence in the promotion and tenure (P&T) process.

At the University level of MSU, the importance of these attributes *is* recognized in research, teaching, and service, and the University provides much support in terms of awards for excellence, intramural funding, and workshops to hone these skills. However, if faculty themselves in a single department do not value these five (or more) areas of innovation, faculty in that particular department are penalized for achieving excellence in these areas. The whole P&T process then effectively stops and that faculty member is not tenured and/or promoted. A clear signal is sent to new faculty to not engage in this type of innovation.

By definition, each Land Grant University has a College of Agriculture and a program in Food Science. How then does a Land Grant University become comprehensively internationalized or able to train students to solve complex global, but local problems, particularly in areas of food security and health? Poverty has no disciplinary home, neither do food security and health. They are each quite complex issues. This is an urgent question. What can an individual faculty member do in a department which has spawned an SFBS-like curricula, but remains contained in a department with a majority of their faculty still teaching with, conducting research and teaching without incorporating cultures' role or even recognizing and valuing culture at all?

CASE STUDY 3: GRAPHIC ARTS, THE HUMANITIES, AND ENTOMOLOGY

In this case study, we explore the initiation of humanities-related teaching and outreach by an entomology professor and a ceramic artist. Their leadership in demonstrating a visual method of how incorporating cultures' role in the food and agricultural sciences could be related to the humanities stands as a reminder that these interconnections are possible. Not only are these interconnections of the two cultures, humanities and food sciences, possible to incorporate into teaching and outreach, but the fusion can be inspiring beyond academe.

Diane Uhlman, entomology professor at University of California Davis and former Associate Dean of the College of Agriculture and Environmental Sciences cofounded and codirected the UC Davis Art/Science Fusion Program with the rock artist, Donna Billick (Fig. 10.3). Diane and Donna provided

FIGURE 10.3
Diane Ullman (left) and Donna Billick (right), cofounders of the Art/Science Fusion Model for providing a visual linkage of the arts, humanities, and agricultural sciences.

leadership for the development of two new undergraduate courses and also led the development and implementation of a year-long colloquium: "The Consilience of Art and Science." A bevy of graduate students were mentored in using an art/science paradigm for teaching which was nationally recognized by the Entomological Society of America. The most noteworthy part of this program development was the visible and permanent beautification of the campus by the students and the community who came together and continue to come together to visually create and celebrate the interconnections of food, the earth, plants, people, nutrition, and biological processes related to insects.

Guided by Diane Uhlman and Donna Billick students chose various media such as wall paintings, mosaics, and silk banners which they hand-painted and displayed to beautify specific plain or even unsightly locations on campus (Table 10.1). The process of designing and painting the murals and banners gave students opportunities to learn specific science knowledge, e.g., the life cycle of the lady bird beetle, or the ecosystems surrounding the UC Davis campus, or bees' role in pollination of food crops. Since this process created a permanent statement of art and food and agricultural science, students had to consider communication of their art and the agricultural science they portrayed to the public.

Table 10.1 Creative Works Associated With Teaching Entomology 001 and Freshman Seminars at University of California Davis

Permanent Installations	Location
1. *Insects and Agricultural Ecosystems of Northern California*—Painted mural that illustrates agricultural ecosystems in the region surrounding the UC Davis campus. Insects found as pests or biological control agents are featured in the ecosystems.	*Briggs Hall, Room 158, Classroom* used by Entomology and associated undergraduate and graduate programs.
2. *Meeting the Challenge*—Painted mural illustrates global environmental challenges from air/water pollution to food production and global warming. Mural incorporates insects, plants, and human endeavors at all levels of the ecosystems illustrated. Students sought to inform the viewer about challenges to the environment while conveying a sense of hope. The quote that best described their thinking is inscribed on the mural: "If we are to survive, our loyalties must be broadened further, to include the whole human community, the entire planet Earth."—Carl Sagan	*Science and Society Conference Room* used for meeting and small classes primarily by the Science and Society Program and the Department of Plant Pathology.
3. *The Where's Waldo of Biological Control*—Painted mural that illustrates the insect pests and their natural enemies in the context of a sustainable garden. Insects and natural enemies are shown highly magnified on one wall and closer to "life size" and in action on the other wall.	*Student Farm*, UC Davis campus, two walls of the *tool shed*. Mural is used extensively for the children in the Garden Program.

Continued...

Table 10.1 Creative Works Associated With Teaching Entomology 001 and Freshman Seminars at University of California Davis *Continued*

Permanent Installations	Location
4. *California's Agricultural, Land, Air, and Water*—Painted mural that illustrates agricultural ecosystems in California (rice to peaches), water systems and associated organisms (e.g., salmon), soil and soil-borne animals (e.g., nematodes), vernal pools, oak forests, and the birds and insects one finds in all these environments.	*Plant Environmental Science Building*. This mural spans the staircase and landing from the first to second floor of this building. It is enjoyed by all those working in the building and visiting the building.
5. *The Redwood Grove from a Lady Bird Beetles Perspective*—This 60-foot-long mural illustrates the life cycle of the lady bird beetle, as well as the many birds and insects common to the Redwood Grove on campus.	*Tunnel under the A Street Bridge*.
6. *Gardens, Food, and Insects*—This painted mural illustrates gardens and how people use gardens, as well as the insects one finds in gardens. It covers two large outdoor walls facing one of the playgrounds and is used as an educational tool for the children and families at the center.	*Center for Child and Family Studies*, first and A Street, UC Davis.
7. *Insect Camouflage*—This ceramic mosaic mural illustrates how insects use camouflage to protect themselves from predators and other natural enemies. It is installed above the discovery garden and there is an educational piece that was prepared by the students who created the mural. It is used as an educational tool for the children and families at the center.	*Center for Child and Family Studies*, first and A Street, UC Davis.
8. *The Insect Collection*—This ceramic mosaic mural illustrates the orders of insects and the importance of insect collections. There is an educational piece that was prepared by the students who created the mural. It is used as an educational tool for the children and families at the center.	*Center for Child and Family Studies*, first and A Street, UC Davis.
9. *California's Environments and Insect Biodiversity*—This is a series of hand-painted and dyed silk banners that illustrate six key environments in California and the insect biodiversity within them. Created by students in the textile section of ENT 001, there is an educational piece that was prepared by Kaino Hopper (TA) and is used by the Bohart Museum for their educational outreach program.	*Hallway Outside the Bohart Museum*, Academic Surge.
10. *The Queen Bee and Her Retinue*—This ceramic mosaic mural illustrates the interactions between the honey bee queen and her retinue (the bees that respond to her queen pheromone and care for the queen). Other honey bee castes are also illustrated, e.g., nurse bees.	*Harry Laidlaw Honey Bee Research Facility*, one block off Hopkins Road, west side of UC Davis Campus.
11. *Pollinators*—This ceramic mosaic mural illustrates a diversity of insects that are known to contribute to pollination of plants in California.	*Harry Laidlaw Honey Bee Research Facility*, one block off Hopkins Road, west side of UC Davis Campus.
12. *The Swarm*—This ceramic mosaic mural illustrates how bees swarm.	*Harry Laidlaw Honey Bee Research Facility*, one block off Hopkins Road, west side of UC Davis Campus.
13. *The Insect Collection*—This ceramic mosaic mural illustrates the major insect orders and highlights the importance of insect systematic and taxonomy to the science of entomology.	*Briggs Hall, Hallway*, Outside Department of Entomology Business Office, Room 376.

Continued...

Table 10.1 Creative Works Associated With Teaching Entomology 001 and Freshman Seminars at University of California Davis *Continued*

Permanent Installations	Location
14. *Insect Biodiversity*—This ceramic mosaic illustrates the wide diversity of insects on the planet and highlights the importance of conserving insects.	*Briggs Hall, Hallway*, Outside Room 383.
15. *Insects and Planet Earth*—This series of ceramic tiles create a portal surround that illustrates the diversity of insects on the planet.	*Briggs Hall*, Portal surround on the main door to Room 383.
16. *Tree of Life*—This 17′ × 11′ ceramic mural illustrates the Valley Oak Tree and the insects and other animals that flourish in the oak ecosystem.	*Comfort Station*, Front Wall, West side of campus (also known as "Hinshaw Hall").
17. *Valley Wise Visions*—This 17′ × 11′ ceramic mural illustrates the Ruth Storer garden and the drought-resistant plants it features. Insects and other animals commonly found in the garden are also illustrated.	*Comfort Station*, Front Wall, West side of campus (also known as "Hinshaw Hall"), side facing the Ruth Storer Garden.
18. *Oak Family Tree*—This 17′ × 11′ ceramic mural illustrates 29 oak species found in the Shields Oak Grove and their evolutionary relationships. Scientific names of the oak species and their groupings are highlighted along with the insects and other animals found in the oak forest ecosystem.	*Comfort Station*, Front Wall, West side of campus (also known as "Hinshaw Hall"), side facing the oak grove and the creek (featured on cover of *Proceedings of the National Academy of Sciences*, October 2009).
19. *Nature's Gallery*—This is mosaic mural celebrates drought-resistant plants in the Ruth Storer Garden along with the pollinators and other insects that are essential to the plants' ecological success.	*US Botanic Garden* 2007, the California State Fair 2008 and is now a permanent installation between the *Ruth Storer Garden and Arboretum Teaching Nursery*. This installation was recently touted by the Sacramento Bee as the "new icon for UC Davis—to replace the water tower."
20. *California's Gold*—This ceramic mosaic mural highlights the fascinating geology of the California land mass and the crops, animals, and insects for which the state is renowned.	Buehler Alumni and Visitor Center and the California State Fair 2008. Awaiting installation (Canon Building, Washington, DC).
21. *The Face of Darwin*—This tile mosaic captures the essence of Charles Darwin's life and the experiences that led him to propose "On the Origin of the Species." It was produced by students in a freshman seminar and community members in collaboration with Mau Stanton and other faculty in Evolution and Ecology.	*Buehler Alumni and Visitor Center*, in the Dean's Office CA&ES lobby. *Currently installed in the Storer Hall Lobby*
23. *Honey Bee Haven Entry Pillars: Pollination and Life Inside the Hive* (work of the ENT 001 Fall 2009 Graphics Studio, led by Sarah Dalrymple).	*Honey Bee Haven*, installed December 2009, a project funded by Häagen-Dazs and facilitated by the Department of Entomology and the California Center for Urban Horticulture.
24. *Interactive Hive Exhibit*—Honey Bee Biology (work of the ENT 001 2009 Fall Textiles Studio, led by Kaino Hopper).	*Honey Bee Haven*, installed December 2009.
25. *Honey Bee Haven Entry Benches* (work of DNC Freshman Seminar Spring 2010, the community and Billick Rock Art).	*Honey Bee Haven*, installed May 2010.
26. *Alternative Pollinators*—wall mural (work of ENT 001 2010 Graphics Studio Sarah Dalrymple).	*Honey Bee Haven*, installed December 2010.

The process, developed in the Department of Entomology and Nematology, is described as "an innovative teaching program that crosses college boundaries and uses experiential learning to enhance scientific literacy for students from all disciplines." The Art/Science Fusion Program was designed to promote environmental literacy with three undergraduate courses, a robust community outreach program, and sponsorship of the Leonardo Art Science Evening Rendezvous. Beyond its role in environmental literacy, the Art/Science Fusion program also serves as a model for a way to specifically fuse the two cultures of the humanities and food sciences through art.

The ENT 001 classes, taught by Ullman and Billick, are part of the UC Davis Department of Entomology and Nematology and affiliated with the Art/Science Fusion Program. One of the most visible and "wow!" projects is the 2500 pound mosaic art, Nature's Gallery in the Storer Garden, UC Davis Arboretum. Showcasing the interaction of insects and plants, it's a product of the ENT 001 class and community outreach. It was initially displayed at the US Botanical Garden in Washington, DC, and at the California State Fair.

Another project that draws much attention and acclaim is the ENT 001 art in the Häagen-Dazs Honey Bee Haven, a half-acre bee garden on Bee Biology Road, west of the central campus. Billick created "Miss Bee Haven," a 6-foot-long honey bee sculpture that anchors the garden. "I like to play with words," said Billick. She also is the artist behind the colorful Harry H. Laidlaw Jr. Honey Bee Research Facility's ceramic sign that features DNA symbols and almond blossoms. A hole drilled in the sign is the portal to a bee hive.

Billick toyed with a scientific career before opting for a career that fuses art with science. She received her bachelor of science degree in genetics in 1973 and her master's degree in fine arts in 1977. This dual "academic citizenship" seems to be, as with C.P. Snow, a good preparation for taking a second look at the two cultures: the humanities and food-related sciences. Billick traces her interest in an art career to the mid-1970s when then Gov. Jerry Brown supported the arts and offered the necessary resources to encourage the growth of art. He reorganized the California Arts Council, boosting its funding by 1300%.

We have shared this example with you as a possibility for taking a second look at the interconnections between the two cultures: humanities and the food sciences. Do the humanities and food sciences have important and urgent messages for each other? Do these details of output from a collaboration of artists, an entomologist, students, and at times the campus and near-campus community provide fertile grounds for ways to link the food sciences and the humanities in the future? This campus beautification project posed questions about the human condition and its intimate relationship with pollinating insect, plants, and their biodiversity, and the earth and water, essential for all of these. This Art/Science Fusion seems to be a successful model for exploring the human relation to food and food science.

CASE STUDY 4: HUMANITARIAN ENGINEERING AND SOCIAL ENTREPRENEURSHIP PROGRAM: THE PSU

For the past 8 years, Khanjan Mehta has directed the Humanitarian Engineering and Social Entrepreneurship (HESE) Program in PSU's College of Engineering and has just assumed the position of inaugural Vice Provost for Creative Activity and Director of the Mountaintop Initiative at Lehigh University. His success in recruiting PSU students from freshman to graduate students and involving faculty and students from many disciplines has created an engaged scholarship model being employed in rural East Africa (principally Kenya) to address local agricultural and health issues.

Introduction

As the founding director of Penn State's HESE program, I, Kkhanjan Mehta, have spent the last 12 years working shoulder-to-shoulder with my students and external partners to design new technologies and build social ventures in several developing (material-resource-poor) countries. These ventures are specifically designed to address endemic challenges like food insecurity, energy poverty, and poor health care access in underserved communities with market-centric approaches. It is essential that these technologies and ventures meet what we regard as the four hallmarks of sustainability—they must be technologically appropriate, culturally acceptable, environmentally benign, and economically sustainable.

Rather than construct individual projects like a greenhouse at a school or a community windmill, HESE's philosophy is to rigorously build scalable ventures—new technologies that can be commercialized into "multimillion smile" enterprises in underserved communities. This sort of sustainability and scalability is understandably very difficult to achieve. Instead of a "let's help the world's poor," it requires robust cooperation with diverse local stakeholders. Our experiences have taught us that such programs and emergent ventures can significantly benefit from a multifaceted convergence of (1) concepts, disciplines, and epistemologies; (2) cultures and countries; (3) teaching, research, and engagement; and (4) multisectoral partners who share a common vision and purpose. Finally, it means that HESE's academic pedagogy must both educate and inspire the next generation of changemakers on technical, professional, and practical grounds. Students must become both social problem-solvers and entrepreneurial global citizens. The core philosophies driving HESE's pedagogies, operations, and ventures are empathy, equity, and ecosystems.

"Building empathy" is about developing true, whole-picture understandings of partner cultures and communities, as well as other global stakeholders and team members. As a Penn State-wide program, HESE uses a progression of courses and fieldwork combined with research projects, commissioned assignments, and volunteer efforts reaching a wide variety of university members. The program encompasses

undergraduate and graduate students from all colleges including the medical and law schools as well as faculty and staff from diverse departments and campuses. Students spend multiple semesters on multidisciplinary teams progressing through a five-course series. Within these courses, their venture teams partner with subject-matter experts, local entrepreneurs, and businesses, and changemakers around the world to co-create a wide variety of technology-based social ventures. This radical collaboration model in HESE is nicknamed "eplum," a portmanteau of "E Plurbius Unum" (out of many, one). Eplum is about reaching out to a wide variety of stakeholders in ways and for reasons that are relevant to them: weaving global awareness, interdisciplinary engagement, and social innovation through a wide variety of connections. The resulting social ventures range from low-cost greenhouses and solar food dryers to telemedicine systems, ruggedized biomedical devices, cell-phone applications, and informal education systems. Every venture can be traced back to a set of ideas and/or needs expressed by partners in the target communities, and HESE participants are taught day one to engage every stakeholder's "story." Also to this end, all students conduct research during their fieldwork (venture-related or not) that strives to provide greater insight into community viewpoints.

We pause here to pose some questions. Exactly what are the details on how this was accomplished? How does the HESE student's *"multiple semesters on multidisciplinary teams"* provide greater insight into community viewpoints? Does this *"five-course series"* include culture general and culture-specific training in the humanities, particularly language, literature, worldviews, ethics, and history? Are the changemakers around the world from the material resource communities where there is food insecurity, energy insecurity, and poor access to clean water? Are the material-resource-poor country team members taking the lead or just follow-along folks?

Research questions have ranged from vendor and consumer opinions of street food to feedback on specific mobile applications to broad questions of how global youth perceive innovation. These efforts have resulted in the publication of over 125 peer-reviewed articles, the vast majority with undergraduate lead authors. Gaining—and appreciating—these cocreated insights and discussions are critical to developing culturally and socioeconomically viable ventures and educating the next generation social innovators and problem-solvers.

"Building equity" is another critical aspect to venture development, with focus on implementation rather than initial understanding. Each phase of each venture affects each stakeholder differently, and it can be a lot to track, much less balance. Empathy is required to understand these potential impacts, but equity is necessary to the successful implementation of each phase. All stakeholders must feel invested, i.e., have "skin in the game," but not overtaxed vis-à-vis their returns from the venture. When developing a venture, teams must also design so as to balance time, sweat, and other equity types (knowledge, credibility, social capital, etc.) as well as money.

Social ventures that aim to help the most disadvantaged entrepreneurs and consu-mers in society cannot afford mistrust, and ventures cannot neglect stakeholders who stand to lose under proposed solutions. The goal is always to cocreate win—win—win situations and to implement them with the equity and trust that requires. This is far from an easy task however, and HESE students are taught to internalize the need for emphasizing equity and trust from the beginning through numerous case studies (many from HESE itself), roleplaying, and venture engagement.

"Building ecosystems" is critical in HESE because no venture can stand up, scale up, or sustain alone. HESE itself is a broad ecosystem of undergraduate, graduate, and professional degree students; staff and faculty; public and private sector partners; and local champions. Each venture operates within its own ecosystem, starting with the initiating community, encompassing the venture team, and then extending to all of the start-up funding, sweat equity, and other equity partners and stakeholders. Each venture ecosystem continues to constantly evolve. One HESE venture stands out above the rest in this regard. The venture has successfully scaled using all the ele-ments just discussed and is now empowering local entrepreneurs in multiple coun-tries: the agricultural venture of low-cost greenhouses.

The Journey to Affordable Greenhouses

Large greenhouses are already a major phenomenon in some Sub-Saharan African countries, where large commercial farms can use multiple 8 m × 15 m industrial-grade greenhouses for growing export crops. These metal structures cost upward of US$3000 (in East Africa) or US$6000 (in West Africa). This price range is far beyond the capability of local smallholder farmers and agro-enterprises, but the struc-tures have, none-the-less, led smallholder farmers, particularly in Kenya, to aspire to have such greenhouses. For this reason, several farming cooperatives and agri-research institutions within HESE did reach out with a request to HESE to build an affordable greenhouse for this target market. Designing and selling such an accessi-ble version of this technology could have a major impact on smallholder farmers. Previous research into similar-sized greenhouses demonstrated that annual yields could increase by 10 times while reducing water consumption 30%—70%. Greenhouses achieve this success both by increasing single-cycle yields and by extend-ing the growing season to additional crop cycles, as well as providing protection for weather and pests. In the end, commercialized HESE greenhouses do all of the above—while also achieving full return-on-investment for the smallholder farmers in under 9 months. As of 2016, over 200 greenhouses have been sold by HESE partners with another 250 projected through the end of the year due to highly nonlinear growth.

Based on this community input, the first critical decision was to select a customer segment of lead users. The venture team determined that there were enough poten-tial customers with their own start-up capital to begin to sustain the venture.

With this initial customer base set, the venture team started a long and arduous process of trial, error, and feedback to develop and validate the correct greenhouse size, materials, and design. Pre-commercialization fieldwork involved construction and long-term testing in difference climatic conditions with different supply chains across three different countries (Kenya, Tanzania, and Rwanda). It was not until this extensive field testing fully developed and validated the venture's design and implementation that the first for-profit company licensed the technology. Mavuuno Greenhouses licensed the HESE design for sale in 10 East African countries in 2012 (this contract has since been modified to focus on Kenya alone). They now sell a variety of different greenhouse sizes with scalable options and drip irrigation systems. In 2013 the HESE technology was again licensed, this time to a social enterprise in Cameroon for sale in six neighboring countries. And most recently in 2014, funding from the USAID Securing Water for Food program and collaboration with World Hope International allowed another social enterprise, GRO Greenhouses, to establish itself in Sierra Leone, Mozambique, Zambia, and other countries. HESE and its ecosystem are now in the midst of further efforts to establish similar manufacturing and vending operations in Cambodia, Burkina Faso, and Senegal. The HESE program also continues to collaborate with its licensees to improve practices like data management and agricultural extension.

"Closing reflections on Case Study 4." In social entrepreneurship and agricultural change-making, my experience directing the HESE program leads me to highlight three key program requirements. Above all, cocreation necessitates building robust and sustainable empathy, equity, and ecosystems among all stakeholders. Empathy means listening to each member individually while observing through their words and into their actions to identify their latent needs and daily choices. Empathy also means connecting and cocreating in order to reach a shared understanding of those needs. Balancing equity means building trust and cohesion between stakeholders who invest with both monetary and nonmonetary input. Every stakeholder must have skin in the game. Finally, every venture requires a sustained but ever-evolving partner ecosystem that is self-aware and responsive to sustainability and scalability needs.

If empathy, equity, and ecosystems are the HESE philosophy, the real driver of success is "convergence." Both HESE's academic teams and its overarching ecosystem of public and private sector partners are highly interdisciplinary, and solutions are constantly found at the convergence of cultures, concepts, and disciplines. There is no "us" and "them" in cocreation, not between cultures or organizations or disciplines. Engineers learn to embrace Indigenous knowledge, businesspeople learn to engage in engineering design, and everyone from full professors and government officials to local champions with minimal education work to collaborate and innovate to find real solutions and deliver impact.

The final key to HESE is its "extremely practical approach." The goal of HESE's social ventures is not for undergraduates to go dig a community well or build a new irrigation system. The goal of every effort that receives HESE energy must be nothing less than a million smiles. And the only way to reach a "million smile" business is to commit to a triple bottom line of economic, social, and environmental sustainability and scalability. HESE emphasizes socioculturally appropriate, market-centric solutions backed by both robust research and Indigenous knowledge, and pushes for big changes through realistic, scalable means. While several academic programs and student club work with communities, five things that set HESE apart are (1) extremely multidisciplinary student and faculty teams, (2) emphasis on sustainable and scalable solutions, (3) a market-centric implementation approach, (4) integration of scholarly research and publication, and (5) focus on execution—getting the job done in the field. The emphasis on execution and sustainable impact continue to strengthen over time. While many faculty members and administrators view HESE, and similar entities, as a mechanism to develop soft skills, that is not the raison d'être of the program. HESE exists, and strives, to address global development challenges while preparing a cadre of social innovators and sustainable development professionals.

"Good intentions and passion are not enough." Any student traveling to a developing country for the first time is going to have a life-changing experience. If "delivering" a prototype or a PowerPoint presentation to a group of farmers actually solved the problem, the world would be a very different place. Practitioners, academic, and otherwise, who have spent decades in the field, know that the technology is a small component of the solution. A wide range of political, cultural, social, economic, and human factors need to be overcome before a technology product has a tangible, measurable social impact. The technology is 10%, the other 90% is implementation, monitoring, evaluation, and pivoting, pivoting, pivoting. The real challenge is in the execution, getting the job done in the field in a harmonious manner. And this is exactly what we need to teach students—how to get things done. Anything less is less than what the world needs.

We close Case Study 4 with the following comments. Although the humanities may not be involved as an academic area in this engineering/entrepreneurship program, the place-based, culture-based approach worked. It may have been the original empathy training and intrinsic focus on the goals and involvement of just "cheetahs" (social entrepreneurs who have the well-being of their community in mind). Maybe this is the other side of the coin of the two cultures. This is the Ayittey's (2005) approach: skip the humanities and aim for the cheetahs. Collaborate in an egalitarian way across cultures—cheetah to cheetah.

CLOSING REFLECTIONS

And now we close this "second look" at the humanities and the food sciences with these observations. Our four case studies were chosen from programs initiated in the United States. Europeans are beginning to respond to the connection they see between food and agriculture and the humanities

(Manouselis, 2016). In 2016 Agroknow carried out a large European project called Europeana Cloud. The project focused on how Europeana could be a research resource to scientists working on areas related to the humanities, i.e., people working on topics such as history, philosophy, theology, and culture. This project was aimed at determining how Europe's cultural heritage library, the Common Agricultural Policy, and agricultural cooperatives could be mutually helpful. The Agriculture Economic and Policy Research Institute, the national agricultural economics, and policy research institute of Greece are heavily dependent on ethnographic studies and other social science methods and tools. In many cases, humanities historical resources and other humanities archives are used as a research resource.

The connection of culture to sciences, including the agricultural and food sciences, is now established in the United States as an institution-wide policy under the national Institutional Review Board. While the European and US food and agricultural sciences sectors are beginning to recognize their connection to the humanities, simultaneously in the United States, the university system is penalizing those faculty who act on this newly recognized connection.[3] The mission statement of MSU, similar to most 1862 Land Grant Universities, is to educate students, create knowledge and art, and *serve communities by integrating learning, discovery, and engagement.* It may be difficult for those who work in more traditional areas to assess the contributions made in a different realm. But American universities are respected worldwide because of the breadth of the research undertaken by their faculty, their training of students from diverse backgrounds, their outreach, and *their ability to be innovative.*

How do the humanities relate to the agricultural and food sciences? Some say they have an essential relationship, and without the humanities, there are only inhumanities. Socioeconomics and policy research and other

---

[3]Remember the folks back home. Administrators, fellow faculty members, your family who are not with you in the field or at the negotiating table, or in the subsistence farm family courtyard with you when the challenges are met, and the stakeholders are pleased. The folks back home cannot just walk across the hall or see you at seminar or drop in on your class. The whole picture, the physical hurdles of temperature, an Indigenous language, and unfamiliar food are not fully appreciated by the people you left at home. During the P&T process, it is especially important to have low expectations for any of this work being rewarded in a professor's dossier. We hope this situation will be changed and that community engagement with Indigenous people, experiential learning, and transdisciplinary research harmoniously solving problems together in material-resource-poor countries is rewarded in the "marketplace" of P&T. The problem may be that this work recognizing culture is so different from the more traditional one of amassing research or field data and publishing it at regular intervals without connection to the overall goals of specific communities within this state or beyond. Certainly there must be places in a comprehensive research university to recognize contributions to the mission of the university that deviate from this pattern of industrial agriculture but nonetheless solve complex problems of our society. By definition a Land Grant University is aimed at teaching, research, and outreach that serve communities that make up our diverse societies. All of these approaches are necessary—the detailed disciplinary approach *and* the approach that accounts for the community itself.

dimensions of the social sciences are an integral part of agricultural research, innovation, and decision making, and form a bridge to the humanities. Humanities help to bring in the human dimension, historical perspectives, and issues in the research process, contributing critically to the analysis and conduct of research in crosscutting themes such as poverty, food security, nutritional imbalances, and natural resource management. The humanities provide research methods and tools for scarce resource allocations, identification of research agenda, research priority setting, adoption and impact assessment studies; and humanities can play important roles by identifying policy gaps and constraints to technology development, uptake, and scaling-up.

For a university to become comprehensively internationalized and able to train students to solve complex global, but local, problems, particularly in areas of food security and health, students must have specific technical training. This includes systematics, physiology, ecology, biochemistry, and other specific areas. But, part of this technical training must teach students how to understand cultures' role in the food and agricultural sciences and the conscious and subconscious role it can play in food choices and health management. This technical training begins with reflective or contemplative listening (Cavanaugh, 2017) and includes the humanities, the arts, ethics, history, language, and literature. Technology can be defined as the entire collection of techniques skills, methods, and processes used in the production of goods or services or in the accomplishment of objectives such as in scientific investigation (Borgmann, 2006).

As such, we consider the recognizing of and incorporating of culture-based, community-based approaches into scientific research, particularly involving food and health, a new technology. We propose that training students to use this new technology along with their specific training in molecular biology, immunology, cell biology, nutritional sciences, and toxicology be recognized as an essential learning outcome.

References

Ayittey, G.B.N., 2005. Africa Unchained: The Blueprint for Africa's Future. Palgrave, Macmillan, NY, 483 pp.

Bylenga, S.A., Clarke, S.A., Hammond, J.W., Morey, R.V., Kayumba, J., 1987. Evaluation of Applying Quality Standards to Bean and Sorghum Markets in Rwanda. Miscellaneous Publication #50. Minnesota Agricultural Experiment Station, University of Minnesota, St. Paul, MN, 165 pp.

Borgmann, A., 2006. Technology as a cultural force: for Alena and Griffin. Can. J. Sociol. 31 (3), 351−360.

Cavanaugh, J.C., 2017. Point of view: you talkin' to me? Chronicle Higher Educ. January 27, A48.

Dunkel, F., Giusti, A., 2012. French students collaborate with Malian villagers in their fight against malaria. In: Thomas, J. (Ed.), Etudiants sans Frontières (Students without Borders): Concepts and Models for Service-Learning in French. Amer. Assoc. of Teachers of French., pp. 135−150.

Dunkel, F., Wittenberger, T., Read, N., Munyarushoka, E., 1986. National storage survey of beans and sorghum in Rwanda. Miscellaneous Publication. Minnesota Experiment Station, University of Minnesota, St. Paul, MN, 205 pp.

Dunkel, F., Clarke, S., Kayinamura, P., 1988. Storage of beans and sorghum in Rwanda: Synthesis of research, recommendations and prospects for the future. Miscellaneous Publication #51. Minnesota Experiment Station, University of Minnesota, St. Paul, MN, 74 pp.

Edmister, J.A., Breene, W.M., Vickers, Z., Serugendo, A., 1986. Changes in the cookability and sensory preferences of Rwandan beans during storage. University of Minnesota - USAID - Govt. of Rwanda. Minnesota Agricultural Experiment Station Miscellaneous. Publication 47-1986. 390 pp.

Hanegreefs, P., Kayinamura, P., Clarke, S., 1988. Alternative Storage Methods for Beans and Sorghum in Rwanda.

Lamb, E. M., Dunkel, F., 1987. Studies on the genetic resistance of local bean varieties to storage insects in Rwanda. Miscellaneous Publication. Minnesota Experiment Station, University of Minnesota. St. Paul, MN, 122 pp.

Lamb, E.M., Hardman, L.L., 1986a. Survey of Bean Varieties Grown in Rwanda, Miscellaneous Publication 45-1986, Minnesota Agricultural Experiment Station, University of Minnesota.

Lamb, E.M., Hardman, L.L., 1986b. Catalog of Bean Varieties Grown in Rwanda, Misc. Pub. 44-1986, Minnesota Agric. Exp. Sta., University of Minnesota, 35 pp.

Jacobsen, K.L., Niewolny, K.L., Schroeder-Moreno, M.S., Van Horn, M., Harmon, A.H., Chen Fanslow, Y.H., et al., 2012. Sustainable agriculture undergraduate degree programs: a land-grant university mission. J. Agric. Food Syst. Commun. Develop. <http://dx.doi.org/10.5304/jafscd.2012.023.004>, pp. 1–14. Available from: <http://www.agdevjournal.com/current-issue/252-sustainable-agriculture-undergraduate-degree-programs.html?catid=101%3Afood-systems-and-higher-education-papers>.

James, J., 2016. Word of mouth. Goethe Institute. <http://www.goethe.de/ins/za/prj/wom/osm/en9606618.htm>.

Jordan, N., Grossman, J., Lawrence, P., Harmon, A., Dyer, W., Maxwell, B., et al., 2014. New curricula for undergraduate food-systems education: a sustainable agriculture education perspective. NACTA J. 2014;58(4) (302-310).

Malone, K., Harmon, A., Dyer, W., Maxwell, B., Perillo, C., 2013. Development and evaluation of an introductory course in sustainable food & bioenergy systems. J. Agric. Food Syst. Commun. Develop. February (1), 1–13. <http://dx.doi.org/10.5304/jafscd.2014.042.002>. Available from: <http://www.agdevjournal.com/volume-4-issue-2/412-intro-course-sfbs.html?catid=155%3Aopen-call-papers> (Published online February 8, 2014).

Manouselis, N., 2016. Digital Humanities and Agricultural Cooperatives: is there a link? Our solutions, projects. Agroknow blog. April 27, 2016. <http://research.europeana.eu/blogpost/digital-humanities-agricultural-cooperatives-is-there-a-link>.

Merriam-Webster, 2017. Definition of culture. #5. <https://www.merriam-webster.com/dictionary/culture>.

Mestenhauser, J., Dunkel, F., Paige, M. 1984. How to pack your parachute to hit the ground running: new concepts in predeparture training for faculty. University of Minnesota-Twin Cities, 60 min.

Ruttan, V., 1982. Agricultural Research Policy. University of Minnesota Press, Minneapolis, MN, p. 369.

Snow, C.P., 1963. The Two Cultures: And a Second Look. Cambridge University Press, New York, NY, p. 92.

Stanford Humanities Center, 2016. Home of the human experience. <http://shc.stanford.edu/what-app.re-the-humanities> (accessed 17.09.16.).

Wharton, C., Harmon, A.H., 2009. University engagement through local food enterprise: community supported agriculture on campus. J. Hunger Environ. Nutr. 4 (2), 112–128.

Couples Counseling: Native Science and Western Science

Florence V. Dunkel[1], Jason Baldes[2], Clifford Montagne[3], and Audrey Maretzki[4]

[1]Department of Plant Sciences and Plant Pathology, Montana State University, Bozeman, MT, United States, [2]Wind River Native Advocacy Center, Washakie, WY, United States, [3]Department of Land Resources and Environmental Sciences (retired), Montana State University, Bozeman, MT, United States, [4]Department of Food Science (retired), Interinstitutional Center for Indigenous Knowledge, The Pennsylvania State University, State College, PA, United States

CONTENTS

Case Study #1: A Change of Thought About Nutrition Education: Kenyan Women's Indigenous Nutrition Knowledge 284

Case Study #2: The Entomologist's Contribution to Early Childhood Nutrition 287

Case Study #3: Recognizing Indigenous Knowledge in the Academy 288

Strategies to Support Academics Interested in Indigenous Knowledge 290

Conclusion 293

References 293

Further Reading 295

In the 1930s British extension agents recognized that corn provided a more dense production of seeds than the millet traditionally used in Western Province, Kenya (Watt, 1936). The corn would yield a higher number of bushels of edible corn per acre than millet, and in the mid-1930s it was mandated by the colonial structure in Kenya that farmers replace the Indigenous crop millet (a nearly complete protein with substantial amounts of the amino acid lysine and a minimal, but acceptable level of tryptophan) with corn that was missing two of the essential amino acids (lysine and tryptophan) and was therefore a sure pathway to kwashiorkor or stunting, both physical and cognitive.

Corn was required by the colonial infrastructure as a quick fix for the increasing population and the decreasing levels of soil fertility; however, the switch to corn introduced malnutrition and nutrient-depleted soils. What was forgotten was the "concept of limiting factor" with essential amino acids, some of the most important nutrients we obtain from grains. These essential amino acids are building blocks. Each of the nine (10 for children) are essential to be available on a daily basis for the body to create important proteins. Simply put, the body can use amino acids only up to the limit of the amino acid that is in lowest concentration, and therefore least available. Corn was so much easier to harvest in the compact ear in contrast to the splayed inflorescence of traditional millet varieties that corn seemed like the "perfect new crop." The connection of this corn to the nutritional biochemistry of the

277

Incorporating Cultures' Role in the Food and Agricultural Sciences. DOI: http://dx.doi.org/10.1016/B978-0-12-803955-7.00011-0

human body was not understood, forgotten, or ignored. Within a decade, corn transformed into a pudding or "ugali" soon became a staple in the Kenyan meal system (Dunkel et al., 2016). Culturally established foods that evolve through a Native Science process, and not a colonial or even a Western Science process, are often the wisest choice for human nutrition as well as for the local environment. Either understanding of how amino acids worked or respect for Traditional Ecological Knowledge (TEK) and, likely, the process of Native science was absent. The result? At least 80 years of malnutrition and depleted soils in Western Kenya.

The message is urgent: Native Science and Western Science need to go into marriage counseling as soon as possible. Native Science needs to learn self-respect and self-empowerment. On the other hand, Western Science needs to learn patience, respect for other ways of knowing, including a rigorous use of a continuous feedback loop incorporated into a time frame of multigenerational repetition.

The key variable, human decision-making, is one of the few variables, which humans can actually control to some degree. It is still impractical to control weather and other natural variables, but decision-making has more promise of being controllable. The colonial model, top-down decision-making, usually led by outsiders, has at times resulted in loss of human dignity and self-sufficiency, along with ecological simplification and unintended consequences such as presence of invasive species or less healthy food choices than the traditional diet. We invite readers to offer examples wherein the Western Culture has replaced a component of the traditional diet and improved the total nutritional content.

In this chapter, we compare and contrast Native Science and Western Culture Science and explore the contributions that Native Science can make to Western Culture Science. Case Study #1 is an example of the intertwining of understanding Native Science with Western culture scientific training on the part of food scientists. Case Study #2 is about Indigenous knowledge (IK) that is well known among entomologists, but little known among early childhood nutrition scientists. Case study #3 explores perceptions of higher education, the academe, about the products of Native Science, Indigenous Knowledge.

The following ideas about Native Science were provided by Jason Baldes, an Enrolled member of the Eastern Shoshone, whom we met in Chapter 9, Listening With Students, and who has coauthored this chapter:

> The group of people generally referred to as Native Americans or American Indians are in fact much older than these terms given to them through colonization. These people are older than America and have been on the North and South American continents for millennia. There are currently 567

federally recognized tribes in the United States and at one time there were over 300 distinct languages spoken. Today, people that come from one of these nations will most certainly specify as to which tribe they belong and is the most correct way to refer to someone's identity. Tribal students that leave their reservations to attend a university have a set of culturally distinct differences from other tribal students of different tribal ties, in addition to the majority of students on campus. On campus' nationwide, it is not uncommon for American Indian or Native Americans of different tribes to have a tendency to relate to one another significantly as indigenous peoples. This inherent relationship is rooted in a concept of cultural commonality or knowledge. Though disregarded previously, it has since been recognized that indigenousness has application in conservation, biological understanding, landuse and many other areas.

In 2000 Berkes, Colding, and Folke defined TEK as "a cumulative body of knowledge, practice, and belief, evolving by adaptive processes and handed down through generations of cultural transmission, about the relationship of living beings (including humans) with one another and with their environments." Berkes et al. (2000) further recognized that "Traditional Ecological Knowledge is an attribute of societies with historical continuity in resource use practice" (p. 1252). This TEK concept can be applied to Indigenous people throughout the world, but this idea can be narrowed down more specifically to be compatible with the concept of Native Science as viewed by the Indigenous peoples of the North American continent.

In his book, *Native Science: Natural Laws of Interdependence*, published the same year as Berkes et al., Cajete (2000) states, "Native Science stems from a deeply held philosophy of proper relationship with the natural world that is transferred through direct experience with a landscape, and through social and ceremonial situations that help members of a tribe learn the key relationships through participation" (p. 67). Cajete, perhaps the foremost author on the subject, has identified the tenets of Native philosophy with the following areas of consideration:

- Native Science integrates a spiritual orientation.
- All human knowledge is related to the creation of the world and the emergence of humans; therefore, human knowledge is based on human cosmology.
- Dynamic multidimensional harmony is a perpetual state of the universe.
- Humanity has an important role in the perpetuation of the natural processes of the world.
- There is significance to each natural place because each place reflects the whole order of nature.

- The history of relationship must be respected with regard to places, plants, animals, and natural phenomena.
- Technology should be appropriate and reflect-balanced relationships to the natural world.
- There are stages of initiation to knowledge.
- Elders are relied upon as the keepers of essential knowledge.
- Acting in the world must be sanctioned through ritual and ceremony.
- Properly fashioned artifacts contain the energy of thoughts, materials, and contexts in which they are fashioned and therefore become symbols of those thoughts, entities, or processes.
- Every "thing" is animate and has a spirit.
- Dreams are gateways to creative possibilities if used wisely and practically.

It is important for students from Indigenous backgrounds to have the opportunity to conduct research that is directly applicable to their home community. It is also important for students from non-Indigenous backgrounds to be able to join in research with these students. When this research begins in the design phase, these tenets need to be in place. These tenets especially need to be in place when any students or researcher is asked to be a part of questionable research. These morals are often within, but not often easy to describe when questions arise. Having them in place, articulated in writing makes it easier to abide by these moral understandings. The Institutional Review Board (IRB) system is gaining in their assistance in this process, but there is still a distance to navigate before the gap is bridged between the practice of Western Science and Native Science.

When departing my own community, the Eastern Shoshone Tribe on the Wind River Indian Reservation in Wyoming, I was told by my elders to be sure to maintain my identity, and do not become what the University wants you to become. I struggled at first to understand what this meant, however after some time in an academic setting it became apparent. My first major was fine art, and while nearly completing a bachelor's program, I was dissatisfied that I was not contributing in some way to my home community in a culturally appropriate way. It was then that I switched majors to Environmental Science and began to address issues that had been underlying my most desired interests particularly around water rights and bison restoration. It is my belief that for many indigenous students, this is a common similarity. Many of these indigenous students that come from a reservation or from a tight knit community of extended family, intend to return home after education to improve conditions. This idea is ingrained from an early age. Go away to school and return home to help your people. If institutions understand the importance of TEK or Native Science and are ready and willing to facilitate indigenous philosophy, the student body and institutions benefit mutually.

The following five underlined insights were gleaned by Clifford Montagne from presentations and conversations at the particularly inspiring meeting of the American Indigenous Research Association, October 2016, Salish Kootenai College, Pablo, MT. We have expanded upon each of the five statements that emerged at the meeting and have drawn from the literature to develop the conversation in this chapter.

Recognition of indigenous values is increasing globally. Fragmentation of ecosystems and indigenous cultures continues around our globe. But, amid the specters of rising human population numbers concentrated in urban centers, loss of habitat, and growing economic inequalities, our remaining indigenous cultures are drawing on knowledge and wisdom of centuries to thread through these challenges, and to even share their ways with other more dominant cultures. These are tribal values (Lambert, 2014) honed by people living closely within contexts of the landscapes where their cultures have evolved. Within the context that life is sacred, and that the entire universe is alive, these are examples of cultures that are particularly good at approaching life challenges with grace, humility, and humor. Since life is universal, respect is likewise a key to nurturing relationships with other people and with all of Nature (Wilson, 2009). Around the world are numerous organizations and meetings evolving from Indigenous cultures as they empower themselves and reach out to other Indigenous cultures to share, network, and collaborate (National Indigenous Research and Knowledges Network, www.nirakn.edu. au; American Indigenous Research Association, americanindigenousresearch-association.org/; International Indigenous Research Conference (annually), http://www.indigenousresearch2016.ac.nz/). This also becomes the opportunity for those from nonindigenous cultures to listen to the conversation and learn at a deepening level about Indigenous cultures and Native Science.

Indigenous values start with the paradigm of people being part of Nature, and with Nature having spirit which extends to inanimate beings (Euro-view) beings such as a stone. In 2016 Clifford Montagne and BioRegions Mongolian Health Team Coordinator Badamsetseg Jargalsaikhan were hiking high on a limestone ridge above the Delger River. They stopped and Cliff, the geologist, took time to discuss the limestone ridges with Badamjargal, explaining how the limestone was formed from calcium carbonate in a seafloor environment and contributed to a soil pH of 8. Badamsetseg replied back that this was interesting, but to her Mongolian mind, the limestone had spirit and was alive.

Similarly, Jason has noted, the Shoshone people and many other Plains tribes lived close to the earth.

In life and ceremony, all elements are held in relation. For instance in sweat ceremonies, participants are reminded that the rocks used to hold the heat are referred to as grandfathers, and they are to be handled with great care.

They are not to be spit on or disrespected in any way. If dropped when being carried that the rocks must be returned to the fire. The most important point, as Jason reminds us, is that the rocks are not seen as inanimate but rather the opposite, they are seen as old men or grandfathers and are to be highly respected. Fire is cleansing and returning rocks to the flames if they are dropped restores the stone so that it may again be used appropriately. There is structure in the beliefs and understandings, reciprocity, and interconnectedness, resulting in specific ways of handling ceremony. Life is to be carried out in this way also and these ceremonies teach about how to live. To be grounded in ceremony with deep cultural understanding provides a way to live that has roots in ties to ancestors from time immemorial. These ways of thinking were nearly taken from us.

Adams (2016) reminds us that Indigenous epistemologies center on the concept that life is everything and everywhere in contrast to Western science, which has used reductionism to describe biological life as centered in living cells (2014) and their assemblages. This could lead to Western conclusions that inanimate objects are not alive and therefore have no spirit. Western religions have developed spiritual values and practices, which extend to the non-living components. But the ongoing dissonance between Western religions and Western science continues as the argument between creationists and evolutionists. Indigenous research methodologies then become suspicious to Western scientists because they also, like Western religions, use dreams, stories, and other ways to express reality.

Duane Champagne (2014) introduces the term "cultural holism" because Indigenous cultures include nature along with the details of human life and its organization, and to Western views this seem like religion. Champagne mentions that Western Culture centers on the individual, and often looks toward controlling nature from an outside view leading to fragmentation of natural processes and boundaries.

Indigenous practitioners, especially Elders, may have insights into how Nature works and how humanity can, and must, work in today's global change situations. These insights may not be obvious to people in the mainstream material culture paradigm. Indigenous, place-based cultures tend to provide a nurturing environment for children to grow up in, close to Nature, physically active, and surrounded by loving relatives and relationships. Ideally, decisions are deliberated among the whole group and are based on consensus, ensuring buy-in by all. These Indigenous groups (tribes) are resilient to much change, although current technologies and outside powers often run rough shod over them, exploiting, fragmenting, and destroying. Nevertheless, the ancestry-based IK nurtures children to learn, and later teach, with respect, to use kinetic (physical feeling) and emotive (performance, visual, and oral arts) expression in stories confirming respect for Nature and lead toward sustainable pathways (Wilson, 2009). These children eventually emerge as respected elders, providing decision-making guidance for younger generations as they face contemporary challenges.

Indigenous research is still downplayed by the dominant culture, which continues to insist that "science" must be objective, measurable, and repeatable, and often does not recognize subjective expression as being important. Although one may like to consider science as unbiased, Lori Lambert, founder of the American Indigenous Research Association, reminds us on the website that although the data collected by Indigenous research methodologies can be analyzed quantitatively as well as qualitatively, it is the acknowledgment of the relationship between researcher and data that fundamentally challenges Western research paradigms (www.americanindigenousresearchassociation.org). In her book (Lambert, 2014), Lori immediately confronts the reader with a realization, looking outward from Indigenous experiences, that the colonialism, which turned many homelands into colonies, still exists as neocolonialisma concept of which many Western scientists and citizens are unaware. Much Western research about Indigenous landscapes and cultures has been funded and directed from outside the communities and peoples of study, often with subsequent exploitation and degradation. Adams (2016) reminds us that Western scientists often wonder why they are not always well received when they come to "do research on" Indigenous peoples. In Chapter 3, Decolonization and the Holistic Process, we have provided some basics on decolonizing methodologies for Western scientists. To reconcile in one's own mind the value of both Indigenous and Western approaches to research is difficult, takes patience over time, but it is worth the journey. Indigenous research methodologies differ from the Western approach because they flow from tribal knowledge. Information is gained through relationship with people in a specific place. In Western academic models, the researcher is just an onlooker in the research project and separated from the data.

Challenges still exist within the scientific academy for indigenous scholars, but progress is being made as more academicians recognize the value of place-based experience, which eventually emerges as usable knowledge and wisdom. Although much Western science has focused on the reductionist, dualistic paradigm based on objective, and measurable and repeatable methods, we are reminded by Wilson (2009) how visionary Western scientists like Einstein have dreamed and used subjective knowledge through story to reveal theories and principles, which were initially dismissed, but later have become foundational natural laws when technologies like microscopy evolved to eventually "prove" them. Science can be unbiased and function across the spectrum from subjective experience to objective measure, both have their places in understanding realities. However, the epistemological gap between lives as depicted by living cells and an alive universe remains, as does the gap between creationism and evolutional theory.

Wilson (2009) reminds us that "caring leads to sharing" as more Indigenous scholars take on activist roles to support the validity and worth of Native knowledge. Adams (2016) ends her paper with the Indigenous advice to learn from the land, even as it changes in response to the colonialism and other global change processes while Mack et al. (2012) promote the importance of Informal Science Education to create inquiry-based learning activities, which help people relate to their native place, where the communities provide the instruction, and where local Native languages provide communication with rich context. These approaches build the human capabilities for all peoples to collaborate on bridging the ecological, social, and economic gaps, which threaten to fragment our world.

How do these five concepts of recognition of TEK and Native Science become incorporated into a university setting, particularly a Land Grant university focused by a US congressional mandate on the land and its people? In a study conducted in 2004 by Semali, Grim, and Maretzki (JHEOE) at the 1862 Land Grant university, Pennsylvania State University, faculty and extension staff throughout the commonwealth were asked about their engagement with indigenous/traditional/local knowledge in their teaching, research, and outreach activities. Those at the flagship research campus (University Park) were less likely to utilize indigenous/traditional/local knowledge than were those in outlying campuses. Faculty in the "hard" sciences were less likely to utilize indigenous/traditional/local knowledge than those in the social sciences and humanities, and faculty in the higher academic ranks were less likely to utilize indigenous/traditional/local knowledge than those in the lower ranks. The single factor that was found in this study to statistically significantly encourage faculty and staff to utilize indigenous/traditional/local knowledge was support from peers to encourage this activity. Audrey Maretzki, an author of this study, has been one of those peers who has never stopped wanting to learn new ideas or new ways of knowing. We meet her both in Case Study #1 and #3 in this chapter through her journey in learning respect for IK.

CASE STUDY #1: A CHANGE OF THOUGHT ABOUT NUTRITION EDUCATION: KENYAN WOMEN'S INDIGENOUS NUTRITION KNOWLEDGE

In 1988, I, Audrey Maretzki, had the opportunity to travel to Kenya to visit the University of Nairobi that had established a Memorandum of Understanding with Penn State University. As a community nutritionist in the Department of Food Science in the PSU College of Agricultural Sciences, my interest was in the health of women and children in rural areas. It was my good fortune to meet Dr. Gabriel Maritim, a Kenyan, who had recently taken on the role of

Head of the Department of Community Nutrition in the College of Agriculture on the Kabete Campus of the University of Nairobi. Gabe, a member of the Kipsigis, offered to take me to his home village of Gele-Gele, in the Rift Valley, a full day's drive from Nairobi. Residents in the Rift Valley site, I learned, were primarily Nilotic Kipsigis, while those in the Central Province were primarily Bantu Kikuyu.

Note that Audrey was, in preparing for her first visit to this area, already recognizing that Gabe belonged to a specific tribe in a specific place. She also learned in her active listening and appreciative inquiry what the origin of Gabe's people were. At this point in the story, we do not know if this will lead to success, but she has at least laid the groundwork to establish respect and trust that could lead to decolonized research methodologies.

I soon learned that Gabe had initiated a number of small-scale women's group income-generating projects that included a bakery and a communal mill for grinding grains. Gabe was idolized by the older women who, years before, had collected money to enable him to travel to the United States where he attended college in Oregon and was subsequently accepted to Johns Hopkins University where he received a graduate degree in Community Nutrition. Before I left on my return flight to the United States, we decided to attempt a community nutrition project involving the women in Gele-Gele and surrounding villages.

In 2002 I became the Principal Investigator for a $750,000 Kenya NutriBusiness Project with collaborators from Tuskegee University and the University of Nairobi. Our novel proposal was to develop NutriBusiness project sites in two distinctly different locations in Kenya, the Rift Valley Province and the Central Province. The Central Province site was less than an hour's drive from Nairobi, while it was a 6- to 8-hour drive to the Rift Valley Province location. In both of the rural locations, there were active women's groups that were undertaking small-scale projects to enhance their household incomes. In both sites, the women were responsible for maintaining their "shambas" on which they grew subsistence crops such as maize, millet, beans, tomatoes, squash, onions, potatoes, and sukuma wiki, a leafy green vegetable similar to kale. Women hoped to be able to harvest sufficient grain and other storable items for their household as well as to be able to take seasonable items to the local market to generate a modest profit. The difficulty facing these women farmers was that most of their vegetable crops came into season at the same time and the local market was temporarily flooded. The price of the crops dropped and nutrient-rich vegetables spoiled and were discarded (Muroki et al., 1997).

The objective of the NutriBusiness initiative was to encourage the small women's groups in each site to form a cooperative that would develop, process, and market regionally a nutritious, culturally acceptable, shelf-stable weaning food, utilizing crops they could grow in their shambas. As a nutritionist, I was especially interested in how the women would decide what ingredients to include in a weaning product for children from 3 to 36 months of age. The observations of two women taught me how they used their indigenous and contemporary knowledge.

Several women in the Central Province site suggested the inclusion of tomatoes in their weaning food product. They had learned from an extension agent that tomatoes were a source of Vitamin C that was needed in the diet, and with the application of a government-recommended pesticide, tomatoes would be free of insect damage. The women began talking about the pleasing flavor of tomatoes and tomato was about to be added to the list of ingredients for the weaning product. At this point a middle-aged woman stood up and said, "We should not use anything in our weaning food that needs pesticide. You know we can't grow tomatoes without spraying them." She sat down and tomato was not added to the proposed ingredient list.

In the Rift Valley Province, an elderly Kipsigis woman was listening intently to a discussion of the use of maize or finger millet as a grain in their infant weaning product. Maize is less expensive and easier to cultivate than finger millet, a tiny grain that is broadcast, time consuming to weed and likely to be eaten by birds. Some of the women argued that only small plots of finger millet are grown to be used for making beer and preparing ugali for special occasions. Infants and children today are fed maize gruel because maize is easier to grow and cheaper to buy at the market than finger millet. "Finger millet," they said, "is too expensive to use for porridge." At this point, the elderly Kipsigis woman said she would tell a story. For others to object would have been very rude, indeed. It was a chilly day and the old woman stood and wrapped a faded, all-purpose cloth around her shoulders. "Once there were two little boys of the same age. One of them was stocky and the other was wiry. The stocky boy's mother fed him maize gruel and the other boy was fed finger millet gruel. When the boys played they sometimes got into a fight about something, like all boys will do. When the two boys began pushing each other to the ground, people thought the stocky boy would knock the wiry little boy to the ground, but that was not what happened. Every time they had a fight the wiry boy pushed the stocky boy to the ground. The boy whose mother gave him finger millet gruel watched her son work hard and was proud that he was strong and healthy while the stocky boy's mother had to urge her son to run errands and she often had to take him to the clinic for medicine." When she sat down, the women said "We must put finger millet in our porridge." I compared the nutritional values of maize and finger millet and was not surprised to find that the old grandmother's knowledge of healthful food and her ability to tell a meaningful story based on her IK was a better nutrition education lesson than I could have taught with my academic knowledge (Maretzki, 2007, 2009).

CASE STUDY #2: THE ENTOMOLOGIST'S CONTRIBUTION TO EARLY CHILDHOOD NUTRITION

Another example of the contribution of IK to childhood nutrition is the use of soft-bodied caterpillars as weaning foods in sub-Saharan Africa. Ecosystems of sub-Saharan Africa have numerous places that support rearing of moths and butterflies. The most well known of these are the mopane or mopane caterpillar/moth, *Imbrasia belina*, that feeds on the mopane tree until it pupates. Boiled in salt water and sun-dried, the mopani can be kept for months without refrigeration. Texture, ease in handling, no legs with spurs on their exoskeleton that need to be removed, and the creamy insides of the fourth instar mopani caterpillar is a perfect weaning food. In addition, the mopani caterpillar provides remarkable nutrition as its iron content is almost six times that of the same quantity of beef steak, in addition to being a good source of potassium, sodium, calcium, phosphorous, magnesium, zinc, manganese, vitamin B12, and copper (van Huis et al., 2013; Williams et al., 2016).

Why has the search for a weaning food that is safe, low cost, locally available, and easily consumed not included mopane in sub-Saharan Africa? I, Florence Dunkel, first became aware of the urgent search for weaning foods in material resource-poor countries in 1984 during a brief consultancy at FAO-Rome. At that time, I was oblivious to the existence of certain species of lepidopteran larvae being the perfect weaning foods that are safe, low cost, locally available, and easily consumed, such as mopane in Zambia, Botswana, and Zimbabwe. The year 1984 was only my first year working in sub-Saharan Africa and fifth year in China, but that was not an excuse. Once again it is likely the disconnect and disrespect between the scientific couple, Native Science, and Western Science, that resulted in my not knowing about this perfect weaning food in 1984. I did have some suggestions for children's malnutrition when I met with FAO's children's nutrition team based on the research of my bean team in Rwanda (Dunkel et al., 1986, 1988; Edmister et al., 1987), but I did not have the best suggestion, insects. I was an entomologist and should have had the best suggestion. Edible insects, as far as I am aware, were not mentioned in entomology courses at University of Wisconsin (UW) when I was a student. Later, in the 1980s, UW became a world leader in this area (Defoliart et al., 2009). It was the Dark Ages of edible insects. Edible insects were almost a taboo topic among professional research entomologists and professors from Western cultures for at least a century. This value judgment was transmitted to their students, to aid agencies with whom they were invited to consult, and, most damaging, to the rural populations in these material resource-poor countries where these scientists were called in to assist with food security issues. Traditional knowledge was suppressed, pushed into the shadows.

When Audrey Marezki, as a food science professor, began awakening to the dangers of suppressing traditional knowledge such as in these food stories, she began working tirelessly over a decade into retirement to bridge the gaps between Western Science and Native Science related to food. She launched and continues to support with her colleagues at Penn State, a reservoir for collecting IK as well as to provide an active campus conversation to learn from this reservoir. Audrey brings us the next case study.

CASE STUDY #3: RECOGNIZING INDIGENOUS KNOWLEDGE IN THE ACADEMY

The view of the Interinstitutional Center for Indigenous Knowledge (ICIK) is that both indigenous and academic knowledge are forms of scholarship that can combine through *thoughtful discovery, transmission and application of knowledge*. Both indigenous and academic knowledge have developed over time, but in different cultural contexts and employing different tools and techniques. From ICIK's perspective, indigenous and academic knowledge can be complementary if the "scholars" are able to communicate and willing to respect the historico-cultural differences in their selective ways of knowing.

In 2000 promotion and tenure guidelines relating to this broad spectrum of scholarship that includes IK were released as part of the report titled "Beyond Boyer: UniSCOPE 2000: A Multidimensional Model of Scholarship for the 21st Century" (Hyman et al., 2001). UniSCOPE guidelines emphasized the integration of teaching, research and service, and defined scholarship as thoughtful discovery, transmission, and application of knowledge. When departmental committees at PSU began to implement the UniSCOPE guidelines, internal reports and interviews with faculty members revealed that (1) the system of rewarding scholarship was biased toward basic research and resident teaching, (2) academic culture favored quantitative over qualitative methods of inquiry, and (3) new faculty members were discouraged from undertaking in their pretenure years, any scholarship that extended to off-campus or to nontraditional audiences.

In 2004 ICIK, Penn State's Interinstitutional Center for Indigenous Knowledge (then a consortium), sponsored a successful international conference on IK and its role in the academy. During a full-day invitational session following the conference, a lively discussion arose about the barriers that prevent IK from being embraced by universities and how those barriers might be overcome. In an effort to objectively describe the academic perspective on IK, the codirectors of ICIK and a Senior Research Fellow in Penn State's Survey Research Center undertook a university-wide survey of faculty, staff, and Cooperative Extension agents on the main (University Park) campus,

Penn State's 23 commonwealth campuses and its 67 county-based Cooperative Extension offices (Semali and Maretzki, 2004; Semali et al., 2006). It should be noted that the ICIK survey took place 4 years after a PSU committee had released promotion and tenure guidelines contained in a report titled "Beyond Boyer: UniSCOPE 2000: A Multidimensional Model of Scholarship for the 21st Century" (Hyman et al., 2001). UniSCOPE guidelines emphasized the integration of teaching, research, and service, and defined scholarship as thoughtful discovery, transmission, and application of knowledge. The ICIK survey also coincided with the World Bank's release in 2004 of a publication titled "Indigenous Knowledge: Local pathways to global development" marking the 5-year anniversary of the World Bank's Indigenous Knowledge for Development Program (World Bank, 2004).

The ICIK survey, implemented in 2004, was conceived as a scholarly effort to identify barriers to the inclusion of "indigenous scholarship" (knowledge generated outside the academy) at a large, research-based, Land Grant university that defines itself as being "engaged in teaching, research and service."

A survey form was emailed to all PSU faculty and extension agents, a total of 6548 individuals. One thousand random individuals also received a survey form by mail. One thousand nine hundred and thirty-two individuals (29.7%) of the PSU combined faculty and extension agents responded. Structural equation modeling was applied in analyzing the data.

The statistically significant findings of the ICIK survey were informative and verified departmental committees' responses to UniSCOPE 2000.

1. Individuals employed at locations away from University Park (UP), Penn State's flagship campus, are *more* likely than those at UP to incorporate IK (identified as local, traditional, and/or folk knowledge or ways of knowing that are grounded in the experience of a local community) into their teaching, research, or outreach.
2. Individuals at higher academic ranks are less likely than those at lower academic ranks to incorporate IK concepts into their teaching, research, or outreach.
3. Individuals in technical disciplines are less likely than those in social sciences, arts, and the humanities to incorporate place-based or IK concepts into their research and teaching activities.
4. Being in a technical discipline is negatively associated with receiving peer support for the incorporation of IK into teaching, research, or outreach.
5. Receiving support or advice from peers, regardless of discipline, is the strongest predictor of whether a faculty member or extension educator

will actually incorporate IK concepts into their professional activities (Grim et al., 2005).

STRATEGIES TO SUPPORT ACADEMICS INTERESTED IN INDIGENOUS KNOWLEDGE

In the decade since the findings of the survey were published, ICIK has designed and implemented initiatives to bring IK to the attention of PSU faculty and students and has employed strategies to engage globally with individuals interested in knowledge generated outside the walls of the academy. These strategies include:

AcademIK Connections is a set of 12 short (5−10 minutes) YouTube videos featuring PSU faculty and graduate students that have embraced IK in various ways. Each video has related discussion questions that a faculty member could choose to expand upon the presenters' content in a 50-minute class period. AcademIK Connections was a collaborative effort between ICIK and the Humanitarian Engineering and Social Entrepreneurship Program at PSU. AcademIK Connections was an effort to provide peer support for faculty interested in bringing IK into their classrooms, laboratories, and extension programs (Mehta et al., 2013).

Endowment focused on IK. In 2008 Penn State received an endowment from the M.G. Whiting Center to "enhance indigenous knowledge at Penn State." With an endowment, it was possible to offer scholarships for student research and seminar speakers could be brought in for the seminar series as well as a myriad of other forms of on-campus visibility could be offered.

The Interinstitutional Center for Indigenous Knowledge. Center designation also improves visibility as does location of the Center outside any defining academic disciplinary group. In 2013 the PSU Board of Trustees approved a transfer of the M.G. Whiting Endowment from the College of Agricultural Sciences to the Penn State University Libraries, an academic unit serving the entire university as well as the local community. In 2014 ICIK was redefined as a center, The Interinstitutional Center for Indigenous Knowledge by the Dean of the PSU Libraries and Scholarly Publications, retaining its ICIK identity as the only global IK resource center in the United States.

Peer-refereed journal. With the official renaming of ICIK, the Libraries also announced the release of "Indigenous Knowledge: Other Ways of Knowing," an open access, peer-reviewed journal issued semiannually. See <journals.psu.edu/IK>. A unique feature of the journal is that authors may submit an article for publication in their indigenous language if they provide an English version of the manuscript for peer review.

Monthly seminar series that is easily accessible provides a forum for face-to-face discussion. Being located in the PSU Libraries enabled ICIK's monthly seminar series to be accessible globally in real time or to be viewed on the ICIK website, see <icik.libraries.psu.edu>, where the seminars are archived. Promotion of ICIK seminars is provided by the Libraries through PSU and community media as well as through the very active ICIK listserv that reaches ~1000 people around the globe with frequent and informative postings on indigenous topics (l-icik@lists.psu.edu).

Collaborative connections with local museums. ICIK has initiated a collaborative IK relationship with the Smithsonian Institution's National Museum of the American Indian, National Museum of Natural History, and National Museum of African Art as well as the Smithsonian Libraries. A number of PSU faculty have established connections with Smithsonian colleagues and ICIK has invited Smithsonian researchers to present seminars and workshops using the Libraries' facilities and promotional expertise.

Student research awards encourage students to understand the importance of IK and learn these research methodologies as a pathway to graduate school, to returning with new skill to their own Indigenous communities, or to gain special skills for professions servicing diverse populations. Interest accrues annually from the M.G. Whiting Endowment supports three to four PSU student (undergraduate or graduate) IK research awards. A team of PSU faculty reviewers evaluates the competitive student applications. Student awardees are required to present ICIK seminars when they have completed their research projects, and in addition, must submit either a peer-reviewed or editor-reviewed article for *Indigenous Knowledge: Other Ways of Knowing*. A student's research advisor is invited to introduce his/her advisee at the ICIK seminar. On occasion, family members travel to the PSU campus to attend their child's seminar or can view the seminar in real time on Media Site Live or on the ICIK website.

Edible insects interest group fits well with IK and follows student-driven interests and entrepreneurial activity that begins on campus. With support from the PSU Life Sciences Librarian, an informal entomophagy group has developed that includes townspeople as well as faculty, staff, and students from Entomology, Biological Sciences, Food Science, Animal Science, and Ecosystem Sciences and Management. Subsequent to an ICIK Co-Director's attendance at the 2014 international conference in the Netherlands on *Insects to Feed the World*, ICIK conducted an entomophagy seminar for the Food Science and Entomology

Departments. The seminar was followed by a Skype session with the Edible Insects Program at the International Center of Insect Physiology and Ecology (ICIPE) in Nairobi, Kenya. Interest in entomophagy led to a PSU library exhibit on edible insects and insect-containing foods designed by a PhD student in biology. The same student organized a well-attended expert panel on edible insects that included Dr. Florence Dunkel as a speaker. When our entomophagy team learned about a 2016 edible insect conference at Wayne State University in Detroit, Michigan, graduate students submitted oral and poster presentations and received financial support from their academic departments to attend the conference. Penn State's annual Great Insect Fair is a free event that draws several thousand people of all ages. Among many other activities, attendees learned about indigenous insect-eating cultures around the world while waiting in line to sample insect dishes prepared by the Entomophagy Team.

Student Society for Indigenous Knowledge (SSIK) captures student enthusiasm for social action. In 2010 several enthusiastic PSU students decided to organize a SSIK. SSIK is now an official student organization and has joined CORED, Penn State's Commission on Racial and Ethnic Diversity, that reports to the PSU President. SSIK's current goal is to engage with PSU students from underrepresented cultures that have too few students to organize and maintain a cultural organization of their own. Native Americans are one such underrepresented group at Penn State. Many SSIK members take part in Penn State's nationally recognized Ojibwe cultural engagement experience that is offered through the Community Environment and Development Program in the College of Agricultural Sciences (<agsci.psu.edu/ojibwe>) (Glenna et al., 2015; College of Agricultural Sciences, 2015). With ICIK's support for more than a decade, PSU students have learned about the Native Americans of the Great Lakes Region who moved west when Pennsylvania was colonized. Students spend 3 weeks learning from Ojibwe elders, leaders, artists, scientists, story-tellers, medicine men, authors, and craft-persons. They also take part in ceremonies, spend an over-night in Ojibwe homes, and observe and listen carefully before offering tobacco (kin nikinnick) to approach a knowledge-holder and respectfully ask a question. After taking the spring semester Ojibwe orientation course and returning from their cultural engagement experience in Northern Minnesota, PSU students in SSIK employ their intercultural skills to engage with underrepresented peers to learn and share informally about each other's cultures and ways of knowing in a safe, supportive academic environment (Glenna et al., 2015).

All of these educational initiatives are examples of how Penn State has structured activities to encourage respect for and interest in IK. The support and assistance provided by Penn State for faculty and students to bring IK into their classrooms, laboratories, and outreach activities serves as a good model to inspire other Land Grant Universities.

CONCLUSION

There is room for both objective and subjective measures and descriptions, and they exist along a continuum of "knowing." If this brief introduction to the interplay of Native Science and Western Culture Science was interesting to you can find more information and examples on the Website developed to accompany this book. Holistic Management (HM) and other forms of systems thinking, like Adaptive Management, provides a framework to accept both at a critical mixing point where synergies can occur. I believe that there is danger in using just a few parts of the HM process without the entire context. For example, monitoring for a sole objective without considering the values of the stakeholders and decision makers is a formula for misunderstanding. So it becomes even more important to build the relationships before getting embroiled in the tools. The HM mantra "Towards a Definite End" may be dangerous if taken out of context without a complete use of the HM process (model).

We believe that we all have opportunity and responsibility to become "indigenous" in our relationships and interactions with our place (landscape, Earth) and culture. Indigenous Research Methodologies, Native Science, and TEK are overlapping and interrelated approaches, which respect and honor being Indigenous.

References

Adams, D.H., 2016. In service to the land: indigenous research methods in the natural sciences. Tapestry Institute Occasional Papers, 2(1b). <http://tapestryinstitute.org/occasional-papers/in-service-to-the-landindigenous-research-methods-in-the-natural-sciences-vol-2-no-1b-october-2016>.

Berkes, F., Colding, J., Folke, C., 2000. Rediscovery of traditional ecological knowledge as adaptive management. Ecol. Appl. 10 (5), 1251–1262.

Cajete, G., 2000. Native Science: Natural Laws of Interdependence. Clear Light Publishers, Santa Fe, NM, 339 pp.

Champagne, D., 2014. <http://indiancountrytodaymedianetwork.com/2014/10/18/understanding-holistic-indigenous-cultures-157339> (accessed 18.10.14.).

College of Agricultural Sciences, The Pennsylvania State University, 2015. No one way of knowing: agricultural science student's perspective changed by ojibwe field experience. IK Other Ways Knowing 1 (2), 179–181, <https://journals.psu.edu/ik/article/view/59887>.

Defoliart, G., Dunkel, F., Gracer, D., 2009. The Food Insects Newsletter: Chronicle of a Changing Culture. Aardvark Global Publishing, Salt Lake City, UT, 414 pp.

Dunkel, F., Clarke, S., Kayinamura, P., 1988. Storage of Beans and Sorghum in Rwanda: Synthesis of Research, Recommendations and Prospects for the Future. Miscellaneous Publication # 51-1988. Minnesota Experiment Station, University of Minnesota: St. Paul No. 51, 74.

Dunkel, F., Hansen, L., Halvorson, S., Bangert, A., 2016. Women's perceptions of health, quality of life, and malaria management in Kakamega County, Western Province, Kenya. GeoJournal. <http://dx.doi.org/10.1007/s10708-016-9701-7>. Web published open source 4 May 2016. 81, 25 pp..

Dunkel, F., Wittenberger, T., Read, N., Munyarushoka, E., 1986. National Storage Survey of Beans and Sorghum in Rwanda. Miscellaneous Publication #46-1986. Minnesota Agricultural Experiment Station, University of Minnesota: St. Paul 205.

Edmister, J.A., Breene, W.M., Vickers, Z., Serugendo, A., 1988. Changes in the Cookability and Sensory Preferences of Rwandan Beans During Storage. University of Minnesota-USAID-Govt. of Rwanda. Minnesota Agricultural Experiment Station Miscellaneous Publication # 47-1986, 390 pp.

Glenna, L., Maretzki, A.N., Martin, B., O'Connor, J., Schlotzhauer, N., Sheehy, H.M., 2015. No easy task: making permanent an indigenous knowledge engagement course that changes lives. IK: Other Ways of Knowing 1 (2), 165–178, <https://journals.psu.edu/ik/article/view/59886>.

Grim, B.J., Semali, L., Maretzki, A., 2005. Checking for nonresponse bias in Web-only surveys of special populations using a mixed-mode (Web-with-mail) design. Working paper. <http://www.ed.psu.edu/icik/%20Working_Paper.pdf> (accessed 08..16.).

Hyman, D., Gurgevich, E., Alter, T., Gold, D., Herrmann, R., Jurs, P., et al., 2001. Beyond boyer: the UniSCOPE model of scholarship for the 21st century. J. High. Educat. Outreach Engage. 7 (1&2), 41–65.

Lambert, L., 2014. Research Methodologies for Indigenous Survival: Indigenous Research Methodologies in the Behavioral Sciences. Distributed by the University of Nebraska Press for the Salish Kootenai College Press, Pablo, MT.

Mack, E., Augare, H., Different Cloud-Jones, L., Davíd, D., Quiver Gaddie, H., Honey, R.E., et al., 2012. Effective practices for creating transformative informal science education programs grounded in Native ways of knowing. Cult. Stud. Sci. Educat. 7, 49–70. Available from: http://dx.doi.org/10.1007/s11422-011-9374-y.

Maretzki, A.N., 2007. Women's NutriBusiness cooperatives in Kenya: an integrated strategy for sustaining rural livelihoods. J. Nutr. Educ. Behav. 39 (6), 327–334 (Reprinted in the *African Journal of Food, Agriculture, Nutrition and Development*: 9(6) September 2009.).

Maretzki, A.N., 2009. Commentary: women's NutriBusiness cooperatives in Kenya—a prologue. Afr. J. Food Agric. Nutr. Develop. 9 (8), November 2009.

Mehta, K., Alter, T.R., Semali, L.M., Maretzki, A., 2013. AcademIK connections: bringing indigenous knowledge and perspectives into the classroom. J. Commun. Engage. Scholar. 6 (2), 83–91.

Muroki, N.M., Maritim, G.K., Karuri, E.G., Tolong, H.K., Imungi, J.K., Kogi-Makau, W., et al., 1997. Involving rural Kenyan women in the development of nutritionally improved weaning foods: nutribusiness strategy. J. Nutr. Educ. 29 (6), 335–342.

Semali, L.M., Grim, B.J., Maretzki, A.N., 2006. Barriers to the inclusion of indigenous knowledge concepts in teaching, research and outreach. J. Higher Educ. Outreach Engage. 11 (2), 73.

Semali, L.M., Maretzki, A., 2004. Valuing indigenous knowledges: strategies for engaging communities and transforming the academy. J. Higher Educ. Outreach Engage. 10 (1), 91–106.

Van Huis, A., Van Itterbeeck, J., Klunder, H., Mertens, E., Halloran, A., Muir, G., et al., 2013. Nutritional value of insects for human consumption. In: van Huis, A., Van Itterbeeck, J., Klunder, H., Mertens, E., Halloran, A., Muir, G., et al.,Edible Insects: Future Prospects for Food and Feed Security. Food and Agricultural Organization of the United Nations, Washington, DC, pp. 67−88. FAO Forestry Paper #171, 187 pp.

Watt, W.L., 1936. Control of striga weed in Nyanza Province, Kenya. East Afr. Agric. J. 1 (4), 320−322.

Williams, J.P., Williams, J.R., Kirabo, A., Chester, D., Peterson, M., 2016. Nutrient content and health benefits of insects. In: Dossey, A.T., Morales-Ramos, J.A., Rojas, M.G. (Eds.), Insects as Sustainable Food Ingredients: Production, Processing and Food Applications. Academic Press, San Diego, CA, pp. 61−84.

Wilson, S., 2009. Research is ceremony: indigenous research methods. ISBN: 9781552662816.

World Bank, 2004. Indigenous Knowledge: Local Pathways to Global Development; Marking Five Years of the World Bank Indigenous Knowledge for Development Program. Knowledge and Learning Group, Africa Region, World Bank, Washington, DC, <http://www.worldbank.org./afr/ik/ikcomplete.pdf>.

Further Reading

Aikenhead, G., Michell, M., 2011. Bridging Cultures: Indigenous and Scientific Ways of Knowing Nature. Pearson Canada Inc, Toronto, Ontario.

Apthorp, H.S., D'Amato, E.D., Richardson, A., 2003. Effective Standards-Based Practices for Native American Students: A Review of Research Literature. Mid-Continent Research for Education and Learning, Aurora, CO.

Archibald, J., Barnhardt, R., Cajete, G., Cochran, P., McKinley, E., Merculieff, L., 2007. The work of angayuqaq Oscar kawagley. Cult. Stud. Sci. Educ. 2, 11−17.

Augare, H., Sachatello-Sawyer, B., 2011. Native science field centers: integrating traditional knowledge, Native language, and science. Dimensions, November−December 38−40.

Bang, M., Medin, D., Atran, S., 2007. Cultural mosaics and mental models of nature. Proc. Natl. Acad. Sci. U.S.A. 104, 13868−13874.

Bartlett, C.M., 2011. Ta'n Wetapeksi'k: understanding from where we come. Proceedings of the 2005 Debert research workshop, Debert, Nova Scotia, Canada. In: Bernard, T., Rosenmeier, L., Farrell, S.L. (Eds.), Integrative Science/Toqwa'tu'kl Kjijitaqnn: The Story of Our Journey in Bringing Together Indigenous and Western Scientific Knowledges. Eastern Woodland Print Communications, Truro, Nova Scotia, pp. 1−8.

Mack, E., Augare, H., Different Cloud-Jones, L., et al., 2012. Effective practices for creating transformative informal science education programs grounded in Native ways of knowing. Cult. Stud. Sci. Educ. 7, 49−70. Available from: http://dx.doi.org/10.1007/s11422-011-9374-y.

CHAPTER 12

Putting It Together: The Way Forward*

Florence V. Dunkel[1] and Hiram Larew[2]

[1]Department of Plant Sciences and Plant Pathology, Montana State University, Bozeman, MT, United States, [2]U.S. Department of Agriculture (retired), National Institute of Food and Agriculture, Center for International Programs, Washington, DC, United States

Story: Picnic at the Great Wall

I, Florence Dunkel, had waited my whole lifetime to visit the Great Wall of China. My first visit to the People's Republic of China was an extensive professional tour to China in 1979 but somehow the Great Wall did not get on the agenda. I was now living in southern China (Guangzhou) as a US National Academy of Sciences Visiting Scholar. Beijing Agricultural University had invited me to give a seminar. I was only an hour's drive from my childhood dream. I shared my desire with the two young professors who were helping me get the slide projector set up for my seminar. A knowing smile passed between them. After my seminar and all the questions that I haltingly tried to answer in Chinese without the translation support I had for the formal part of the seminar, all logistics seemed to be in place.

The car with a driver was waiting. With a picnic basket in the back of the car, and I still in my blue business suit, nylons, and heels was whisked off with two of the professors at Beijing Agricultural University to my dream-come-true visit to the Great Wall of China. It was a beautiful spring day. Not many cars on the road. None as I recall after we left the edge of the city. When we arrived at the Great Wall, there were actually no tourists. It was 1981. China had only opened its doors to the West for the past several years following the Cultural Revolution. Looking forward to that first step on the wall, I bit my tongue as my hosts suggested we first spread out a blanket and try out the contents of the picnic basket. Two hours later I realized I was not on a tourist visit to this 2000-year-old structure. It was an opportunity away from the listening ears of the omni-present communist staff member who was part of every professional gathering and who organized the weekly meetings to

CONTENTS

Introduction 299

The Chapul Story 302

Other Young People Tell Their Stories 303

The Quiet Revolution 305

View From 30,000 Miles Above Academia 308

References 308

---

*With contributions from Cameron Ehrlich and Jordan Richards.

Incorporating Cultures' Role in the Food and Agricultural Sciences. DOI: http://dx.doi.org/10.1016/B978-0-12-803955-7.00012-2

discuss one of the chapters of the little red book. It was a chance for these two young professors with halting English skills to explain to me, the scientist from the West, what it was like to be a professor during the Cultural Revolution.

I learned about the hope in the hearts as these young professors as they had returned a few months earlier to their university buildings and to their laboratories they had been forced to leave behind when they were exiled along with most other scientists in Chinese universities to the countryside in a western province of China. For 2 years (1978–80), they had helped the subsistence farmers of a commune in Qinghai Province raise pigs and other agricultural products, the doors of their laboratories had been locked. When they were allowed to return to their laboratories, they found all the glassware smashed, overturned chairs and tables. Textbooks and laboratory notebooks were gone, likely burned by the Red Guard. I saw in the eyes of these young professors the passion for science and the tears for their struggle. Two hours later when the stories were over, tears dried, they asked if I still wanted to walk on the Great Wall of China. "Yes," I said, "this was a childhood dream."

Walking alone up on the wall, I could feel the layers of history that passed as the Wall had stood through two millennia. I could see the parts that had been rebuilt so that it was safe for tourists to walk on. I could see for miles and miles to the West, no structure or road. It was as if I was transported back into history for 2000 years. But it was not the exhilaration I had imagined I would feel. This was an amazing series of civilizations that had literally built a wall to keep out those that were "different." Of course, this wall also protected those inside the wall against those who would potentially have overtaken their culture, as the European influence on Native Americans accomplished. As a final stroke, 2000 years later, in the 1970s, descendants of those who built the Great Wall conducted a Cultural Revolution and cleansed from within, most vestiges of intellectual or Western scientific thought.

"Culture" is a complex term. In 2014 it was the most looked-up word in the electronic Webster's Dictionary. Why? In our polyglot of cultures in the United States, is there a rising fear of invaders from the West or the East or South? Is culture, now, an even more important word than it appeared to be in 2014? Food is essential to our health. Health and well-being are essential to our lives. What is the value of incorporating the cultural dimension in all of our decisions about food and agriculture? Are we suggesting that the cultural dimension is essential to our personal health? Culture is essential to our environmental health. Is recognizing culture and incorporating its role in the food and agricultural sciences essential to the continuation of our human life on Earth?

INTRODUCTION

First they ignore you, then they laugh at you, then they fight you, then you win.
— **Mahatma Gandhi**

This chapter helps the reader put together all the innovative, somewhat radical ideas that were laid out in the previous chapters. We suspect, certainly hope, that readers had a cognitive dissonance during the process. This is positive. Recognizing one's cognitive dissonance provides an opportunity for learning to take place, and we support your journey. As you close the book, we want it to be perfectly clear what you can do right now to incorporate the eleven concepts and associated skills illustrated in this book into your daily work, be it research, teaching, policy-making, outreach, and/or other forms of leadership. We have proposed some basic tools that have worked for us:

- Recognize and understand one's own culture.
- Be open to failure and how to learn from it.
- Decolonize (use language of those with whom you work and live).
- Strive to have an ethnorelative worldview.
- Use the holistic process in one's own life and facilitate its use in learning and sharing in a community.
- Value immersion experiences.
- Listen to Indigenous peoples.
- Listen across power lines.
- Instructors, listen to students, students listen to each other.
- Listen across campus.
- Include Traditional Ecological Knowledge and the doing of Native Science as an important basis and on-going process.

These are the basic components of a comprehensively and inclusively internationalized university or college or other educational or policy-making unit, one that can recognize and incorporate cultures' role in the food and agricultural sciences. These ways of listening and use of these additional basic tools create an environment where students, faculty, administrators, and staff create a rich teaching and learning environment in which local and global grand challenges can be addressed and solved sustainably. These ways of listening and basic tools can also create a place where Native American students and all other students regardless of heritage feel valued, ready to learn, and do not need to be "crossing the border" every day they attend class.

These tools are not the purview of any single academic discipline or profession or technology. These tools need to be used by everyone from staff and students to professors and administrators in a comprehensively, inclusively internationalized institution of higher education. And, they can and should be used to communicate across power lines. Hiram Larew has helped us understand the dynamic use of the holistic process to "Speak to Power." He shares with us these insights.

> While our own personal and professional growth and creativity, as a faculty member, student or peer, depends on our ability to deeply listen to others—especially to those who are very different from us—such skills can also help us breakthrough the seemingly impenetrable barriers that we face when speaking to power. Influencing key decision makers, steering them toward broader and bolder understanding of positive differences and the holistic process, and calling them to heed the perils of "cultural deafness" are all important outcomes of this book's lesson-in-a-word: Listen.
>
> Recent analyses and reports from organizations such as Association of Public and Land Grant Universities (APLU) and Association for International Agriculture and Rural Development (AIARD) have highlighted a crucial crossroads in fighting hunger at home in the United States and abroad, a crossroads that would have the US Government, college campuses, private sector and foundations step more intentionally forward with commitments to address hunger and thus, health. The recent reports (AIARD, 2017) echo many of the points made in this book, i.e., trans-disciplinary approaches are crucial, cultural awareness is the key, and listening is the precursor to addressing the scourge of hunger, and ultimately, world peace, itself. This report has opened the conversation. With this book, we are hereby adding to the conversation an additional approach to attaining food security—recognizing and incorporating cultures' role.
>
> Reaching those at the top—be they College leaders, professional icons, eminent scientists, budget gurus, political wizards, business moguls, religious forces or just good, solid officials—is so often a fearful challenge. Not only are such responsible people extraordinarily busy and hard to reach, but their passion for what they do may make them seem narrowly focused or make it difficult to capture their attention. In fact, those in power often are a distinct and distant culture all their own. While different approaches are more or less effective, in engaging those who are members of the power culture, usually being brave, upbeat, completely honest, clear, and intentional are good basic ingredients in the mix for approaching a leader. So is practicing the very core of participatory learning: *Listening is key*. In fact, never forget the point that echoes throughout this book's chapters: Listening is the most important cultural and communication tool we enjoy, and it is our most powerful tool as we engage with those in power. Listen to what is on powerful people's minds and, to the extent possible, speak to what they are saying. Understand their culture, and as you can, consider their background and interests. Learning what moves a decision maker—what she or he loves or hates—provides your grounding for speaking up.
>
> Speaking from your personal experience—"I learned the hard way how foolish I was to not listen to others"—is also a powerful way to affect the way powerful people think and feel. Testimonies from the heart and head

are hard to ignore. Messages that are short, sweet and bravely true are also a must. As demonstrated in this book, short personal stories that illustrate the difference that cultural insight has made in your own life is one of the best ways to catch decision makers attention and to sway their thinking.

Keep in mind, too, that when Speaking to Power, shyness is best overcome; being a little nervous is natural, but you should never shrink back because the powerful person seems too powerful—they are human just like everyone else. Said slightly differently, respecting power—like respecting differences—is a norm in so many cultures. But, respecting power is not the same as fearing it: Letting ideas rule simply because they are held by those in power is counterproductive. Speaking up and speaking truth as you know and feel it is your right and your responsibility.

One of the best ways to build skills in Speaking to Power is to practice with a few others to make a point. A small group can be especially effective in sharing a new idea or approach with power brokers. And, group members can learn from one another about what works well, what resonates, what convinces.

Like learning to listen across cultures, Speaking to Power is a long-haul effort. Persistence is usually needed; do not give up if the powerful person at first seems preoccupied or uninterested. Be confident in your message and your motives. Even if your first visit with the CEO or Chairperson goes terrifically well, even if you click, and even if she promises to change this or, to look into that, follow up will probably be needed. The nature of change is by definition slow, and leaders are often the slowest to shift opinions because of their deep experience, self-confidence, or vested interests. Changing anyone's mind requires more than a single dose of insight, and leaders can be especially "dug in" to the status quo. So, be thankful for quick breakthroughs, but be ready to try again, follow through and circle back as needed.

Speaking to Power is also, of course, done in many formats. In-person visits may be the favored mode, but letters to editors, opinion pieces on the web, or speaking up in meetings are often very effective ways to Speak to Power. Good ideas presented in compelling terms—spoken, written, or in any other way presented—will eventually reach the intended powerful person.

Lastly, you should be prepared to change as you Speak to Power. Powerful people are powerful for a reason—they have a proven track record of accomplishment and they know the ropes. They may, in fact, have good powerful life lessons and ideas to share—ones that will sway and influence you. The holistic process is all about positive change and how those who practice it can both promote change and be changed themselves through the very act of engagement. By listening carefully, you remain open to powerful ideas whatever their source.

Appreciating colleagues origins, being alert to invisible border crossings, striving to be ethnorelative, using the holistic process, merging STEM and Humanities disciplines, and the acceptance of Native Science in the Western

Science community are essential tools in recognizing culture in food and agricultural sciences. On the website associated with this book are a variety of tools, audio, video, on-line, and in hard-copy that await the reader to implement these concepts. What can the confident and inspired reader do, immediately? Contributions are made by an encouragingly wide age/education/ethic range of people. From the Millennial Generation, for example, there is Patrick Crowley, founding CEO of Chapul, Salt Lake City, UT. He is a successful young business man, who has been trained as a hydrologist, able to make a decent income as a white water rafting guide in the Western United States.

THE CHAPUL STORY

Pat is a curious combination of being intensely connected to the landscape and having a keen business sense. While rafting, Pat began to notice that the waters in the West were declining, dewatering and it bothered him. Pat is a quiet, soft spoken activist. He began to draw on his disciplinary training coupled with his passion for water and his deep understanding of interconnectedness of all of the living and nonliving components of the ecosystem. Pat explains the rational for his company as follows:

> We are children of the arid Southwest, and passionate advocates for a more environmentally sustainable footprint on our fragile environment, particularly with respect to freshwater usage. Personally, I grew up camping and exploring the Colorado River basin, and now work as a professional river guide in the Southwest and around the world. As we searched for ways to reduce humanity's destructive, even reckless use of our precious and limited freshwater resources, we came to two fundamental facts: 70—90% of all the water we use flows into agriculture, and 70% of all agricultural land exists entirely to feed industrial livestock production. Our meat consumption is literally draining the planet dry; already, the majestic Colorado River—the lifeblood of an entire region—no longer reaches the sea. Change must begin at home, and it starts with what we eat.

Pat consulted local Native Americans, the Ute people. He learned the stories of the caves where the local katydids banded together and how the traditional knowledge of the Ute regarding these katydids actually provided emergency nutrition for Western settlers who came to the area around the Great Salt Lake during a drought. Their families are still there. The katydids were later renamed the Mormon crickets (Dunkel, 2013; Dunkel et al., 2013). Pat, fascinated by this story did some research on energy requirements and other environmental costs of these edible insects in comparison to other crops introduced by the settlers. He also checked out the nutritional benefit these insects provided for humans. It all now made sense to him. Other crickets, the common house cricket, that is mass produced in the United States for pet food and are farmed in Thailand might be reared in Salt Lake City. A modern "prairie cake" like the Ute made 140 years ago, might solve both issues: humans search for a near perfectly nutritious food and the growing water scarcity in the West.

A successful session on the TV program, Shark Tank, provided the start-up funding for his idea, and Chapul, Inc., launched in 2012 in Salt Lake City, UT. Now the Chapul bars are available in five flavors to match taste preferences of five ethnic groups worldwide to whom Pat markets his product.

The Chapul bar is a protein bar with a date paste base. Pat purchases human food quality crickets live, roasts them in his facility and creates his own cricket powder for the protein bars. Since 2012, Pat speaks about feeding the revolution and uses that as his slogan for the protein bar. Two new definitive books, one on Sustainable Protein Sources (Nadathur et al., 2016) and one on the science of using insects as a food ingredient (Dossey et al., 2016), were published in 2016. Both books used the Chapul bar as an example of incorporating cultures' role in the food and agricultural sciences. Pat Crowley, is an example of "The Way Forward." The path that Pat has laid out for us is an example of marrying traditional cultural practices with environmental stewardship to create a highly nutritious human food product while minimizing the environmental footprint. This is a powerful model for the way forward. Indeed, the launching of Chapul paved the way for a plethora of other human food companies that recognized the environmental message in food choices. The floodgate of companies that subsequently opened in the United States included wholesale suppliers of cricket powder and a variety of consumer products, such as chips (Chirps), other energy bars, smoothie and shake mixes, gluten-free powder, snack food, and trail mix. In Montana, the first food insect farm, Cowboy Cricket Farms, has launched and their gluten-free, chocolate chip cookies are a hot item on the local and the Web-based markets.

The revolution has begun. It is with the Millennials and future generations that we can watch for where the revolution will go from here.

OTHER YOUNG PEOPLE TELL THEIR STORIES

Listen carefully to the young people around you. They may have more ideas for you about the Way Forward. Cameron Ehrlich, for example, comes from a Sustainable Foods disciplinary background overlain with various immersion experiences on several continents. Cameron participated in a set of role plays in AGSC 465R at Montana State University to help students and us faculty understand the decision-making process the people of Sanambele, Mali, may be going through as the women of the village begin the planning phase for a 1000 hen chicken coop. Cameron's student predecessors had been successful in explaining to the women in the village the nutritional biochemistry underlying physical and cognitive stunting in village children. Now the women knew what village foods would provide the lacking vitamin B12 and sufficient calcium, and the essential amino acids, lysine, and tryptophan in the children's diet. Two eggs a day per child supply these missing nutrients. The women have tested the acceptability of eggs in the daily diet of their children, and it was a success. The plan of the women villagers is bold. Cameron studied carefully for his classroom role as chief and then his role as a young, innovative farmer. He has shared a reflection about empathy and role playing with us during his early moments of getting to know the Sanambeleans.[1]

---

[1]Start the way forward with building empathy through role playing. "The role play exercise on Thursday not only helped to challenge my understanding of the Bambara culture, but also my understanding of my own culture and perspectives. For me, it seemed the experience was divided into two distinguishable stages: preparation and presentation. Preparing for the presentation was a process of gradually ascending a spectrum of understanding and empathy towards the Bambara culture. Our studies of the Bambara through readings and videos does help to build a certain level of understanding, and beliefs. However, when we learn through such materials, we do so through the lens of our own perspective. We process the information we receive in the context of our own worldview, and then apply those ideas in our own environment. The value of this process cannot be overlooked, as it can act as a catalyst to expand the potential of individuals as well as that of communities and societies. Still, such an understanding creates only a superficial image of the true nature of others. What is lacking in the process is the formation of empathy. While it is extremely difficult to establish, empathy allows us to step outside of our own perspective and into that of the other. Throughout the preparation for the role play exercise, I was forced to revisit the readings and videos we had used to study the Bambara, but this time was doing so from the perspective of a Bambara person. As I did so, I realized that I was absorbing that information in an entirely different way. I was internalizing the material in a way that actually placed me in the village, and in the shoes of the people there. I began to listen to Malian music and as I did so the music reinforced that mindset, and made every attempt I could to actually find myself in Sanambele. Then, as I applied that mindset to my research topics, I began to ask questions that I hadn't even considered before.". "The second stage of the role play, the actual presentation, was almost as valuable as the preparation itself. During the preparation stage, I was in control of the information I was receiving and was able to process that information at my own pace. However, during the presentation, this was not so, and I was forced to think on my toes to develop both questions and responses. This was even more exaggerated by the presence of guests from outside our class, who did not have the same understanding and background knowledge as that of the members of our class. I was forced to ground myself in that empathetic mindset that I had begun to develop during the preparation. When questions of ideas were presented, I had to address them from the Bambaran perspective, which required me to challenge my preconceived beliefs, step back, and formulate an answer that may have not been my own. Surprisingly, I didn't have as much difficulty with that as I initially anticipated. I think wearing the traditional clothes and creating the set really acted as the embodiment of all of the empathy I had built throughout the preparation process. Our outer appearance has a large effect on our individuality, and we change that appearance to resemble someone else, we instantly gain a new understanding of the person we are emulating. During my preparation, I felt I had begun to put one foot in the Bambara shoe, but still kept the other in my own shoes. However, once I put the traditional clothes on, I felt as though I was fully immersed in the culture and perspective of the characters I was playing. The unscripted conversation that took place raised a number of questions about our research and the approach we are implementing. Particularly, I found myself reflecting on our site mentor, Ibrahima's suggestion of asking the women's group for their vision of this chicken coop project. I realized that I have mostly been looking at the research from my own perspective, but in order to develop meaningful and empathetic solutions, we first need to understand their perspective. Hopefully doing so will create a shared vision that can be successful and beneficial for the Sanambeleans."

Jordan Richards completed a BS degree in cell biology and neuroscience and a MS degree in health sciences MSU. She shared her story on putting the concepts of decolonizing methodologies and the holistic process together with her background in immunology to help us understand the significance of a sound nutritional state and sound functioning of a child's immune system especially at the beginning of the malaria season that coincided with the hunger season in Mali. Using this experience in working with the Sanambeleans, she took a job after graduation with Americorps at a family health center in Santa Fe, NM. Her story of taking a small step as a junior member of a team to introduce the holistic process to the team of non-Hispanics who were making plans to introduce nutritional changes into low-income, elderly Hispanic communities in Santa Fe are shared in the footnote.[2]

THE QUIET REVOLUTION

There is a quiet revolution underway now in higher education. We are recognizing that to solve our wicked problems and to prepare the young people of our world to solve new grand challenges there must be a systemic change that takes place in higher education. This systemic change must take place horizontally and vertically, in every nook and cranny of the playing field of higher education, a grassy plain where academic silos are not allowed according to the "building code." This systemic change must take place in all colleges and departments of a university and at all levels from the custodians to the financial aid advisors to the President herself or himself.

Poverty and the whole area of health and food security is one of our global grand challenges right now and is predicted to be even more of a challenge in the future. Nobel Prize recipient, Professor Muhammed Yunus (2007), is positive and upbeat about poverty in the future. His advice is to work like poverty will only be found in museums by the year 2050. Right now my students say, "Oh this is an impossible dream." If higher education does its job in teaching and learning about what is poverty and what is not poverty, poverty may well be found only in museums by 2050. If higher education can reform the curriculum and retrain the faculty to engage with communities and let the community drive the learning action, and if higher education can

---

[2]As an AmeriCorps member working at La Familia Medical Center in Santa Fe, New Mexico, I was able to introduce the holistic process into the clinical setting. The Centro Educativo Naciones Unidas (CENA) program, a health education program for over weight and obese kids, was the perfect framework to reconstruct a concise model to better engage and assist these patients. Through focus groups and analyzed data of the CENA program, I was able to develop a program structure that was structured around the desires and needs to the CENA participants. When a program listens and is developed around the desires of the patients is when the greatest change and results will occur.

teach the holistic process to all students and professors, poverty may be in museums by 2050. Not all the answers rest in higher education. The Quiet Revolution is taking place in diverse places outside the halls of higher education. In the State of Montana Department of Agriculture offices we have noticed, edible insects are taken seriously as an alternative cropping system for Montana.. The US Department of Agriculture National Institute of Food and Agriculture (USDA NIFA) ELI has recognized the inseparability of culture and food in some of their competitive funding programs.

It is urgent that the listening process be intensified everywhere. To address issues of food security completely, those passing through higher education and those teaching and administrating in higher education will need to broaden their Western culture concept of food and listen carefully to other cultures' concepts of food and let the curriculum reflect this new attitude. For example, we need to be more open to what is truly edible and nutritious and what is not edible and detrimental to health. Ignoring as food important protein sources of one-third of the world's population by US managers of International development and those who manage funding for research and outreach is taking a step backward in addressing food security. There are glimmers of positive examples coming from World Bank and the IMF. The USDA NIFA, though, stands out as a bold trend-setter in this area. It was the chance that the peer review panel took in 2002 and the continual encouragement of the program officer, Gregory Smith to whom this book is dedicated, that began this quiet revolution of letting a Land Grant, College of Agriculture professor bring intercultural competency into a science classroom and place value on learning from Indigenous peoples about food and agricultural sciences. Thanks especially go to the vertical listening and inspiring conversations that have happened at the USDA NIFA Waterfront Center and in the Whitten Building on the Mall this past decade that have helped this book to come to be.

One of my (Dunkel's) favorite stories about inclusive and ethnorelative thinking I learned from Daniella Martin (2014). She and I have, on several occasions, shared stories with each other about how people from non-Western cultures find us Western culture folks so amusing in our adverse reactions to edible insects and to other interesting, but edible materials. One of her fun-to-listen to stories goes like this. Five hundred years ago, pests like insects living in corn kernels, called corn earworms, were part of people's diet, rather than deducted from it (Martin, 2014, pp. 89–90). Adult humans cannot produce and must ingest daily nine essential amino acids and children need 10. Amino acids are used by our bodies to build protein. Protein is needed by our bodies, muscles, and minds, to do what they need to do. When food has all 10 of these amino acids or building blocks, we say it is a complete protein food. Western culture folks often think insects in anything, except a silk factory or a bee hive, are intruders. The Aztecs, through their Native Science process, developed the perspective that rather than an intruder, the corn earworm was a bonus. In other words, an ear of corn with a worm in it was better than one without. When Western Science technology was finally developed enough to do the amino acid analysis, scientists found out that the Aztecs were right on the money. Corn by itself is usually lacking in lysine and tryptophan, two essential amino acids. Insects, specifically the corn earworm, provides both of those missing amino acids, in addition to nutrients corn does not supply in great quantities needed daily by humans, like calcium, iron, riboflavin, thiamine, copper, iodine, selenium, and zinc. So corn plus "worms," actually insect larvae, are a complete protein and even more. These moth larvae are soft bodied and when boiled or roasted provide an additional boost of flavor as well as nutrients. Another piece of traditional knowledge is practiced by Western cultures and Western science today, as Daniella reminds us, "An ear of sweet corn neglected by these insects is of lesser taste and quality [i.e., less nutrients]. Insects know their corn."

Just in case we have not made our point, consider corn smut or *huitlacoche*. Corn smut is a bluish black tumor-like fungus that I (Dunkel) often observed as I wandered through our corn field in Wisconsin as a child. Smut was hard to miss as it stood out on the plant in stark contrast to the green background. I was told it was a blight by my parents, but Danielle shared with me that she also learned from those in Mexico it is sold as a delicacy with a subtle, earthy, mushroomy taste. Here again, Western Science has reaffirmed the traditional wisdom of Mexican farmers. *Huitlachoche* is lysine-rich. The original discovery, though, was from the process of doing Native Science.

As revealed in the book, I am an entomologist by training, so there is a relevance for me to become involved in issues like malaria and kwashiorkor. But, what does a professional entomologist have to do with topics like diabetes management, climate change, and particularly culture? There is a simple answer. Once I adjusted my "entomology eye-shades," I could see that insects and other arthropods are part of everyone's life no matter who they are or where they live in the world.

Perhaps this is the key to comprehensively and inclusively internationalizing higher education, to use our discipline eye-shades, to enhance our vision of the world as humans living in communities within ecosystems within biospheres within a solar system. The disciplines are places to grab specific knowledge when we have identified a specific problem. It is all interconnected. In our ultra-fragmentation of knowledge in higher education and the research that takes place in the STEM areas within higher education, we have somehow lost our ability to see the human, the community, the ecosystem, and the biosphere. Comprehensive internationalization that is inclusive recognizes equally all cultures, peoples, living within various ecosystems and biomes.

We began this book looking at failures and how they are a fertile ground for learning to take place. We hope that this book, therefore, is a failure. We hope this book creates concern, fear, disgust, and a fair share of criticism. That will be the seedbed for creation of a new way of thinking about higher education. When the entire university system and the policy makers of governments and nongovernment organizations who disperse resources to others can recognize the fundamental importance of culture, of intercultural competency and that it is not only the realm of the "soft sciences" and the humanities, but part of all of what it means to be educated, then, only then, can we move into the 21st century with the tools to navigate with our 9 billion fellow human beings, dwindling clean water and air, and complete protein for everyone and all of their great, great grandchildren that will be born in this century.

VIEW FROM 30,000 MILES ABOVE ACADEMIA

The scientific world, the peer-refereed publication system, the promotion and tenure process, the forces inside one's self, and, in some respects, the US Land Grant system itself, try to push us into a narrow focus and to remain sustainably narrow. Herein lies the conundrum. How can we avert failures similar to those described in Chapter 2, Failures? If we use the holistic process and incorporate cultures' role in the food and agricultural sciences, is this enough? Those at the top or 30,000 miles above our daily narrowness do understand. If there is to be a systemic change of this magnitude in higher education, we must all step out of our bubbles of homogeneous ideas, and listen.

References

AIARD, 2017. SMART Investments in International Agriculture and Rural Development. Association for International Agriculture and Rural Development White Paper. 55 pp. http://www.aiard.org/uploads/1/6/9/4/16941550/smart_investments_final.pdf.

Dossey, A.T., Morales-Ramos, J.A., Rojas, M.G. (Eds.), 2016. Insects as Sustainable Food Ingredients: Production, Processing and Food Applications. Academic Press, San Diego, CA.

Dunkel, F.V., Coulibaly, K., Montagne, C., Luong, K.P., Giusti, A., Coulibaly, H., Coulibaly, B., 2013. Sustainable integrated malaria management by villagers in collaboration with a transformed classroom using the holistic process: Sanambele, Mali and Montana State University, USA. Am. Entomol. 59, 15−24.

Dunkel, F. 2013. Examples from Mali and the United States. In: van Huis et al. (Eds.), Edible Insects: Future Prospects for Food and Feed Security. Food and Agricultural Organization of the United Nations. FAO Forestry Paper #171, pp. 38-39.

Martin, D., 2014. Edible. New Harvest. Houghton Mifflin Harcourt, New York, 250 pp.

Nadathur, S., Wanasundara, J., Scanlin, L. (Eds.), 2016. Sustainable Protein Sources. Elsevier Publ. Co, Boston, MA.

Index

Note: Page numbers followed by "*f*", "*t*" and "*b*" refer to figures, tables and boxes, respectively.

A

Academia. *See* Western Science
AcademIK Connections, 290
Acanthoscelides obtectus, 26
Acetylcholine, 30−31
Acheta domesticus, 123
Achillea millefolium, 58
Adams, D.H., 282
Addams, Jane, 9−10
Africa Unchained: Blueprint for Africa's Future (Ayittey), 36, 56, 224
African Commission on Human and Peoples' Rights, 11−12
African culture, 5−6
Agency for International Development (USAID), US, 28, 135−136
  Bamako headquarters of, 203−204
  in classroom learning, 189, 203−204
  CRSP of, 221−222
  locust/grasshopper management program and, 29−34, 99−100
  in Mali food and health, 75−76, 78
  Securing Water for Food Program of, 272
The Agricultural Education, 236−237
The Agricultural Magazine, 236−237
Agricultural Science
  in Darhad Valley, 179−181
  EC Model in, 63−66, 64*f*
  ethno-relativity and, 13−17
  Holistic Process in, 62−66
  role of culture in, 13−17, 19−20
Agriculture and Human Values journal, 236−237
Agriculture Economic and Policy Research Institute, 273−274
Agroecology, in SFBS, 261
AIARD. *See* Association for International Agriculture and Rural Development
Aikido, 159−160, 165, 181
Akagera National Park, 3−4
Akihiru, Kuriki, 226−227
Aldrin insecticide, 30−31
Alexander, Sierra, 140−141, 147, 155, 211
Alfalfa, 35
Algae, 209−210
American Indigenous Research Association, 281, 283
American Prairie Reserve, 182
Amino acid deficiency, 32
Amskii Pikuni, 228−229
Ancient Futures: Learning From Ladakh (Norberg-Hodge), 80−81, 229−230
Anderson, James C., 237−238
Anderson, Rachel, 92−93
Animal and Range Sciences, 260
Animal herding, 49, 162−163, 175−177, 203
Anopheles gambiae, 116−117
Antibiotics, 58, 171
Apsaalooke, 12−13
  beginning relationship with, 141−143
  children in, 141−142
  eating habits in, 142
  Elder of, 4−5, 142
  grasshoppers and, 142−143
  language of, 95
  LBHC in, 142
  "Let's Pick Berries" Project in, 95−98, 143−146
  listening to, 141−143
  with students in, 211, 218−221, 234
  mini-immersion and, 90
  spears of, 90−91
Aragona, Sicily, 11
Arapaho, 36
Art and humanities, 268
Art/Science Fusion Model, 264*f*
Association for International Agriculture and Rural Development (AIARD), 222−223
Atinga, 56
Ayittey, George, 36, 56, 224
Azadirachta indica, 26, 59, 202−203
Aztec knowledge, 306−307

B

Baldes, Jason, 98, 214−215, 224, 233, 278−279
Bambara ethnic group, 110−115
  culture of, 213
  farmers of, 11, 13−14
  farming villages, 25−26
  language of, 136−137
  tutor of, 23
Bayanhangai Valley, 175−176
Bean storage, in Rwanda, 254−255
Beartusk, Janelle, 134
Beaver, 152
Beijing Agricultural University, 297
Bennett, Milton, 14−15, 101*b*
Berries
  cultural meaning of, 210

Berries (*Continued*)
  elders and, 144
  picking of, 15
  in Tlingit culture, 209−210
Big Horn River, 214−215
Bilateral communication, 23. *See also*
    Communication
Billick, Donna, 259*f*, 264−265
BioRegions International (BRI),
    17−18, 49
  annual work visits in, 168−176
  art in, 172−173
  environment of Mongolia in,
    169−170
  festival of Darhad Blue Valley in,
    173−174
  field clinics in, 171−172
  formation and evolution of,
    161−167
  Ger schools in, 168−169
  health in food and agriculture of,
    179−181
  importance of listening and,
    183−184
  learning in, 164−167
  lessons of, 174*b*
  Mongolian students in, 177−179
  Native American students in,
    177−179
  reflections on, 176−177
  water quality measurement in, 172
  whole community and business
    in, 174−176
  Yellowstone connection to, 181−183
BioRegions Mongolian Health Team,
    281
Bison, 214−215
  population reduction of, 57−58,
    138
  reintroduction of, 217
  restoration of, 280−281
Blackfeet Tribal College, 227
Blue Valley Festival, 173−174, 179
Bob Marshall Wilderness, 227
Body language, 4−5
Boojum Expeditions, 161
Bordeau, Amy, 122
Bottle gourd, 117*f*, 124−127
Bougoula, Mali, 11
Bozeman, Montana, 3−4
Bradley, Charles, 45−46

BRI. *See* BioRegions International
Brown, Jerry, 268
Bruchid beetle, 25−26
Buffalo berry, 145, 218−221
Buffalo ribs, 150−151
Building Resilience of Mongolian
    Rangelands: A
    Transdisciplinary Research
    Conference, 176−177
Burkina Faso, 272
Busby, Montana, 149
Butterfield, Jody, 45−46, 229−230

C
Cajete, Gregory, 197−198, 279−280
Calcium, 303−304
  in chicken eggs, 122−123, 123*t*
Calcium carbonate, 202−203
California Arts Council, 268
California biodiversity, 265*t*
California Davis, University of, 265*t*
Callabasse gourd, 117*f*, 124−127
Callosobruchus maculatus, 25−26
Camara, Abdoulaye, 137
Cambodia, 272
Cameroon, 224
Cape Town, South Africa, 30
Case study
  of failure
    Chicken award, 25−29
    fry bread in, 34−35
    introduction to, 23−25, 24*f*
    reflections on, 36−37
    use of river systems and, 35−36
    war on locusts and
      grasshoppers, 29−34
  of humanities and food science
    closing reflections on, 273−275
    graphic arts and entomology in,
      259*f*, 264, 264*f*, 265*t*
    HESE Program in, 269−273
    introduction to, 249, 253
    SFBS in, 252*f*, 257−264
  of immersion
    "Let's Pick Berries" Project in,
      95−98
    mini-immersion in, 89−92, 91*b*
    native food in, 85−99, 86*b*
    Northern Cheyenne Reservation
      in, 92−95

  summary of, 98−99
  of intercultural science, 194−205
  of listening to students
    classroom and policy-making in,
      222−233
    communication patterns in,
      211−214
    connecting reality with action
      in, 218−222
    unique backgrounds in,
      214−218
  of USDA in classroom, 188−194
Cashews, 52−54
Caterpillars, as weaning food,
    287−288
Cave Junction, Oregon, 230−233
CDKC. *See* Chief Dull Knife College
Celiac disease, 34−35
Chaikin, E., 6−7, 136−137
Champagne, Duane, 282
Chapul story, 302−303
Cheyenne culture, 6−7. *See also*
    Northern Cheyenne
Cheyenne language, 134
Chicago Council on Global Affairs,
    222−223
Chicken
  The Award of, 25−29
  feed of, 123−124
  in Mali, 303−304
  nutrition of eggs, 122−123, 123*t*
Chief Dull Knife College (CDKC),
    6−7, 133, 195
  Dean of Academic Affairs at, 212
  immersion in, 85
Children. *See also* Stunting
  corn in diet, 120*t*
  development of, 29−34
  finger millet and, 285−287
  malaria and, 115−118
  of Mali, 32
  of Rwanda, 30
  Sanambelean story of, 137
  stunting in Mali, 109−110
China, Cultural Revolution and,
    297−298
Chokecherry, 95−96, 145, 151−152,
    218−221
  pudding of, 145−146
Christensen, Scott, 182−183
Cicero, Marcus Tullius, 8

Classroom, 189−190
  and policy-making, 222−233
  USAID in, 189, 203−204
  USDA and, 188−194
Cognitive dissonance, 299
  in immersion, 89
Coliform bacteria, 172
Collaborative Research Support
      Program (CRSP), 25,
      221−222
Colonization. *See also*
      Decolonization
  of Cameroon, 225
  history of, 56−57
Committee on Scholarly
      Communication with People's
      Republic of China (CSCPRC),
      254−255
Common Agricultural Policy, of
      Europe, 273−274
Communication. *See also* Listening
  bilateral, 23
  CSCPRC in, 254−255
  former colonization in, 26−27
  gap of, 23−25, 24*f*
  nonverbal, 254−255
  self-confidence in, 214
  SFBS and, 259−260
  students in, 211−214
  in subsistence farming, 110−115
  village Chief in, 112
  women in, 27−28, 110−115
Community
  driven learning of, 305−306
  EC Model in, 63−66, 64*f*
Consensus-based decision-making,
      112, 213
Consumer culture, 8−9
Cooperative Extension offices,
      288−289
Corn
  in child's diet, 120*t*
  earworm and, 306−307
  finger millet nutrition and,
      285−287
  smut and, 307
Cotton, 33
Coulibaly, Bourama, 54−55
Coulibaly, Hawa, 28, 110, 117*f*,
      118−119, 127−128
Coulibaly, Keriba, 13−14, 28, 112

Cowboy Cricket Farm, 303
Cowpea, 29, 202−203
  nutrition and, 28−29
  postharvest of, 59
  study of, 26
Cowpea weevil, 25−26
Creationism, 283
Crickets
  callabasse and, 117*f*, 124−127
  farming of, 123
Crop production, in SFBS, 261
Cross-cultural education, 220.
      See also Culture
Crow Mercantile, 142−143
Crowlyey, Patrick, 302−303
CRSP. *See* Collaborative Research
      Support Program
CSCPRC. *See* Committee on
      Scholarly Communication
      with People's Republic of
      China
Cultural holism, 282
Cultural identity, 6−7
  awareness of, 14
  tectonic plate concept in, 6−7
Culture
  appreciation of others in, 210
  "center of origin" in, 10−11
  definitions for, 8
  inclusivity in, 14
  in Mali food and health, 77−80
  personal origins of, 3−4, 8
  of transparency, 8−9
Cures Without Care (Stanhope),
      10−11
Curlee, Michele, 6−7, 212

D

Dagget, Dan, 58
Dairy cows, 35
Dancing Across the Gap (Chaikin),
      6−7, 136−139
Darhad Valley, Mongolia, 19. *See also*
      Mongolia
  animal agriculture in, 177−178
  Bioregional Management Plan,
      166
  Elders of, 178−179
  festival of, 173−174
  GIS in, 178

  initial visits to, 161−164
  learning through crisis in,
      164−167
  people of, 164
  Renchinlumbe School in,
      172−173
  sugar consumption of, 167
  water quality in, 172
Darija, 83−84
Dartmouth College, 46−47
Decolonization
  colonization history and, 56−57
  conclusions on, 66−68
  connection to land in, 45−46
  Holistic Process introduction in,
      46−49
  introductions of, 43−45
  language in, 51−54, 53*f*
  in Mali, 55*t*
  methodologies of, 49−59, 221
  ownership of data or products in,
      54−55
  reductionist adjustment in, 57−58
  TEK in, 58−59
*Decolonizing Methodologies: Research
      and Indigenous Peoples* (Smith),
      56, 97
Deer, 149
Department of Agriculture (USDA),
      US
  in the classroom, 188−194
  CSREES Secondary Education
      Grant of, 194
  Higher Education Challenge grant
      program, 134, 262
  NIFA of, 203−204, 305−306
Descartes, René, 56−57
Diabetes, 34−35, 139
Dialakoroba, Mali, 11
Dieldrin, 30−31
Diggs, Robert, 137−138
"Disgust factor", 15, 34
Disney, 217−218
Dovchin, Badamgarav, 61−62, 176,
      180
Dreams, 280
Dunkel, Florence, 131−132, 287,
      291−292
  China and, 297−298
  immersion and, 83
  students and, 215−217, 238−239

E

Easterly, William, 80−81
Eastern Shoshone, 36, 80−81, 138,
    278−281
    grasshoppers and, 142−143
    in listening with students,
        214−215, 217
EC. *See* Expansive Collaborative
Echinacea, 87−88
Ecology Department of the
    National University,
    Mongolia, 166
Ecosystem fragmentation, 281
Edible Entrails booklet, 92−93
The Edible Entrails Cookbook, 151
Edible Entrails semester, 149
Edible insects
    Aztec knowledge, 306−307
    interest group of, 291−292
    mopani caterpillars as, 287−288
    moth larvae as, 306−307
Egg, chicken, 122−123, 123*t*
Ehrlich, Cameron, 303−304
El Salvador, 44−45
Elders
    of Apsaalooke, 4−5, 142
        berry picking and, 144
    consensus and, 282−283
    of Darhad Valley, 178−179
    in farmer information, 59−60
    interruption of, 212−213
    of Japan, 226−227
    as keepers of wisdom, 280
    of Kipsigis, 285−287
    Mali council of, 77
    of Northern Cheyenne, 148,
        150−151
    of Ojibwe, 292
    wisdom of, 3−4, 7−8
    youth-connection and, 91−92
Elk, 149
Empathy
    between cultures, 269−270, 272
    training, 273
Endangered Wildlife Trust, 30
'Enebla,', 50
Engineers Without
    Borders, 238−239
Entomological Society of America,
    264−265

Entomology
    early childhood nutrition and,
        287−288
    graphic arts and, 259*f*, 264, 264*f*,
        265*t*
    Humanities and, 264−268
    International Congress of,
        155−156
"Eplum", 269−270
Equity, in venture development,
    270−271
Ethiopian food, 50
Ethno-relativity, 13−14
    in decolonization, 41, 43
European culture
    colonization build up of, 56−57
    Common Agricultural Policy of,
        273−274
    stork as symbol in, 3−4
Europeana Cloud, 273−274
Evolutionary theory, 283
Expansive Collaborative (EC) Model,
    63−66, 64*f*, 233

F

Faculté des Lettres, Arts, et Sciences
    Humaines de Bamako
    (FLASH), 76
Failure
    averting of, 308
    cases of
        Chicken award in, 25−29
        fry bread in, 34−35
        introduction to, 23−25, 24*f*
        reflections on, 36−37
        use of river systems and, 35−36
        war on locusts and
            grasshoppers, 29−34
Falcon, Dawn, 227
FAO. *See* Food and Agricultural
    Organization
Federal Commodity program,
    138−139
Federal Indian Law, 89
FFA. *See* Future Farmers of America
Finger millet, 285−287
Fire, in land management, 46−49
The First One Thousand Days
    (Thurow), 80
Fish population, river systems and, 36

FLASH. *See* Faculté des Lettres, Arts,
    et Sciences Humaines de
    Bamako
Food
    algae as, 209−210
    Darhad Valley health of, 179−181
    deserts of, 85−99
    importance of sharing in, 50
    insecurity of, 223
    limpets as, 209−210
    taboos of, 152
Food and Agricultural Organization
    (FAO), 16−17, 28, 33, 222
    standards of, 52−54
Food and Agricultural Science
    EC Model in, 63−66, 64*f*
    Holistic Process in, 62−63
    National Food Quality Laboratory
        of Rwanda in, 253−257
Food Insects Newsletter, 16−17, 30
Foreign students, 222−233
Fort Robinson, Nebraska, 93−94, 134
Francophone Africa, 73, 75
French language, 63−64
Friends of Local Food, 260
Fry bread, 139
    case study of, 34−35
Future Farmers of America (FFA),
    95−96, 145
Future Leaders Forum, 222−223

G

Gandhi, Mahatma, 299
Genocide, in Rwanda, 255−256
Geographic Information Systems
    (GIS), 96, 145, 178
    at MSU, 218−221
Ger, 161, 168−169
Gibson, Susan, 171−172, 180
GIS. *See* Geographic Information
    Systems
Giusti, Ada, 63−64, 73, 75,
    258−259
Glacier National Park, 227
Global College of Long Island
    University (LIU Global),
    218−221
Global Positioning Systems (GPS),
    178

Global Research Symposium, 203–204
Gourd, in cricket, 117*f*, 124–127
GPS. *See* Global Positioning Systems
Graf, Birgit, 142
Grand Teton National Park, 182
Grasshoppers. *See also* Insects
    Apsaalooke and, 142–143
    war on, 29–34, 99–100
Great Insect Fair, 291–292
Great Lakes, 292
Great Salt Lake, Ute people of, 302–303
Great Wall of China, 297–298
Greater Yellowstone Coalition, 163–164, 182–183
Greenhouses, HESE and, 271–273
GRENARWA. *See* National Grainary of Rwanda
Grimm, Carly, 51, 114, 258–259
GRO Greenhouses, 272
Guatemala, 44–45, 225–226
Guest lectures, 187

H

Häagen-Dazs Honey Bee Haven, 268
Haiti, recover of, 36
Hansen, Kathleen, 238–239
HBCUs. *See* Historically Black Colleges and Universities (HBCUs)
Health, Poverty, Agriculture: Concepts and Action Research class, 4–5, 188–189, 229, 258
    listening with students and, 209–210, 237–238
    mini-immersion and, 90
Heart Butte, Montana, 227
Herbal knowledge, 9–10
Herding families, 162–163, 175–177, 203
Higher Education Challenge grant program, 134
Himalaya, 229–230
Historically Black Colleges and Universities (HBCUs), 251–252
History of Aboriginal People course, 218–221
HMM. *See* Holistic Management Mandala

Holistic Management (HM), 293
    connection to land and, 45–46
    listening with students in, 230–231
    origins of, 45–46
Holistic Management: A New Framework for Decision Making (Savory and Butterfield), 229–230
Holistic Management Mandala (HMM), 61–62, 61*f*
Holistic Process, 18, 59–62
    conclusions on, 66–68
    goal statement of, 65–66, 66*f*
    teaching of, 62–63
Holistic Resource Management, 46–49
Holistic Thought and Management, 17–18, 258
Homo sapiens, 11–12
Hoonah, Alaska, 15, 209–210
Huitlachoche, 307
Hull House Sicilian Day, 9–10
Human cosmology, 279
Humanitarian Engineering and Social Entrepreneurship (HESE)
    affordable greenhouses and, 271–273
    closing reflections in, 273–275
    "eplum" in, 269–270
    introduction of, 269–271
    philosophy of, 269
Humanitarian Engineering and Social Entrepreneurship Program, 290
Humanities, 18
    Art/Science Fusion Model in, 264*f*
    entomology and, 264–268
    food science and
        introduction to, 249, 253
        National Food Quality Laboratory of Rwanda in, 253–257
    HESE Program and, 269–273
    integration in SFBS of, 263–264
    merging with STEM of, 301–302
    in socioeconomics and policy, 274–275
Hunt, Mary, 258–259
Hutu, 255–256
Hyptis spicigera, 28

I

IAV. *See* l'Institut Agronomique et Vétérinaire
ICIK. *See* Interinstitutional Center for Indigenous Knowledge
ICIPE. *See* International Center of Insect Physiology and Ecology
IEFA. *See* Indian Education for All
IER. *See* l'Institut d'Economie Rurale
IK. *See* Indigenous knowledge
Imbrasia belina, 287
IMF. *See* International Monetary Fund
Immersion
    background to, 73–74
    concluding thoughts on, 100–102
    definition of, 74–75
    food deserts in, 85–99, 86*b*
    at institutional level
        evaluating results of, 83–85
        introduction to, 80–81
        undergraduates in, 81–83
    Malian food and health project in
        applying cultural knowledge in, 77–80
        growing awareness in, 76–77
        introduction to, 75
        naïve state in, 76
    MSU in, 73–74
    native foods in, 85–99, 86*b*
        "Let's Pick Berries" Project and, 95–98
        mini-immersion in, 89–90
        mini-immersion process of, 90–92, 91*b*
        Northern Cheyenne Reservation and, 92–95
        summary of, 98–99
    omission in eduction of, 99–100
    student view of, 101*b*
Inclusivity, 14
Indian Education for All (IEFA), 99
Indian frybread. *See* Fry bread
Indian tradition, HMM in, 61*f*
Indigenous, 10–13
    culture of
        fragmentation of ecosystems and, 281
        subjugation of, 11–12
        transfer of, 11–12

Indigenous (*Continued*)
  language, 9–10
  People's Day, 12*f*
  relationship with land of, 11–12
Indigenous knowledge (IK), 278
  and engineering design, 272
  ICIK in, 288–290
  strategies to support, 290–293
Indigenous Knowledge for
    Development Program,
    288–289
*Indigenous Knowledge: Other Ways of
    Knowing*, 291
Information dissemination,
    203–204
Insects
  as disease vector, 23
  as food, 13–17
  in Mali food and health
    cultural knowledge in, 77–80
    growing awareness in, 76–77
    introduction to, 75
    naïve state in, 76
    neurotoxins for, 30–31
  as traditional snack food
    in Mali, 32
    in Rwanda, 29–30
Insects to Feed the World conference,
    33–34, 291–292
l'Institut Agronomique et Vétérinaire
    (IAV), 81, 84
l'Institut d'Economie Rurale (IER),
    26, 54–55, 112–113
  in Mali food and health, 76
Institutional Review Board (IRB)
    system, 279
Integrated Pest Management (IPM),
    25
Integrated Pest Management
    Collaborative Research
    Support Program (IPM CRSP),
    64–65, 75
Intercultural competency, 14–15
Interinstitutional Center for
    Indigenous Knowledge (ICIK),
    288–290
  website of, 291
International Center of Insect
    Physiology and Ecology
    (ICIPE), 291–292

International Congress of
    Entomology, 155–156
International Institute for
    Environment and
    Development, 25
International Monetary Fund (IMF),
    306
International School of Bamako, 27
IPM. *See* Integrated Pest Management
IPM CRSP. *See* Integrated Pest
    Management Collaborative
    Research Support Program
IRB. *See* Institutional Review Board
Iron content, 287
ISAR. *See* National Research Institute
    of Rwanda
Issues of Insects and Human Societies,
    13–14

J

Jamaica, 223
Japan, 226–227
Jargalsaikhan, Badamsetseg, 281
Johns Hopkins University, 285–287
Johnstad, Mark, 180–181
Jones, Warren, 46–47
June berries, 145, 218–221

K

Kaba, Lansine, 255–256
Kansas State University, 254
Kayinamura, Phocus, 255
Kenya
  meal system of, 277–278
  women's IK of, 284–286
Kibungo Prefecture, Rwanda, 3–4
Kinyarwandan language, 254
Kipsigis Elders, 285–287
Klein, Robyn, 132–133
Kola nut, 77
Kolb, David, 68
Kroos, Roland, 48–49, 59–62
Kwashiorkor. *See* Stunting

L

Ladakh, 225–226, 231
Lady bird beetle, 264–265

Lambert, Lori, 283
Lame Deer, Montana, 85, 137, 196
  Rezzeria of, 86*b*
Lame Deer High School, 154
Land
  spirit and, 281
  spiritual relations with, 11–12
Land Grant University, 264
  Colleges of Agriculture at, 166
  IK support in, 290–293
  in immersion programs, 84
  mission statement of, 274
  National Food Quality Laboratory
    of Rwanda and, 253–257
  native foods and, 85–86
Land management, 46–49
Language
  in culture, 12–13
  of Darija, 83–84
  in decolonization, 51–54
  French in, 254
  graphic representation and,
    51–52, 53*f*
  immersion and, 74–75
  Kinyarwandan in, 254
  in SFBS, 259–260
  Spanish-speakers in, 214
Larew, Hiram, 188–189, 193, 222,
    233, 299–301
*The Last Hunger Season: A Year in an
    African Farm Community on the
    Brink of Change* (Thurow),
    80–81
LBHC. *See* Little Big Horn Tribal
    Community College
Ledger, John, 30
Lefthand, Fredericka, 155
Leonardo Art Science Evening
    Rendezvous, 268
"Let's Pick Berries" Project, 15–16,
    95–98, 143–146, 234–235
Lewin, Ross, 84–85
Limestone, 281
Limpets, 209–210
Listening, 299–301. *See also*
        BioRegions International;
        Policy leaders and listening
  in cowpea story, 29
  importance of, 37, 183–184
  interrupting speakers in, 212–213

Native Americans and
  Apsaalooke in, 141–143
  awakening in, 132–136
  background to, 131
  indigenous teach and learning
    in, 136–146
  "Let's Pick Berries" project in,
    143–146
  reflections on, 155–156
  shared curricula in, 146–155
"outsiders" and, 183–184
students and
  background in, 209–211
  communication patterns in,
    211–214
  community missions of,
    214–218
  conclusion in, 240
  cross-cultural immersion of,
    236–237
  encouraging passion in,
    237–238
  personal connections in,
    233–234
  policy-making in, 222–233
  reality and action in, 218–222
  response to, 238–239
  teaching methodologies and,
    234–236
  unique backgrounds of,
    214–218
in subsistence farming, 109–110
  bottle gourd and, 117f,
    124–127
  concluding remarks in,
    127–128
  main communicators in,
    110–115
  malnutrition and, 118–124,
    120t, 123t
  role of health in, 115–118
Listeria, 58
Little Big Horn Tribal Community
  College (LBHC), 142, 155
Littlebear, Richard, 134, 146–147,
  154
LIU Global. *See* Global College of
  Long Island University
Livestock in Sustainable Systems
  course, 230–231

Livestock production, in SFBS, 261
Locust. *See also* Insects
  children collecting of, 30
  malaria and, 29–30, 33–34
  population cycles of, 32–33
  war on, 29–34, 99–100
Lodge Grass, Montana, 95, 143,
  218–221
Lodge Grass High School, 144
Long Island University, 218–221
Lysine
  FAO standards of, 52–54
  in stunting, 32

M

Madin, Kent, 161
Madsen, Michelle, 133
Madsen, Robert, 91b, 133
Maize. *See* Corn
Malaria, 198–199
  cowpea nutrition and, 28–29
  epidemic of, 239
  food insecurity and, 41
  locust and, 29–30, 33–34
  in Mali children, 115–118,
    231–232
Malathion, 30–31
Mali
  authors of, 76–78
  The Chicken Award
    and, 25–29
  children's story of, 137
  food and health project in
    cultural knowledge in, 77–80
    growing awareness in, 76–77
    introduction of, 75
    naïve state, 76
    USAID in, 75–76, 78
  HM Model in, 55t
  malaria in, 115–118, 115f
  middle school cultural exchange
    with, 194–205
  Northern Cheyenne exchange
    with, 135–136
  subsistence farming in, 109–110
    bottle gourd and, 117f,
      124–127
    concluding remarks of,
      127–128

  main communicators in,
    110–115
  malnutrition and, 118–124,
    120t, 123t
  role of health in, 115–118
  Women's Association in, 51
Mali Extern Program, 27–28,
  133–134, 221–222
Malian scientists, 17–18
MAP. *See* Montana Apprenticeship
  Program
Maretzki, Audrey, 284, 288
Martin, Daniella, 306–307
Matzo, 150
Mavuuno Greenhouses, 272
McCartney, Heather, 135–136
McFarland U.S.A. (Disney), 217–218
McNulty, Bridget, 94–95
Meat, storage of, 98
Mediterranean, 5–6
Mehta, Khanjan, 269
Melanoplus differentialis, 142–143
Merriam-Webster, 8
Mestenhauser, Josef, 255
Mexico, trade with, 152
M.G. Whiting Endowment, 290
Millennial Generation, 8–9, 19–20,
  240
  in listening with students, 221
Miller, M.M., 36
Mineral rights, 140–141
Mini-immersion, 89–92
Minnesota, University of (U of MN),
  26, 83
  in food storage, 253–257
  and National Food Quality
    Laboratory of Rwanda,
    253–257
Mishig, 162, 168–170
Mongolia
  Bayanhangai Valley of, 175–176
  GPS in, 178
  herding in, 49, 162–163,
    175–177, 203
  initial visits to, 161–164
  learning through crisis in,
    164–167
  middle school cultural exchange
    with, 194–205
  President of, 179–180

Mongolia (*Continued*)
  Russian history and, 161
  seminomadic communities in,
      17–18
Mongolia River Outfitters, 179
Montagne, Cliff, 44–45, 61–62, 95,
    281
  in Darhar Valley, 159
Montagne, Joan, 163–164, 197
Montana, Department of Agriculture
    of, 305–306
Montana Apprenticeship Program
    (MAP), 133
Montana State University (MSU),
    3–5. *See also* BioRegions
    International
  BioRegions Program in, 49
  College of Agriculture of, 75
  College of Liberal Studies of, 15
  Community Service-Learning
      award from, 213
  cowpea study and, 26
  GIS in, 218–221
  holistic process and, 46–49
  Holistic Thought and Management
      in, 211
  Horticultural Farm of, 154–155
  immersion in, 73–74
  Mali Extern program in, 65
  mission statement of, 274
  Native American Studies program,
      218–221
  Soil Conservation class of, 160
  STEM curricula in, 218–221
Montana Wilderness Association,
    44–45
Mopani caterpillar, 287–288
Morocco, 5–6
Morocco, University of, 83
Mosquito
  anopheline larvae of, 59
  malaria and, 28
  in Mali, 115–118, 115f
Moth larvae, 306–307
Mountain Research Center, 161
Mozambique, 272
MSU. *See* Montana State University
Mule deer, 149
Multiple origins, 6–7
Museum collaboration, 291
Muskrat, 152

N
Naadam Festival, 173
NACTA. *See* North American College
    of Teachers of Agriculture
Nadathur, S., 303
Nagoya, University of, 226–227
Nairobi, University of, 283
NAS. *See* National Academy of
    Sciences
National Academy of Sciences
    (NAS), US, 254–255
National Agricultural Research
    Organization, 54–55
National Food Quality Laboratory of
    Rwanda, 253–257
National Forest Service, 48–49
National Geographic, 162–163, 175
National Grainary of Rwanda
    (GRENARWA), 257
National Institute of Food and
    Agriculture (NIFA), UDSA,
    203–204, 305–306
National Office for the Development
    and Marketing of Food and
    Animal Products (OPROVIA),
    257
National Research Institute of
    Rwanda (ISAR), 257
*National Standards for Foreign
    Language Learning*, 259–260
Native American Plains people, 3–4,
    139
  "comfort food" and, 34–35
  disease emergence in, 34–35
  Elder wisdom in, 4–5
  Fry bread and, 34–35
  water utilization and, 35–36
Native Americans
  Art History of, 147
  Blackfeet, 227
  Cheyenne, 6–7, 134
  culture
    interrupting speakers in,
        212–213
  Eastern Shoshone, 36, 80–81,
      138, 142–143, 214–215,
      217, 278–281
  Elders of, 132–133
  First People of, 227
  foods of, 92–95

  identity and, 214–215
  listening with
    Apsaalooke and, 141–143
    awakening in, 132–136
    background to, 131
    "Let's Pick Berries" project in,
        143–146
    reflections on, 155–156
    shared curricula in, 146–155
    teaching and learning in,
        136–146
  Mongolian students and, 177–179
  Northern Arapaho, 36, 138,
      214–215, 217
  Northern Cheyenne, 11, 16,
      135–137, 147–148,
      150–153, 196, 213
  Ojibwe, 292
  poetry reading of, 146–147
  shared curricula in, 146–155
  Shoshone, 281–282
  sovereignty and, 215–217
  sweat ceremony of, 281–282
  Tlingit, 15–16, 141–142,
      209–210
  Ute, 302–303
  Western Shoshone, 80–81
Native Science, 18
  edible insects and, 306–307
  Holistic Process in, 59–62
  recognition of TEK and, 284
  TEK and, 58–59
  and Western Science
    background of, 277, 284
    conclusions of, 293
    early childhood nutrition in,
        287–288
    ICIK in, 288–290
    Kenyan TEK in, 284–286
    strategies to support TEK,
        290–293
*Native Science: Natural Laws of
    Interdependence* (Cajete),
    279–280
Natural Resources Conservation
    Service (NRCS), 48–49
Neal, Ty, 96, 145–146
Neem trees, 26, 31–32, 202–203
Newton, Sir Isaac, 56–57
NGO. *See* Nongovernmental
    organization

NIFA. *See* National Institute of Food and Agriculture
Niger, 31–32
Nightwalker, George, 93–94, 147, 153
Nilotic Kipsigis, 283
Nitizelyayo, Anastase, 257
Nomadacris septimfaciata, 30
Nongovernmental organization (NGO), 122
  in Mali food and health, 76
Nonverbal communication, 254–255
Norberg-Hodge, Helena, 80–81, 229–230
North American College of Teachers of Agriculture (NACTA) Journal, 236–237
Northern Arapaho, 36, 138, 214–215, 217
Northern Cheyenne, 11, 16
  dry meat in, 147, 150
  Elders of, 148, 150–151
  exchange with Malians of, 135–136
  interruption of speech in, 213
  Mexico trade with, 152
  middle school cultural exchange with, 196
  nomadic diets of, 152–153
  women of, 137
Northern Cheyenne Reservation, 86b, 133
  immersion in, 92–95
  natural resources and, 140–141
Northern Cheyenne Tribal College, 85
Northern Cheyenne Tribal Reservation
  middle school cultural exchange with, 194–205
Northern Rockies, 3–4
NRCS. *See* Natural Resources Conservation Service
NutriBusiness Project, 285–287
Nutrition
  of cowpea, 28–29
  in early childhood, 287–288
  of eggs, 122–123, 123t
  of finger millet, 285–287
  malnutrition and, 118–124, 120t, 123t

Obama, Barrack, 36–37
Oedaleus senegalensis. *See* Locust
Ogiman bread, 149
Ojibwe Elders, 292
Om Sta Nah, 150
OPROVIA. *See* National Office for the Development and Marketing of Food and Animal Products
Oral storytelling, 16, 303–305
  Mali children's tale in, 137
  in Northern Cheyenne culture, 92–93
Orchard, Charles, 59–60, 68
Ownership, of original data, 54–55

P
Parathion, 33
Peace Corps, 27, 113–114, 133, 222
Pennsylvania State University (PSU), 284
  HESE program in, 269
  ICIK of, 288–290
  Survey Research Center of, 288–289
The People of the Darhad Blue Valley (Poulsen), 166
Pesticides
  Aldrin, 30–31
  Mali food and health and, 75
  and stunting, 123–125
  in war on locusts/grasshoppers, 30, 33
  weaning food and, 285–287
Peynado, Gizelle, 222–223
Peyou, Aedine Ndi, 224
Phaseolus vulgaris, 26
Phillips, Ronald, 237–238
Pine, Francesca, 141
PLAN Mali, 33, 122
Plasmodium falciparum, 116–117
Plums, 218–221
Poaching, 149–150
Policy leaders and listening
  in the classroom, 188–194
    getting started, 189–190
    impacts of, 193–194
    planning of, 190–193

intercultural science and, 194–205
  data analysis and conclusions in, 201–203
  design and data collection in, 200–201
  global research symposium of, 203–204
  information dissemination in, 203–204
  partners in, 195
  reflections on, 205
  research areas of, 198–199
  research process of, 199–200
  tips and suggestions on, 204
  trunks in, 196–198
  introduction in, 187–188
  student influence on, 222–233
  USDA in, 188–194
Porcupine, 152
Postharvest
  of cowpeas, 59
  of dry edible beans, 256–257
Potatoes, 201–202
Poulsen, W., 166, 172–173
Poverty, Inc. (Miller), 36
"Pregnant silence", 37
Prince, S.J., 166
Promotion and tenure (P&T) process, 263
Protein, cricket powder in, 302–303
PSU. *See* Pennsylvania State University
P&T. *See* Promotion and tenure

Q
Qinghai Province, 298
Queen bee, 265t
The Quiet Revolution, 3–4, 8, 305–308
  culture definition in, 8–20
  Ethno-relativity in, 13–17
  Indigenous peoples in, 10–13
  Millennial Generation and, 17–20

R
Ramaswamy, Sonny, 155–156
Reductionist Science, decolonization in, 57–58

Reflective listening, 275
Reflective writing, 62
Renchindaava, 170
Renchinlumbe School, 172–173
Renewable Resource Universities, 251–252
Resource management, river systems and, 35–36
Respiratory infections, 9–10
The Rezzeria, 86b, 91
Richards, Jordan, 305
Rides, Roylene, 68
Robison, Greta, 96, 218–221
Rome, culture and, 9
Rozin, Paul, 15
Rubus spectabilis, 15–16, 141–142
Ruppel, Kristin, 178
Russia, 228–229
Ruttan, Vern, 251–252
Rwanda, 3–4
    bean storage in, 254–255
    GRENARWA of, 257
    ISAR of, 257
    Minister of Agriculture of, 257
    OPROVIA of, 257
    traditional snack food of, 29–30
    U of MN and, 253–257

S
Salish Kootenai College, 281
Salmon berry, 15–16, 141–142
Sanambele, Mali, 11, 32, 136–137, 303–304
    malaria and, 231–232
    Women's Association in, 213
Savoie, Brent, 80
Savory, A., 229–230
Savory, Allan, 45–46, 48–49, 68
Science, technology, engineering, math (STEM) curricula, 155, 184, 218–221
    merging with Humanities of, 301–302
Science trunks, 196–198
Scientific process, teaching of
    conclusions in, 201–203
    data analysis in, 201–203
    partners in, 195
    research area in, 198–199

research design and data collection in, 200–201
research process in, 199–200
SOP in, 196
Scott, Earl, 256
SCS. See Soil Conservation Service
Seminole, Mina, 150–151
Senegal, 272
Serugendo, Ausumani, 256
Shambas, 285–287
Share-the-Wealth Symposium, 92, 131, 146–147
Shark Tank, 303
Shea butter, 28, 202–203
Shield, Pretty, 132–133
Shoshone people, 281–282
Shyness, 299–301
Sicilian culture, 4–5, 9–10
Sierra Leone, 272
Sindelar, Brian, 59–60, 68
Skunk, 152
Small, Kurrie, 4–5, 90–91, 218–221
Small, Tracie, 4–5, 90–91, 131, 153, 218–221
Smallholder farmers
    greenhouses and, 271–273
    importance of listening to, 183–184
Smith, Gregory, 140, 306
Smith, Linda, 56, 97
Smithsonian Institution, 291
Snell, Alma Hogan, 132–133
Snow, C.P., 264f
Social media, 36–37
Social ventures, in HESE, 270–271
Socioeconomics, humanities and, 274–275
Sociolinguistics, 5–6
Soderquist, Lora, 61–62
Soft Systems Circle, 48–49
Soil Conservation course, 160
Soil Conservation Service (SCS), 48–49
Soil nutrient-depletion, 277–278
SOL. See Standards of Learning
SOP. See Standard operating procedure
Sowell, Jamie, 229–230
Spang, Larissa, 137–138

Spanish-language speaker, 214
Speaking to Power, 299–301
Spiritual orientation, 279
SSIK. See Student Society for Indigenous Knowledge
Standard operating procedure (SOP), 199–200, 205
Standards of Learning (SOL), 194, 197
Stanford University, 251
Stanhope, Michael, 10–11
Staphylococcus, 58
Stearns County, Minnesota, 4–5
Storage, food, 98, 253–257
Story telling
    Mali children's tale in, 137
    young people in, 303–305
Student Society for Indigenous Knowledge (SSIK), 292
Students
    communication patterns in, 211–214
    community missions of, 214–218
    cross-cultural immersion of, 236–237
    encouraging passion in, 237–238
    personal connections and, 233–234
    policy-making in, 222–233
    reality and action in, 218–222
    view of immersion, 101b
Stunting, 29–34, 277
    chicken eggs in, 122–123, 123t
    food security and, 118–124, 120t
    Mali children and, 109–110
    Mali women and, 303–304
Sub-Saharan Africa, 3–4
    Akagera National Park of, 3–4
    greenhouses in, 271–273
    stork story of, 3–4
Subsistence farming
    Atinga in, 56
    importance of listening to, 183–184
    in Mali, 109–110
        bottle gourd and, 117f, 124–127
        concluding remarks in, 127–128
        main communicators in, 110–115

malnutrition and, 118–124, 120t, 123t
role of health in, 115–118
Sugar beets, 35, 195
Summer field clinics, 171–172
Sustainable Cropping Systems course, 232–233
Sustainable Food and Bioenergy Systems (SFBS)
  agroecology in, 261
  crop production in, 261
  curriculum development in, 260
  design team of, 252f, 258–260
  goals of, 258
  integration of humanities and, 263–264
  learning outcomes of, 260–262
  lessons learned from, 262
  livestock production in, 261
  student response to, 262–263
Sustainable Protein Sources (Nadathur), 303
Swanson, Tim, 49, 163
Sweat ceremony, 281–282
Switzerland, 4–5
Systematic behavior, 8–9

T

TA. *See* Teaching Assistant
Tallbull, Linwood, 90–92, 91b, 149
Tallbull, Meredith, 86, 86b, 91, 154
Tanzania, 30, 225–226
TEA. *See* Teaching Excellence and Achievement
Teaching Assistant (TA), 209–210, 215–217
Teaching Excellence and Achievement (TEA), 112
TEK. *See* Traditional ecological knowledge
Testimonials, 299–301
Three Irons, Jade, 141
Thurow, Roger, 80–81
Tlingit people, 15–16, 141–142, 209–210
Tobacco ceremony, 292
Tou, 52–54
Towne's Harvest Garden, 260, 262–263

Traditional ecological knowledge (TEK), 4, 18
  definition of, 279
  of Niger, 31–32
Trans-disciplinary approach, 63
Traore, Ibrahima, 112
Tryptophan
  FAO standards of, 52–54
  Mali sources of, 121
  in stunting, 32
Tsagaan, Teki, 163, 165
Tuareg people, 11–12
Tuskegee University, 283
Tutsi, 255–256
Twodot Land and Livestock Co., 46–47

U

U of MN. *See* Minnesota, University of
Ugali, 285–287
Uhlman, Diane, 259f, 264–265
Ulaanbaartar, Mongolia, 163, 166, 174, 176–177
UN. *See* United Nations
Undergraduate education. *See also* Classroom
  foreign students in, 222–233
  immersion in
    annual activities of, 81–83
    evaluating results of, 83–85
    introduction to, 80–81
  Native Americans
    shared curricula in, 146–155
UniSCOPE guidelines, 288
United Nations (UN)
  FAO of, 16–17, 28, 33, 52–54, 222
United States (US)
  Agency for International Development (USAID), 28, 135–136
    Bamako headquarters of, 203–204
    in classroom learning, 189, 203–204
    CRSP of, 221–222
    locust/grasshopper management program and, 29–34, 99–100

  in Mali food and health, 75–76, 78
  Securing Water for Food Program of, 272
  Botanical Garden of, 268
  Department of Agriculture (USDA)
    in the classroom, 188–194
    CSREES Secondary Education Grant of, 194
    Higher Education Challenge grant program, 134, 262
    NIFA of, 203–204, 305–306
    Federal Commodity program of, 138–139
    federally recognized tribes in, 278–279
    genocide in, 34–35
    Land Grant Universities of, 85–86, 251–252
    National Academy of Sciences (NAS), 254–255
    Northern Rockies, 3–4
  University Park (UP), 284, 288–289
  Uranium, 214–215
  US. *See* United States
  USAID. *See* Agency for International Development
  USDA. *See* Department of Agriculture
  Ute people, 302–303
  UW. *See* Wisconsin, University of

V

Vaccarello people, 11
Venture development, equity in, 270–271
Vigna unguiculata, 25–26
Village Chief, 112
Virginia Tech, 237–238
Visions of the Blue Valley (Poulsen), 172–173
Vitamin B12, 54, 287, 303–304
  in chicken eggs, 122–123, 123t
Vitamin C, 146, 285–287
Voice intonation, 4–5, 214

W

"War on Locusts and Grasshoppers", 29–34, 99–100

Warren Rush, Warren, 30
Water
    quality of, 199
        in Darhad Valley, 172
    rights to, 280–281
    river diversion and, 35–36
    Wind River conflict of, 225
Wayne State University, 291–292
Weaning food
    product development of, 285–287
    soft-bodied caterpillars as,
        287–288
    tomatoes in, 285–287
West Africa
    locusts and, 29–34
    middle school cultural exchange
        with, 194–205
Western Province, Kenya, 277
    TEK of, 277–278
Western Science
    early childhood nutrition in,
        287–288
    Holistic Process and, 59–62
    ICIK in, 288–290
    IK support in, 290–293
    Native Science and
        background of, 277, 284
        conclusions on, 293
        early childhood nutrition in,
            287–288
        ICIK in, 288–290

Kenyan TEK in, 284–286
    strategies to support TEK,
        290–293
Western Shoshone, 80–81
White Man's Burden (Easterly),
    80–81
Wilderness Society, 44–45
Wildfires, 46–49
Wilson, Kathy, 68
Wind River Indian Reservation,
    214–215, 217
Wind River Native Advocacy Center,
    217, 233
Wind River Reservation Liaison
    Committee, 98
Wisconsin, University of (UW), 287
Wisdom. *See also* Elders
    of Elders, 3–5, 7–8, 280
    of past, 9–10
Women
    association in Mali of, 51
    chicken feed and, 123–124
    in communication, 27–28
    as communicators in Mali,
        110–115
    Kipsigis Elders of, 285–287
    Mali stunting and, 303–304
    of Northern Cheyenne, 137
Women's Association, 111–112
Woolbaugh, W., 195
Work visits, to Mongolia, 168–176

environment and, 169–170
    field clinics and, 171–172
World Bank, 222, 288–289, 306
The World Citizen in Sustainable
    Communities: Exploring the
    Geographies of Hope course,
    163
World-Readiness Standards for
    Learning Languages, 252f,
    259–260
Wound healing, 58
Wright, Amy, 15, 209–210
The Written River (Prince), 166
Wyoming, 36, 44–45, 99, 217

Y

Yarrow, 58, 87–88
Yeats, William Bulter, 239
Yellowstone, 44–45
    Coalition of, 163–164, 182–183
    connection to Mongolia of,
        181–183
    wildfires in, 46–49
Yellowtail, Bill, 142
Yunus, Muhammed, 305–306

Z

Zambia, 272

Printed in the United States
By Bookmasters